16th European Workshop on Modern Developments and Applications in Microbeam Analysis (EMAS 2019 Workshop)

IOP Conference Series: Materials Science and Engineering
Volume 891

Trondheim, Norway
19 - 23 May 2019

ISBN: 978-1-7138-2810-5
ISSN: 1757-8981

Printed from e-media with permission by:

Curran Associates, Inc.
57 Morehouse Lane
Red Hook, NY 12571

Some format issues inherent in the e-media version may also appear in this print version.

Institute of Physics
Dirac House, Temple Back
Bristol BS1 6BE UK

Phone: 44 1 17 929 7481
Fax: 44 1 17 920 0979

techtracking@iop.org

Additional copies of this publication are available from:

Curran Associates, Inc.
57 Morehouse Lane
Red Hook, NY 12571 USA
Phone: 845-758-0400
Fax: 845-758-2633
Email: curran@proceedings.com
Web: www.proceedings.com

TABLE OF CONTENTS

IOP Conf. Series: Materials Science and Engineering **891** (2020) 011001 doi:10.1088/1757-899X/891/1/011001

16th European Workshop on Modern Developments and Applications in Microbeam Analysis (EMAS 2019) Trondheim, Norway, 19-23 May 2019

Bjørn E. Sørensen[1], Hans Dijkstra[2], Kristian Drivenes[1], Michael B. Matthews[3] and Silvia Richter[4]

[1] Norwegian University of Science and Technology, Department of Geoscience and Petroleum, 7491 Trondheim, Norway

[2] Thermo Fisher Scientific BV, Breda, The Netherlands

[3] AWE Plc., Aldermaston, Reading RG7 4PR, Great Britain

[4] RWTH Aachen, Gemeinschaftslabor für Elektronenmikroskopie (GFE), 552074 Aachen, Germany

Preface

This volume of the IOP Conference Series: Materials Science and Engineering contains papers from the 16th Workshop of the European Microbeam Analysis Society (EMAS) on Modern Developments and Applications in Microbeam Analysis, which took place from the 19th to the 23rd of May 2019 at the Realfagbygget of the Norwegian University of Science and Technology (NTNU), Trondheim, Norway.

The primary aim of this series of workshops is to assess the state-of-the-art and reliability of microbeam analysis techniques. The workshops also provide a forum where students and young scientists starting out on a career in microbeam analysis can meet and discuss with the established experts. The Trondheim meeting was organised within the usual EMAS format, comprising invited plenary lectures by internationally recognized experts, poster presentations by the participants and round table discussions on the key topics led by specialists in the field. The workshop was organised in collaboration with the Norwegian University of Science and Technology (NTNU). The technical programme included the following topics: Electron probe microanalysis, Electron backscatter diffraction (EBSD), Complementary techniques, Automated mineralogy, μ-X-ray fluorescence, and their applications.

As at previous workshops, there was also a special oral session for young scientists. The best presentation by a young scientist was awarded with an invitation to attend the 2020 Microscopy and Microanalysis meeting in Milwaukee, Wisconsin. The prize went to Grzegorz Cios, AGH - University of Science and Technology, Krakow (PL), for his presentation entitled: "*Local tetragonality of martensite investigated by EBSD*" (by G. Cios, A. Winkelmann, G. Nolze, T. Tokarski and P. Bała).

Seven half- or full-day pre-meeting courses were held covering the following areas of microbeam analysis: Introduction to EBSD, Introduction to quantitative EDS X-ray microanalysis, Quantitative compositional mapping by EPMA - An introduction to XMapTools software, Introduction to WDS X-ray microanalysis, mapping and phase analysis, Monte Carlo simulation, Electron probe maintenance, and Shielded EPMA for use in irradiated nuclear fuel analysis: Problems and opportunities.

The continuing relevance of the EMAS meetings and the high regard in which they are held internationally can be seen from the fact that 71 posters from 20 countries were on display at the meeting and that the participants came from as far away as Australia, Canada, Japan, and USA. A selection of participants with posters was invited to give a short oral presentation of their work in three dedicated sessions. The EMAS Young Scientist Poster Award was awarded with an invitation to participate in the 26th Australian Conference on Microscopy and Microanalysis (ACMM 26) in 2020 in Canberra, Australia. The prize was awarded to Håkon W. Ånes, Norwegian University of Science and Technology (NO), for his contribution entitled: "*Processing and indexing of electron backscatter patterns using open source software*" (by H.W. Ånes, J. Hjelen, A.T.J. van Helvoort and K. Marthinsen). The EMAS Workshop Poster Prize was given to William E. Goosen, Nelson Mandela University, Port Elizabeth (SA), for his contribution entitled: "*Characterisation of precipitates in a 12 % Cr tempered martensite ferritic steel*" (by G. Marx, W.E. Goosen and J.E. Westraadt).

This proceedings volume contains the full texts of 11 of the invited plenary lectures and of 12 papers on related topics originating from the posters presented at the workshop. All the papers have been subjected to peer review by a least two referees.

March 2020

EMAS 2019 Workshop, held at the Realfagbygget of the Norwegian University of Science and Technology (NTNU), Trondheim, Norway, from 19 to 23 May 2019.

Acknowledgements

On behalf of the European Microbeam Analysis Society I would like to thank all the invited speakers, session chairs and members of the discussion panels for making the meeting such a great success. Special thanks go to Bjørn Sørensen and Luc Van 't dack who directed the organisation of the workshop

IOP Conf. Series: Materials Science and Engineering **891** (2020) 011001 doi:10.1088/1757-899X/891/1/011001

giving freely of their time and talents. As was the case for previous workshops, the EMAS Board *in corpore* was responsible for the scientific programme. The Workshop also included a commercial exhibition where many leading instrument suppliers were represented. Several companies that exhibited provided financial support, either by sponsoring an event or by advertising.

Below, in alphabetical order, is a list of exhibiting companies and sponsors of the workshop.

- Ametek GmbH, Edax Business Unit
- Ametek, CAMECA Business Unit
- Bruker Nano GmbH
- European Commission, JRC Directorate G – Nuclear Safety and Security (Germany)
- Jeol (Germany) GmbH
- Micro to Nano
- Norwegian University of Science and Technology (NTNU
- Probe Software
- NordicNano Solutions AB
- Target-Messtechnik
- TESCAN ORSAY HOLDING, a.s.
- Thermo Fisher Scientific

Michael B. Matthews
EMAS President

IOP Conf. Series: Materials Science and Engineering **891** (2020) 011002 doi:10.1088/1757-899X/891/1/011002

Peer review statement

All papers published in this volume of *IOP Conference Series: Materials Science and Engineering* have been peer reviewed through processes administered by the proceedings Editors. Reviews were conducted by expert referees to the professional and scientific standards expected of a proceedings journal published by IOP Publishing.

IOP Conf. Series: Materials Science and Engineering **891** (2020) 012001 doi:10.1088/1757-899X/891/1/012001

U-Th-Pb$_{TOTAL}$ dating of REE-phosphate by electron microprobe: Review and progress

J M Allaz[1], M J Jercinovic[2] and M L Williams[2]

[1] ETH Zürich, Inst. für Geochemie und Petrologie, Department of Earth Sciences, Clausiusstrasse 25, 8092 Zürich, Switzerland
[2] University of Massachusetts Amherst, Department of Geosciences, 611 North Pleasant Street, 233 Morrill Science Center, 01003-9297 Amherst, MA, U.S.A.

E-mail: julien.allaz@erdw.ethz.edu

Abstract. Monazite (Mnz) and xenotime (Xnt) are rare earth element (REE) phosphate minerals that occur in a large variety of rocks. They are known to contain very limited amounts of common Pb, and thus, a date can be calculated by simply measuring the total lead- and actinide-contents. Such a measurement can be done at high spatial resolution with an electron probe microanalyser (EPMA), together with a complete mineral analysis. The U-Th-Pb$_{total}$ dating method was developed almost 30 years ago, to provide an in-situ age for several (U, Th)-bearing minerals such as uraninite, Mnz, and Xnt. This technique permits identification of events that are not recorded by other minerals, allows dates to be related to structural or geochemical features, and ultimately reinforces the age interpretation. The principal challenge is the accurate measurement of Pb, which is commonly present at the trace level (< 1,000 ppm).
This paper reviews the most recent advances in the U-Th-Pb$_{total}$ dating technique of Mnz and Xnt, notably in terms of a) accurate background acquisition using the multipoint background technique, and b) improved peak interference correction for trace and REE elements. This paper also discusses a new method of quantitative element mapping, which helps with the identification of compositional domains in Mnz (or Xnt). Finally, a date cannot be interpreted as an age without the proper context, and several key factors in the interpretation are discussed and illustrated with several case studies.

1. Introduction

Geochronology is a part of the geological sciences that aims to determine the age of a rock, mineral, or domain within a mineral in order to constrain the age of some geological event. For instance, an age can be obtained from the decay of an actinide (^{238}U, ^{235}U, and ^{232}Th) to lead (^{206}Pb, ^{207}Pb, and ^{208}Pb) [1]. An isotopic analysis is commonly necessary in order to distinguish the daughter element of the decay from the initial amount of the element present in the mineral at its time of growth, e.g., the "common lead" in the U-Th-Pb decay series. However, several actinide-bearing minerals incorporate very little, if any, common lead, and a date can be obtained with a simple elemental analysis of U, Th and Pb. This is notably the case for uraninite, and for the REE-phosphates, monazite (Mnz) and xenotime (Xnt), for which the common Pb content is only a few ppm and essentially negligible compared to the Pb produced by the decay of the relatively abundant actinides [2, 3].

The U-Th-Pb_{total} dating technique was first applied on uraninite using an electron probe microanalysis (EPMA) [3, 4], followed in the nineties by its successful application to Mnz and Xnt [5-7] (see review in [8]). The vague term "chemical dating" is commonly used for this technique, and should be dismissed and replaced by "U-Th-Pb_{total} dating" or "electron microprobe dating" if an EPMA is being used, as any isotopic or elemental dating technique is "chemical" in essence. The U-Th-Pb_{total} dating technique of Mnz and Xnt can be subdivided into two distinct approaches. First, the Chemical Isochron Method (CHIME) dating technique [7, 9-11] involves measurement of several Mnz (or Xnt, or zircon) of different compositions in order to obtain a "pseudo-isochron" of (Th,U) versus Pb, in which the slope is proportional to the date. All Mnz grains or domains included in the analysis should be within the error of the same age; otherwise results yield an age mixture or worse a totally spurious age. Whereas sector zoned Mnz found in pegmatites and in some granitic rocks can satisfy this condition, it is rarely the case in metamorphic rocks, as the variation of Mnz composition indicates a different monazite-forming reaction and commonly, a different age. The second approach [5, 6] involves making a high-precision measurement of U, Th, and Pb from a chemically homogeneous domain established by X-ray mapping (e.g., [12-19]). The goal is to link the date, or a population of dates from similar compositional domains, to specific mineral growth reactions. The single domain age technique is applicable to any actinide-bearing phase providing the common Pb content can be shown to be negligible. In this sense, the single-domain dating technique is more appropriate and will be the main focus of this paper.

The goal of this paper is to review the U-Th-Pb_{total} dating technique, and to discuss the recent improvements in the methodology, such as the peak interference corrections on REE and the utilisation of the multipoint background approach for trace element analysis [14]. Novel approaches are also presented, notably a) quantitative element mapping to permit better targeting of a set of homogeneous domains for precise spot analysis, and b) the progress on a combined EDS-WDS analysis to reduce the overall acquisition time or to improve the precision on minor elements analysis, while preserving accuracy of the interpreted age.

2. Monazite and xenotime

Mnz and Xnt are common accessory minerals that occur in a wide variety of geological settings, notably in (meta-)sedimentary rocks, placer deposits, gneisses, high-silica plutonic and volcanic rocks, and carbonatites [20]. They commonly display strong compositional zoning at the micrometre-scale, which can be related to magmatic, metamorphic, or hydrothermal reactions or events. Grains are commonly a few micrometres to 0.5 mm in size, and occasionally reach mm- to cm-size, notably in pegmatites. Mnz has a monoclinic structure and incorporates mostly light REE (La to Eu; LREE), whereas Xnt has a tetragonal structure with mostly Y and heavy REE (Gd to Lu; HREE). Actinides enter the mineral structure through the cheralite or the huttonite/thorite substitutions [21] (Table 1). Due to crystallographic constraints the solid solution towards cheralite or huttonite/thorite is incomplete, and Mnz incorporates up to 31.5 wt% ThO_2, with an average around 6 wt%, and a usually < 1 wt% UO_2 content [20, 22, 23]. The actinide content is generally higher in Mnz compared to Xnt. Monazite tends to incorporate Th over U, whereas Xnt tends to incorporate U over Th [24], although there are exceptions. Low temperature REE-phosphates tend to be poor in actinides and their compositions approach the ideal formulas of Mnz $[LREE]PO_4$ or Xnt $[HREE]PO_4$ [15], and the actinide content increases with temperature [25]. Other major substitutions in Mnz and Xnt are summarized in Table 1 [26-30].

Table 1. Possible substitutions in Mnz and Xnt. M^{2+} = divalent cation: Ca^{2+}, Sr^{2+}, etc.

Name	**Substitution**	**End-member**	**Ref.**
Monazite-(Ce), Xenotime-(Y)	$LREE^{3+} \Leftrightarrow HREE^{3+}$	$REE(PO_4)$	-
Cheralite (brabantite)	$2\ REE^{3+} \Leftrightarrow Ca^{2+} + (U,Th)^{4+}$	$Ca(U,Th)(PO_4)_2$	[21]
Thorite, Huttonite, Coffinite	$P^{5+} + REE^{3+} \Leftrightarrow Si^{4+} + (U,Th)^{4+}$	$(U,Th)SiO_4$	[21]
Gasperite-(Ce), Chernovite-(Y)	$P^{5+} \Leftrightarrow As^{5+}$	$REE(AsO_4)$	[26, 27]
Sulphate (anhydrite, celestine, etc.)	$REE^{3+} + P^{5+} \Leftrightarrow M^{2+} + S^{6+}$	$M^{2+}(SO_4)$	[28-30]

Compositionally homogeneous domains are commonly in the range of a few micrometres, which necessitates the use of a high-spatial resolution analytical instrument. Mass spectrometry techniques, such as the laser-ablation inductively-coupled plasma mass spectrometry (LA-ICP-MS), sensitive high-resolution ion microprobe (SHRIMP), or secondary ion mass spectrometry (SIMS) can provide excellent in situ isotopic analysis of U, Th and Pb. Yet, these techniques are spatially limited by their large ablation pit of 10 - 20 μm. Moreover, it is difficult to obtain a complete compositional analysis along with a date on these instruments, notably due to the lack of appropriate REE-rich standards in many laboratories for accurate measurement and the problem of isobaric interferences from $[REE]O^+$ compounds, which complicates the analysis of HREE (M. Guillong, pers. comm.). Access to these instruments, especially SIMS or SHRIMP, also remains limited. EPMA U-Th-Pb_{total} dating offers an alternative with high spatial resolution down to the submicrometre scale [31]. However, limitations arise from the Pb detection limit, in general limiting its application to samples older than 100 - 200 Ma.

3. The U-Th-Pb_{total} age equations

The very first approximate calculation of U-Th-Pb_{total} age of Parslow *et al.* [4] using a single point analysis involved an iterative calculation of two equations to determine the Pb content that stems from the decay of U and Th:

$$t_U = 3.323 * 4.5 * 10^9 \left(\log\left[1 + \frac{x}{U}\right]\right) \quad (1a)$$

$$t_{Th} = 3.323 * 14 * 10^9 \left(\log\left[1 + \frac{Pb-x}{Th}\right]\right) \quad (1b)$$

where U, Th, Pb and x are atomic or molecular proportions, t_U and t_{Th} is the time in years, and the coefficients of each equation are related to the half-life of U and Th isotopes. The variable x is adjusted until the times t_U and t_{Th} converge. Shortly after, Bowles [3] proposed a single equation to iteratively calculate the date:

$$Pb = U * \left[0.9928\left(e^{\lambda^{238}*t} - 1\right) + 0.0072\left(e^{\lambda^{235}*t} - 1\right)\right] + Th * e^{\lambda^{232}*t} - 1 \quad (2)$$

where U, Pb and Th are atomic proportions, t is the time in years, λ^{238} ($1.55125*10^{-10}$/y), λ^{235} ($9.8485*10^{-10}$/y), and λ^{232} ($4.9475*10^{-11}$/y) are the decay constants of ^{238}U, ^{235}U, and ^{232}Th [32], and the coefficients are the relative abundances of ^{238}U and ^{235}U. Equation 2 is very similar to the equation proposed by Montel *et al.* [5, 6], with the difference that U, Th and Pb are expressed as weight% or ppm:

$$Pb = 206\frac{U*0.9928}{238.04}\left(e^{\lambda^{238}*t} - 1\right) + 207\frac{U*0.0072}{238.04}\left(e^{\lambda^{235}*t} - 1\right) + 208\frac{Th}{232}\left(e^{\lambda^{232}*t} - 1\right) \quad (3)$$

where 238.04 is the average atomic weight of U, and 232, 206, 207 and 208 are the atomic weights of ^{232}Th, ^{206}Pb, ^{207}Pb and ^{208}Pb. In Eqs. 2 and 3, the date is iteratively calculated to match the measured Pb-content (a Microsoft® Excel macro for the iterative calculation of Eq. 3 can be requested by email to JMA).

The approach of Suzuki and Adachi [7, 9, 33] is different and requires multiple analysis of Mnz domains of variable composition. The set of analyses is plotted in a PbO versus ThO_2* (or UO_2* for U-rich samples, e.g., Xnt or zircon) to obtain a "pseudo-isochron". ThO_2* is the total measured ThO_2 plus the equivalent UO_2:

$$ThO_2^* = ThO_2 + \left(\frac{UO_2*W_{ThO2}}{W_{UO2}*\left(e^{\lambda^{232}*t}-1\right)}\right)\left(\frac{e^{\lambda^{235}*t}+137.88*e^{\lambda^{238}*t}}{138.88} - 1\right) \quad (4)$$

A first approximation of t in Eq. 4 is obtained by calculating an apparent date:

$$\frac{PbO}{W_{PbO}} = \frac{ThO_2}{W_{ThO2}}\left(e^{\lambda^{232}*t} - 1\right) + \frac{UO_2}{W_{UO2}}\left(\frac{\left(e^{\lambda^{235}*t} + 137.88*e^{\lambda^{238}*t}\right)}{138.88} - 1\right) \quad (5)$$

where W_{PbO}, W_{ThO2}, and W_{UO2}, are the measured atomic weight of PbO, ThO_2 and UO_2. The date T is then obtained using the slope m of a regression on the PbO versus ThO_2* plot:

$$T = \frac{1}{\lambda^{232}}\ln\left(1 + m * \frac{W_{ThO2}}{W_{PbO}}\right) \quad (6)$$

The newly calculated time T is then inserted back in Eq. 4, and an iteration follows until t and T converge. The intercept of the regression of ThO_2* versus PbO is an indicator of the presence of common Pb. In the following, we will only consider the single-domain U-Th-Pb_{total} dating technique of Montel [5, 6], yet the analytical considerations are applicable to all the above calculations.

4. Analytical considerations for electron microprobe analysis of Mnz and Xnt

The use of EPMA to date Mnz and Xnt is advantageous due to a) the high spatial resolution reached by the EPMA; b) the non-destructive nature of the analysis, c) the in situ character of the analysis, which preserves the textural and structural features, d) the capacity to collect element maps to identify the compositional variations, and e) the ability to simultaneously provide an U-Th-Pb_{total} age and a complete compositional analysis. An acceleration voltage of 15 keV is optimal, as it provides a good overvoltage for the L-family X-rays of light and heavy REE (critical ionisation energy between 5.48 [La-Lα] and 9.25 keV [Lu-Lα]), while keeping the analytical volume small (~1 - 2 μm^3; see extended discussion in [34]). The challenge lies in the precise and accurate measurement of U, Th and especially Pb, as the Pb-content is commonly < 1,000 ppm, which classifies this type of EPMA analysis as a trace element analysis [34, 35]. The numerous peak and background interferences from LREE, Y, Th and S, and the need for an accurate background correction complicate the analysis of Pb in REE-minerals, and the limitation of the technique arises chiefly from the detectability and analytical sensitivity of Pb, which restricts the application of the U Th Pb_{total} method to either old REE-phosphate (> 100 - 200 Ma) or to young and very actinide rich chronometers (Fig. 1). Williams and Jercinovic [16] discussed the numerous pitfalls and ideal conditions for performing precise and accurate quantitative analysis, Jercinovic *et al.* [34] reviewed the optimal analytical conditions for trace element analysis, notably in Mnz, and this paper presents an updated analytical protocol (Table 2).

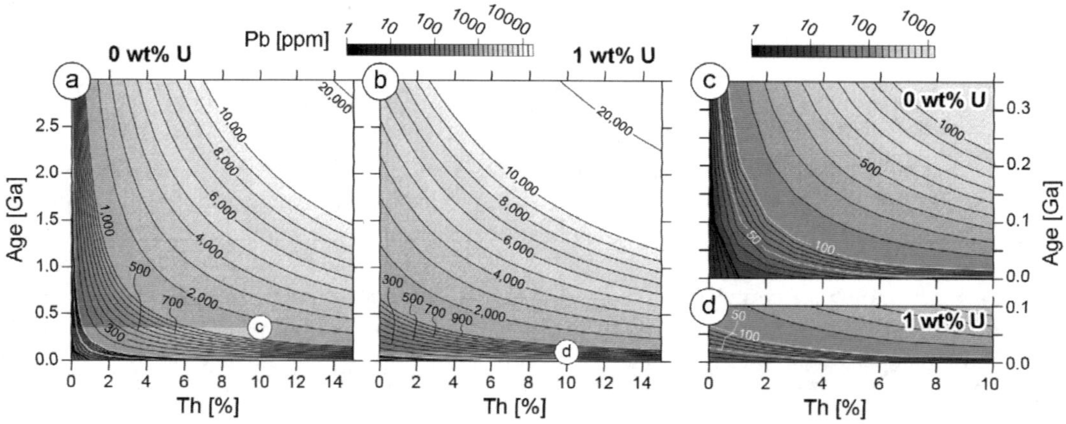

Figure 1. Production of Pb as a function of the age and Th-content assuming a) 0, or b) 1 wt% U. c, d) Enlargement of a, b). Limitations of the technique arise at Pb content < 50 - 100 ppm (white lines).

Table 2. Analytical protocols for Mnz or Xnt analysis by EPMA.

What	Comment	Analytical condition
1) Thin section preparation and coating (15-20 nm Al + 7-8 nm C or 20-30 nm C).	Aluminum coating preferred, thick (>20 nm) carbon coating possible, do not use Au or Ag coating (X-ray lines around Pb-Mα, or U-Mβ and Th-Mα, respectively).	
2) Full thin section map with WDS or EDS phase map to identify accessory phases.	In samples with Ba-mineral, use La-Lα to avoid interference of Ba-Lβ on Ce-Lα. Other element used as a base map, depending on the type of sample (e.g., Si-, Al-, or Mg-Kα).	Stage mapping, 15 keV, 200-300 nA, defocused beam 20-30 µm, ca. 4-5 hours per thin section: Ce or La-Lα (Mnz), Y-Lα (Xnt), Ca- or P-Kα (apatite), Th-Mα (Th-rich phase), other element.
3) Identification of accessory minerals (Mnz, Xnt, zircon, apatite), selection of 10-20 grains.	Selection of grain can be done based on structural considerations, shape of the grains, location of the grain with regards to other minerals, etc.	
4) Element mapping of individual Mnz and Xnt grains, ~10 to 20 grains.	For grains >50-100 µm, prefer either a mosaic image with several beam maps or a stage map to minimize defocussing effect. If Si- or Ca-content is very low, replace it by a REE.	Beam mapping, 15 keV, 200-300 nA, focused beam, <0.5 µm step size, ≥20-30 msec dwell time, usually 5-30 min per grain, Th-Mα, U-Mβ, Y-Lα, Si-Kα, Ca-Kα (or one REE).
5) (optional) Age map of monazite for illustrative purpose.	*Age mapping is a very long analysis, and only useful for high-Pb content and if a large age variation is expected.*	*Beam mapping, 15 keV, ≥300 nA, focused beam, <0.5 µm step size, ≥500 msec dwell time, usually 8-12 hours per grain, Th-Mα, U-Mβ, Pb-Mα, Y-Lα.*
6) Complete quantitative analysis (U, Th, Pb, P, REE, etc; see Table 4).	Alternative option: two-step combined analysis with (a) low current (20 nA) analysis of major element, and (b) high current (200 nA) analysis for minor and trace elements.	15 keV, 200 nA, focused beam, 10-15 min per point (peak only), 5 to 8 points per homogeneous domain, combined with EDS analysis for P (±REE). See details in Table 3.
7) Data treatment and date calculation, and age interpretation	Age calculation following method of Montel et al. [5,6], one age per homogeneous domain (average of 5-8 points), bring together full thin section, grains maps, and age data.	

4.1. Instrumentation and analytical conditions

Most data considered here were obtained on a 5-spectrometer JEOL JXA-8230 EPMA at the University of Colorado-Boulder equipped with a W filament (~1.2 µm focused beam diameter at 15 keV, 200 nA) or a LaB_6 cathode (~0.7 µm). Some data were obtained on a CAMECA SX 100 "Ultrachron" EPMA at the University of Massachusetts-Amherst equipped with a LaB_6 or CeB_6 cathode. Both instruments have a 10 mm^2 EDS detector, yet only the JEOL instrument was used for combined EDS-WDS analysis. Complementary tests were also obtained at ETH Zürich using a JEOL JXA-8200 and a scanning electron microscope JEOL JSM-6390 equipped with a 30 mm^2 Thermo UltraDry SDD EDS detector. A summary of the detector and spectrometer configurations is listed in Table 3. Unless specified otherwise, analytical conditions were 15 keV, 200 nA, and a focussed beam (see also Table 2).

Table 3. EPMA configurations used in this study. "L-..." = large-area monochromator, "VL-PET" = very large area PET monochromator, "H-..." = H-type spectrometer with smaller Rowland circle (JEOL only): 100 mm versus 160 (CAMECA) or 140 mm (JEOL).

	Cameca SX-100		JEOL JXA-8230		JEOL JXA-8200	
1	TAP	P-10, low P	TAP	P-10, low P	PET	P-10, low P
2	L-PET	P-10, high P	L-PET	P-10, low P	H-TAP	P-10, low P
3	VL-PET	P-10, high P	L-PET	Xe sealed	LiF	Xe sealed
4	VL-PET	P-10, high P	L-PET	Xe sealed	H-PET	Xe sealed
5	L-LiF	P-10, high P	L-LiF	Xe sealed	H-PET	Xe sealed
EDS	Bruker XFlash 4010		Thermo UltraDry		JEOL SiLi EDS	

4.2. Monazite identification and element mapping

A highly polished sample surface (e.g., thin section or epoxy mount) is necessary for accurate analysis. After cleaning, the thin section is coated with either a thick C-coating or an Al-coating with a thin overcoat of C to prevent the rapid oxidation of Al. A dual (Al+C)-coating is preferred, as it significantly minimizes the beam damage effect compared to C alone [36].

The first step is to identify all Mnz and Xnt grains in the thin section in order to select a representative set of grains. Identification of accessory phases in the thin section can be done either with a phase analysis map using the energy-dispersive X-ray spectrometer (EDS) or with a full thin-section element map using the wavelength-dispersive X-ray spectrometers (WDS) and scanning for Ce-Lα or La-Lα (Mnz), Y-Lα (Xnt), Ca- or P-Kα (apatite), Zr-Lα (zircon), and a major element map (e.g., Si-, Al-, or Mg-Kα) or a BSE image to be used as a base map (step 2 in Table 2; e.g., [37]). Even with a step size of 20 – 30 μm (and equivalently defocussed beam), Mnz or Xnt grains as small as 5 μm can be identified with the full thin section WDS maps as these phases are usually the only ones to incorporate a large quantity of Ce or Y, whereas an EDS phase analysis map usually requires a smaller pixel size to permit the proper identification of 5 - 10 μm-sized phases. A faster identification by EDS can be performed with a dedicated particle search software that uses a threshold on a mosaic BSE image to highlight dense particles, completed by an EDS identification of each particle, with the caveat that the thresholding must be set to reveal the smallest target grains in the section.

In the second step, a number (10 to 15 minimum) of individual Mnz or Xnt grains per sample are mapped using the WDS to reveal the variation of Th-Mα, U-Mβ, Y-Lα, Ca-Kα and Si-Kα (step 3 in Table 2). EDS mapping is not useful here, as it does not have the required sensitivity to reveal all the compositional variations. Moreover, there is a critical interference between P-Kα and Y-Lα, which will obscure small Y variations. In order to permit an inter-grains and even an inter-samples comparison, the maximum and minimum intensity levels of each map should be treated simultaneously [38], or the maps should be quantified (see example below). The grains are targeted in order to have a representative set of all sizes, shapes, and textural or structural context in order to capture all possible compositional and textural variations in the Mnz and Xnt populations.

High-resolution element mapping of Th-Mα, U-Mβ, Y-Lα, and Pb-Mα at very high current (> 500 nA) and long counting time (0.5 - 1 s per pixel; step 4 in Table 2) can provide low precision age maps for Mnz grains [39]. At these conditions, a single map of a 50 - 100 μm grain can easily take 8 - 12 hours to complete, whereas in the same amount of time multiple grains can be quantitatively analysed and dated with higher precision. High-resolution age mapping is, therefore, recommended mainly for an illustrative purpose, where complexly zoned grains present a strong contrast in age domains. A single-domain quantitative analysis should remain the rule to provide precise data (Table 2).

4.3. Quantitative analysis

The U-Th-Pb_{total} analysis of Mnz or Xnt by EMP must be considered as a "trace element analysis" [34, 35], as Pb, some REE, and sometimes U or Th are commonly found in trace amounts (< 1,000 ppm). The use of a high beam current, long count times, and a set of high count-rate spectrometers is thus, necessary. This type of analysis is challenging due to a) the non-linearity of the background, b) the presence of numerous peak and background interfering X-ray lines, notably from Th and REEs, and c) a high potential for beam damage during the analysis.

The following is a summary of some of the key analytical considerations necessary to obtain an accurate U-Th-Pb_{total} date and a complete compositional analysis (step 4 in Table 2). A generalised analytical setting suitable for both Mnz and Xnt analysis is summarized in Table 4. One or more secondary Mnz (or Xnt) reference materials of known age are measured before, during, and after the analysis of unknowns in order to ensure that the analytical protocol is sound and that the results are consistent from day to day (= "age consistency standard"). Recently Goncalves *et al.* [40] has reviewed several Mnz reference materials from Brazil that are suitable for isotopic or elemental analysis. In this study, the Moacyr Mnz (TIMS age ~507 Ma; equivalent to the Itambé Mnz [40]) and Burnet Mnz (SIMS age ~1,088 Ma) are used as reference materials.

IOP Conf. Series: Materials Science and Engineering **891** (2020) 012001 doi:10.1088/1757-899X/891/1/012001

Table 4. Analytical setting for Mnz and Xnt analysis by EPMA.

SP	El.	Line	Xtal	Peak [mm]	Peak [10^5.sinθ]	Recommanded standard	Time [s] Peak	Time [s] Bkg#	Bkg Type	Major and minor peak interferences (see Table 7)	
1	Y	Lα	TAP	69.70	24890	YPO_4	180	-	MAN	Sr Lβ	(1)
1	*Sr*,***	*Lα*	*TAP*	*74.77*	*26705*	*Strontianite*	*180*	-	*MAN*	*Si Kβ + satellite peaks*	
1	Si	Kα	TAP	77.10	27535	Microcline	180	-	MAN	Nd Lα (III)	
1	*As*,***	*Kα*	*TAP*	*105.36*	*37630*	*As-pyrite, GaAs*	*180*	-	*MAN*	*Tb Mβ; Dy Mα; Sm Mγ*	
2	Th	Mα	L-PET	132.44	47300	$CaTh(PO_4)_2$, ThO_2	550	50##	MPB	-	
2	Ca	Kα	L-PET	107.50	38395	Apatite	60	-	MAN	-	
2	S	Kα	L-PET	172.09	61460	Sulfate or sulide	60		MAN	-	
3	U	Mβ	L-PET	119.04	42470	UO_2	600	50##	MPB	K Kα; Th Mγ; Sm $Lβ_2$ (II)	
3	K	Kβ	L-PET	110.64	39515	K-feldspar	100	-	MAN	U Mγ; Tm Lα (II); Ca Kα	
4	Pb	Mα	L-PET	169.24	60445	Pyromorphite	700	50##	MPB	Y $Lγ_{2,3}$; Th Mζ; La Lα (II)	(2)
5	*La***	*Lα*	*L-LiF*	*185.38*	*66205*	*$LaPO_4$*	*30*	-	*MAN*	*Nd Ll*	*(3)*
5	Ce	Lα	L-LiF	178.12	63615	$CePO_4$	30	-	MAN	-	(3)
5	Nd	Lα	L-LiF	164.82	58864	$NdPO_4$	30	-	MAN	Ce Lβ; La $Lβ_6$	(3)
5	*Pr***	*Lβ*	*L-LiF*	*157.08*	*56100*	*$PrPO_4$*	*30*	-	*MAN*	-	*(3)*
5	Sm	Lβ	L-LiF	138.97	49630	$SmPO_4$	30	-	MAN	Tb Lα	(3)
5	*Tb**	*Lα*	*L-LiF*	*137.41*	*49075*	*$TbPO_4$*	*60*	-	*MAN*	*Sm Lβ*	
5	Eu	Lβ	L-LiF	133.58	47705	$EuPO_4$	60	-	MAN	Dy Lα	(4)
5	Dy	Lα	L-LiF	132.66	47380	$DyPO_4$	60	-	MAN	-	
5	*Er**	*Lα*	*L-LiF*	*123.98*	*44280*	*$ErPO_4$*	*60*	-	*MAN*	*Tb Lβ*	
5	Gd	Lβ	L-LiF	128.32	45830	$GdPO_4$	60	-	MAN	Ho Lα; Sm $Lβ_2$	(4)
5	*Tm**	*Lα*	*L-LiF*	*119.96*	*42845*	*$TmPO_4$*	*60*	-	*MAN*	*Sm Lγ; Dy Lβ; Gd $Lβ_{9,10}$*	(5)
5	Yb	Lα	L-LiF	116.13	41475	$YbPO_4$	60	-	MAN	Tb $Lβ_{2,15}$; Dy $Lβ_{3,6}$; Sm $Lγ_2$	
5	*Ho**	*Lβ*	*L-LiF*	*114.40*	*40855*	*$HoPO_4$*	*60*	-	*MAN*	*Yb Lα,n; Dy $Lβ_{2,15}$*	
5	*Lu**	*Lα*	*L-LiF*	*112.62*	*40220*	*$LuPO_4$*	*60*	-	*MAN*	*Ho $Lβ_{3,6}$; Dy $Lβ_{2,15}$; Eu Lγ*	
-	P	Kα	EDS	2.01 keV		$CePO_4$	300	-	EDS	Y Lβ	(6)

*, **: minor to trace element in Mnz (*) or in Xnt (**).
\# Background measurement is done separately from the peak measurement (see text for discussion).
\#\# The MPB background counting time can be increased to 100 s for higher precision.
MAN: Mean atomic number background correction; MPB: Multipoint background correction.
(1) Consider analysing Y-Lα on a PET (SP 2) if the line is too close to the spectrometer limit.
(2) Use Pb-Mβ for Xnt. If U-Mβ is analysed on the same spectrometer as Th-Mα, two spectrometers can be used for Pb analysis to enhance precision.
(3) To gain time, the main LREE elements can be analysed by EDS, at the cost of a lower precision.
(4) Strong interferences on Eu-Lβ and Gd-Lβ in Xnt; prefer Eu-Lα and Gd-Lα.
(5) Strong interference of Sm-Lγ on Tm-Lα in Mnz; Tm-Lβ could be an alternative option, although it is strongly interfered by Gd-$Lγ_2$ and Tb-Lγ (not tested here).
(6) P-Kα is best analysed in Mnz by EDS in order to minimize beam damage effect. Alternatively, P-Kα can be analysed by WDS with a PET monochromator (required for Xnt due to interference from Y-Lα).

4.3.1. Choice of spectrometer. It is necessary to select a spectrometer that yields a high count rate with a high peak-to-background ratio, in order to lower the detection limit and increase the analytical sensitivity for the analysis of minor and trace elements. Integrating counts from one or more spectrometers measuring the same element, a common feature of modern software, can increase precision substantially. A comparison of the count rate obtained on UO_2, ThO_2, pyromorphite, and a few REE phosphate standards on JEOL or CAMECA instruments (Table 5) demonstrates the large gain in intensity using a spectrometer equipped with large- (L-type crystals) or very large area monochromators (VL-type PET monochromator on CAMECA only), or a spectrometer with a smaller Rowland circle (H-type; on JEOL only) compared to a normal spectrometer (J-type). For the selected elements, a count-rate gain of 2.1x to 3.2x is observed between J-type and L-type spectrometers equipped with Xe counter, and of 2.7x to 4.1x between J-type and H-type spectrometers (Table 5).

A disadvantage of the H-type spectrometer is the loss of spectral resolution compared to J- or L-type spectrometers (Fig. 2a), which complicates the selection of an appropriate background and increases the effect of peak interferences, especially for REE analysis. The H-type spectrometer also has a higher background and a lower peak-to-background ratio compared to the VL- or L-type spectrometers (Table 5; Fig. 2b). This reduces its sensitivity despite the higher count rate. In terms of count rate, the L-type spectrometer appears to be superior compared to the H-type spectrometer for X-ray lines of energy > 6.4 keV (Table 5). In the present study, tests of VL-type spectrometer from CAMECA are only shown for the measurement of Pb-Mα, for which this spectrometer was designed, and U-Mβ. This spectrometer beats all others with a high-count rate, a good peak-to-background ratio, and a similar spectral resolution compared to the L-type spectrometers. The use of a Xe counter or a high-pressure (2 - 3 atm) Ar-counter is recommended for the analysis of medium to high energy X-ray lines, with a small advantage to the high-pressure Ar-counter on intermediate X-ray lines energies (4.0 - 6.4 keV) and to the Xe-counter on lower or higher energies, although the difference is only ±20 - 30 % relative, and is likely to be dependent on the spectrometer alignment.

Table 5. Comparison of count-rate and peak-to-background ratio for J-, L-, H- or VL-type spectrometers.

				Net peak [cps/nA]						Peak-to-background ratio					
				J-type	J-type	L-type	L-type	H-type	VL-type	J-type	J-type	L-type	L-type	H-type	VL-type
	Standard	X-ray	keV	Ar, LP	Xe	Xe	Ar, HP	Xe	Ar, HP	Ar, LP	Xe	Xe	Ar, HP	Xe	Ar, HP
PET	Pyromorph.	Pb Mα	2.35	61.1	66.5	141.2	145.1	239.3	266.4	27.0	124.9	195.7	242.9	116.1	179.8
PET	ThO_2	Th Mα	3.00	72.5	151.6	397.1	-	563.1	-	15.9	82.1	153.7	-	78.3	-
PET	UO_2	U Mβ	3.34	69.3	166.4	336.5	280.4	636.0	420.2	11.4	80.2	83.1	63.0	74.5	56.7
LiF	$LaPO_4$	La Lα	4.65	-	22.2	62.2	88.7	91.8	-	-	101.1	137.8	106.6	78.3	-
LiF	$NdPO_4$	Nd Lα	5.23	-	37.1	98.9	121.0	137.0	-	-	104.2	125.7	100.2	75.7	-
LiF	$TbPO_4$	Tb Lα	6.27	-	54.3	144.7	164.3	172.6	-	-	70.3	82.2	63.0	48.4	-
LiF	$DyPO_4$	Dy Lα	6.49	-	57.0	184.1	168.1	177.2	-	-	61.4	73.2	61.5	42.2	-
LiF	$HoPO_4$	Ho Lβ	7.52	-	36.1	114.0	89.5	98.8	-	-	26.8	25.7	21.5	16.7	-

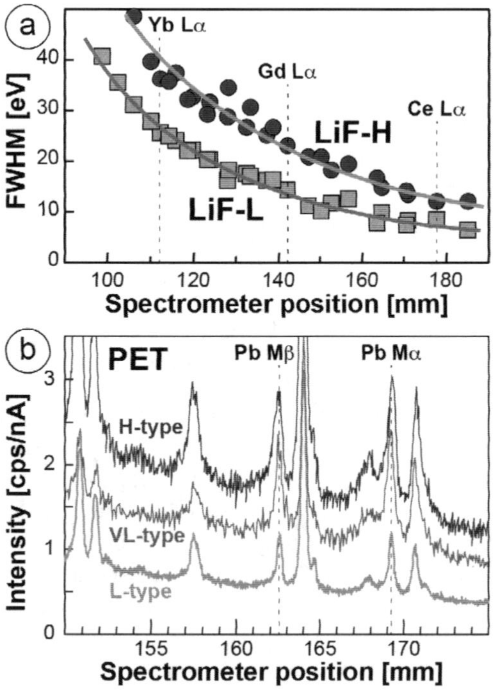

Figure 2. a) Variation of the full-width at half maximum (FWHM) on Lα and Lβ X-ray lines of REE on L-type versus H-type spectrometer (LiF). b) Comparison of scans around Pb-Mα obtained on L-, VL-, and H-type spectrometers (PET) in Burnet Mnz. WDS scan on L-type spectrometer acquired with higher precision.

An optimum instrument for trace element analysis should be equipped with an L-type spectrometer for REE, U, and Th analysis, and with an L-, an H-, or even better a VL-type spectrometer for the analysis of Pb-Mα, with either a high-pressure Ar-counter or a Xe-counter. Figure 3 shows the detection limits obtained on L-type spectrometers for the analysis of U, Th, Pb and REE (see [14, 34] for a detailed discussion on detection limits and sensitivity). In the case of Mnz or Xnt analysis, it is desirable to reach at least a 50 to 30 ppm detection limit on Pb Mα, which is achievable after 5 to 10 min counting time at 200 nA on a L-, VL-, or H-type spectrometer. The detection limit of REE on LiF-L already reaches 130 to 250 ppm on Lα- or Lβ- lines after 30 s counting on the peak at 200 nA, and 90 to 180 ppm after 60 s counting (Fig. 3).

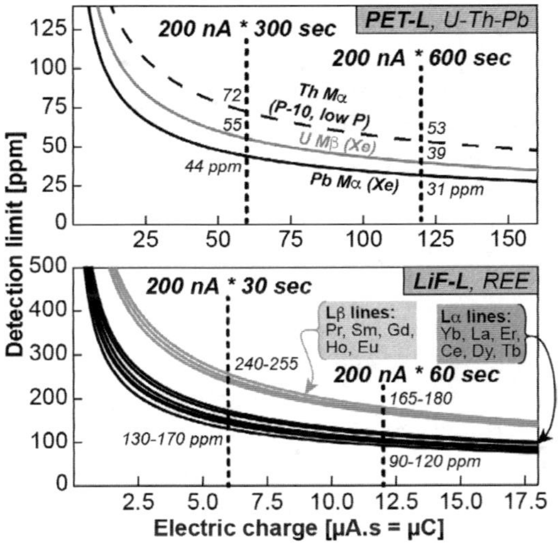

Figure 3. Detection limits on L-type spectrometers for U, Th, Pb (PET-L) and REE (LiF-L).

A danger of using a high count-rate spectrometer is an inaccurate dead time correction and a potential shift of the pulse in the X-ray detector, the latter being significant on JEOL instruments due to the smaller detector size and the absence of an exit window. The deadtime correction should be carefully checked regardless of the detector, especially for the major elements (e.g., La, Ce, Nd, Th) that are analysed at high current with a standardisation done at low current and when the count rates are above ~20 - 30k. It is recommended to keep the PHA setting of those elements either in integral mode or in differential mode with a significantly large window to accommodate a possible pulse shift. Conversely, an optimum PHA window setting for X-ray lines interfered by high-order X-ray lines should be set to minimise this type of interferences (e.g., La-Lα [n = 2] on Pb-Mα)

4.3.2. Background acquisition. An accurate background measurement is absolutely essential for trace element analysis [34]. In fact, one could go so far as to say that trace element EPMA analysis is mostly an exercise in characterising background at high precision and accuracy. Gathering peak counts at high precision is obviously a consideration, but if the background is inaccurate, the result may be meaningless (as P/B nears unity). It is well established that the X-ray continuum is not perfectly linear and is better modelled using an exponential or a polynomial curve [14, 16, 34]. Moreover, the background positions must be chosen in order to avoid any background interference. Numerous interfering X-ray lines are present around Pb-Mα and -Mβ, notably Y-L$\gamma_{2,3}$, Th-Mζ, and several 2nd order lines of LREE (La, Ce, Pr; Fig. 4a). Further complications can arise from additional X-rays from unexpected elements that are either rarely present (e.g., S-Kα,β in Moacyr; Fig. 4a) or induced by secondary fluorescence at distance from a neighbouring phase (e.g., Ti-Kα or K-Kα; [41]).

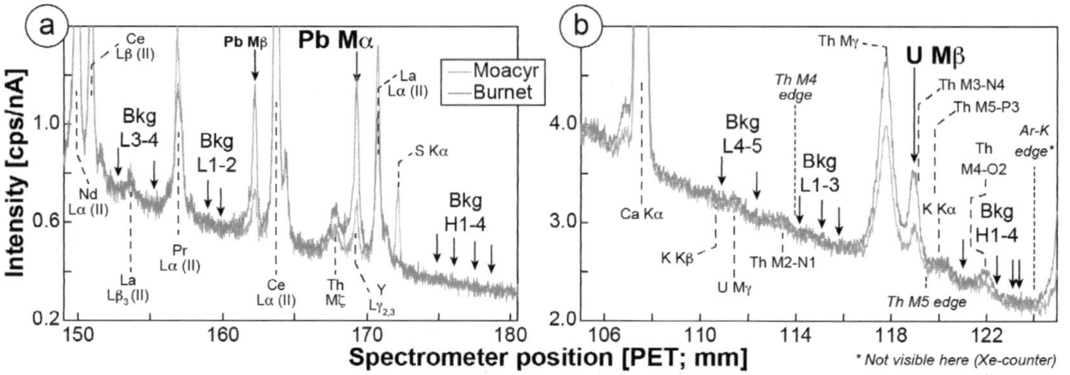

Figure 4. WDS scans acquired on L-type monochromator (PET-L) equipped with Xe-detector in Moacyr and Burnet Mnz around a) Pb-Mα, and b) U-Mβ with multipoint background positions on low (Bkg Lx) and high side (Bkg Hx). When using a P-10 detector, the analyst should be wary of the Ar-K edge at 124.0 mm, which will limit the location of high-side background.

Williams and Jercinovic [16, 42] initially suggested acquiring a detailed quantitative (step, not time-sliced continuous scan ["ROM" scan], scan yielding accurate, dead-time corrected cps/nA) WDS scan at high current (≥ 200 nA) around the X-ray lines of U-Mβ, Th-Mα and Pb-Mα, and to use this scan to accurately determine the background intensity by taking into account the background curvature and avoiding the background interferences. Because the background value is directly dependent on the composition (i.e., the average Z), a single WDS scan is necessary within each homogeneous domain. While this technique remains accurate, it is time-consuming and subjective, as the analyst has to select areas in the scan, which appear to represent the background. Moreover, minor background interferences might be barely visible and lost in the low analytical precision of a WDS scan unless a very long counting time is used.

A better background measurement technique called the "multipoint background acquisition" has recently been developed [14]. This method involves measurement of several background intensities (perhaps 6 or more) at pre-set spectrometer positions around the peak of interest, and the acquired intensities are interpolated with a linear or better an exponential or polynomial fitting in order to accurately and precisely determine the background curvature and intensity. The set of background positions is selected based on a collection of WDS scans in a series of materials similar to the unknowns being analysed (Table 6; Figs. 4a and 4b). This approach is less subjective than the WDS scan technique, and it tolerates a few erroneous background positions that are affected by one or more unexpected background interferences in a specific sample. Erroneous background(s) are removed during the reprocessing either automatically by selecting the points that yield the statistically lowest background (within the precision of the acquisition) or manually by the analyst. The technique has been successfully tested in multiple Mnz reference materials for which an isotopic age exists (see details in [14]).

Background acquisition for the REE is also challenging, as multiple X-ray lines prevent the selection of a "true" background position, especially for the HREE. The "shared background" method [14] can be used to minimise the risk of a background interference and to correct for the effects of absorption edges. In this paper, we consider the use of the mean atomic number (MAN) background correction [43], which consists of calibrating an absorption-corrected background intensity curve as a function of the average atomic number. This calibration is done by measuring the element of interest in several reference materials, which do not contain this element or a peak interference. Once calibrated, the background is not measured, and the total analysis time is either cut in half, or the extra time is used to double the peak counting time and thus improve the analytical sensitivity especially for HREE in Mnz and LREE in Xnt. This technique is appropriate for elements above ~200 ppm [44], and thus, is

Table 6. Multipoint background positions for the analysis of U-Mβ, Th-Mα, Pb-Mα and Pb-Mβ.

		Cameca position [10^5 sinθ]				JEOL position [mm]			
		U Mβ	Th Mα	Pb Mα	Pb Mβ	U Mβ	Th Mα	Pb Mα	Pb Mβ
Bkg Low	1	-1095	-1035	-3200	-375	-3.0	-2.9	-9.0	-1.0
	2	-1355	-1270	-3500	-2800	-3.8	-3.5	-9.8	-7.8
	3	-1685	-1470	-4750	-3760	-4.7	-4.1	-13.3	-10.5
	4	-2330	-	-5460	-4000	-6.5	-	-15.3	-11.2
	5	-2850	-	-	-	-8.0	-	-	-
Bkg High	1	760	1940	1900	430	2.1	5.4	5.3	1.2
	2	1265	2600	2300	935	3.5	7.3	6.4	2.6
	3	1525	3780	2800	1100	4.2	10.6	7.8	3.1
	4	1600	-	3200	1750	4.5	-	9.0	4.9

generally applicable to Mnz or Xnt dating (Table 4), as it remains in the range of the detection limit for most REE elements (Fig. 3). To improve accuracy, the shared background method [14] or the use of a "blank correction" [44] using a secondary reference material of known composition is recommended.

4.2.3. Peak interference corrections. The complex composition and the presence of heavy elements in Mnz and Xnt lead to the production of many potential peak interferences. Of major interest for U-Th-Pb_{total} analysis is the correction for interferences of Y-$L\gamma_{2,3}$, Th-Mζ, and La-Lα (n = 2) on Pb-Mα, or Ce-Lα (n = 2) on Pb-Mβ, and Th-Mγ and K-Kα on U-Mβ (Tables 4, 7). Some interferences are very small, such as the La-Lα (n = 2) on Pb-Mα (~0.04 cps/nA on $LaPO_4$; Table 7), yet, they are essential because the Pb-content is low and the La-content in Mnz is high. An extensive discussion on these correction procedures, notably the problem of Y-$L\gamma_{2,3}$ peak shifting between YPO_4 and Mnz, is found in [16]. In order to improve the accuracy of some crucial peak interference corrections such as those applicable to U-Mβ and Pb-Mα, the multipoint background acquisition method is used (see [14]).

The REEs also mutually interfere, which complicates the analysis of REE minerals in general. Fortunately, an improved correction routine that properly corrects for mutual interferences is used in this study [45]. A summary of all possible interferences and the count rates on primary and interference standards is listed in Tables 4 and 7, along with an approximation of the percentage correction for each interference:

$$\%_{interference} \approx \frac{I^A_{interf}/C^B_{interf}}{I^A_{std}/C^A_{std}} * \frac{C^B_{unk}}{C^A_{unk}} \tag{7}$$

where I^A is the measured intensity at the spectrometer position of the measured element A, C^A and C^B are the concentration in weight% of element A and B, and *std*, *interf*, and *unk* stands respectively for the primary standard for element A, the interference standard, and the unknown. This calculation does not consider the matrix effect, which is variable depending on the exact composition of the unknown, yet it is a first-order approximation for the strength of a possible interference. Based on these approximations, interferences can be classified as important, minor (up to ~2 % correction), or too small to be significant (<0.5 % correction; Table 7).

The choice of the optimal X-ray lines depends on whether Mnz or Xnt is being analysed. For example, Pb-Mα is recommended for Mnz as it provides the highest count rate and thus the lowest detection limit. However, it should be avoided and replaced by the measurement of Pb Mβ in Xnt due to the strong interference from Y-$L\gamma_{2,3}$. Pb-Mβ is only interfered by Ce-Lα (n=2), but this can be ignored in the case of Xnt as the Ce-content is always very low. Two other problematic X-ray lines are Gd-Lβ and Eu-Lβ. These lines are respectively interfered by Ho-Lα and Dy-Lα, which are minor elements in Mnz, and minor to major in Xnt. For Xnt analysis, Gd-Lα and Eu-Lα are preferred as they are only interfered by elements present in trace amounts in Xnt (Nd-$L\beta_{2,15}$ and La-$L\gamma_{2,3}$, and by Pr-$L\beta_{2,15}$, Ce-$L\gamma_5$ and La-Lγ, respectively). In any case, accurate results can be obtained providing that all major interferences are properly corrected.

Table 7. Summary of peak interference corrections, count rates, and estimate of interference correction. Measurement done on J-type (TAP) or L-type (PET-L, LiF-L) spectrometers on a JEOL-8230.

Measured element				Monazite*		Xenotime*		Interfering element			Approx. % interf. corr.			
Element	Xtal	Standard	cps/nA	Avg wt%	Max wt%	Avg wt%	Max wt%	Interference	Interf. Std.	cps/nA	Mnz avg	Mnz max	Xnt avg	Xnt max
Y Lα	TAP	YPO_4	336.6	0.85	4.92	35.44	41.89	Sr $L\beta_x$	$SrCO_3$	4.60	**0.2%**	**9.2%**	*No data for*	
Sr Lα	TAP	$SrSO_4$	362.9	0.10	7.04	?	?	Si Kβ + sat.	Orthoclase	2.98	**0.8%**	**6.0%**	*Sr in Xnt*	
Si Kα	TAP	Orthoclase	581.0	0.43	3.08	0.15	0.61	Nd Lα (III)	$NdPO_4$	0.62	**1.2%**	**3.3%**	*<0.1%*	*0.3%*
As Lα	TAP	As-pyrite	201.3	0.10	25.56	0.15	0.15	Tb Mβ	$TbPO_4$	7.85	*<0.1%*	*0.4%*	**4.9%**	**8.4%**
								Dy Mα	$DyPO_4$	5.67	0.3%	2.1%	**27.5%**	**39.7%**
								Sm Mγ	$SmPO_4$	1.70	0.4%	2.1%	0.8%	2.9%
Th Mα	PET	ThO_2	114.2	5.56	30.71	0.34	7.42	*No significant interference*						
Ca Kα	PET	F-apatite	684.9	0.80	14.94	0.03	0.17	*No significant interference*						
S Kα	PET	As-pyrite	52.3	0.29	1.36	0.02	0.04	*No significant interference*						
U Mβ	PET	UO_2	252.0	0.60	13.79	0.56	5.13	K Kα	Orthoclase	7.40	*Variable (fluorescence)*			
								Th Mγ	ThO_2	2.10	**7.7%**	**42.5%**	**0.5%**	**11.1%**
								Sm $L\beta_{2,15}$ (II)	$SmPO_4$	0.17	0.3%	1.2%	*0.1%*	*0.3%*
K Kβ	PET	Orthoclase	18.5	Nominally 0 wt%, intensity depends on secondary fluorescence effect				U Mγ	UO_2	4.10	*Variable (fluorescence)*			
								Tm Lα (II)	$TmPO_4$	6.70	*Variable (fluorescence)*			
								Ca Kα	F-apatite	0.55	*Variable (fluorescence)*			
Pb Mα	PET	Pyromorphite	141.2	0.23	1.46	0.17	0.37	Y $L\gamma_{2,3}$	YPO_4	0.51	**2.1%**	**12.0%**	***Prefer Pb Mβ***	
								Th Mζ	ThO_2	0.22	**3.2%**	**17.8%**	**0.3%**	**6.0%**
								La Lα (II)	$LaPO_4$	0.04	**1.9%**	**3.9%**	*<0.1%*	*<0.1%*
La Lα	LiF	$LaPO_4$	62.2	12.11	24.86	0.07	0.14	Nd Ll	$NdPO_4$	0.18	0.2%	0.6%	0.8%	2.8%
Ce Lα	LiF	$CePO_4$	78.2	24.67	32.83	0.17	0.46	*No significant interference*						
Nd Lα	LiF	$NdPO_4$	98.9	9.43	26.00	0.20	0.69	Ce Lβ	$CePO_4$	0.45	1.2%	1.6%	0.4%	1.1%
								La $L\beta_6$	$LaPO_4$	0.36	0.5%	1.0%	*0.1%*	*0.3%*
Pr Lβ	LiF	$PrPO_4$	59.2	2.85	10.77	0.05	0.09	*No significant interference*						
Sm Lβ	LiF	$SmPO_4$	78.6	1.63	7.60	0.47	1.65	Tb Lα	$TbPO_4$	0.60	*<0.1%*	*0.2%*	1.0%	1.7%
Tb Lα	LiF	$TbPO_4$	144.7	0.12	0.34	0.61	1.06	Sm Lβ	$SmPO_4$	0.56	**1.2%**	**3.3%**	*0.1%*	*0.5%*
Eu Lβ	LiF	$EuPO_4$	82.8	0.35	1.00	0.21	0.73	Dy Lα	$DyPO_4$	17.90	**23.0%**	**141%**	***Prefer Eu Lα***	
Dy Lα	LiF	$DyPO_4$	184.1	0.39	2.36	4.81	6.95	*No significant interference*						
Er Lα	LiF	$ErPO_4$	199.8	0.15	1.12	2.15	5.77	Tb Lβ	$TbPO_4$	12.30	**5.2%**	**14.6%**	**1.8%**	**3.1%**
Gd Lβ	LiF	$GdPO_4$	156.4	1.26	3.60	2.89	7.95	Ho Lα	$HoPO_4$	126.50	**6.1%**	**14.8%**	***Prefer Gd Lα***	
								Sm $L\beta_{2,x}$	$SmPO_4$	0.23	0.2%	0.9%	*<0.1%*	*0.1%*
Tm Lα	LiF	$TmPO_4$	207.6	0.14	0.42	1.90	4.55	Sm Lγ	$SmPO_4$	20.30	**116%**	**542%**	**2.5%**	**8.7%**
								Dy Lβ	$DyPO_4$	1.73	**2.3%**	**13.9%**	**2.1%**	**3.0%**
								Gd $L\beta_{9,10}$	$GdPO_4$	0.65	**2.8%**	**8.1%**	0.5%	1.3%
Yb Lα	LiF	$YbPO_4$	209.9	0.08	0.39	1.91	10.01	Tb $L\beta_{2,15}$	$TbPO_4$	0.95	0.7%	2.0%	*0.2%*	*0.3%*
								Dy $L\beta_{3,6}$	$DyPO_4$	0.88	**2.1%**	**13.2%**	1.1%	1.6%
								Sm $L\gamma_2$	$SmPO_4$	0.50	**5.3%**	**24.7%**	*<0.1%*	*0.2%*
Ho Lβ	LiF	$HoPO_4$	114.0	0.10	0.24	1.90	7.10	Yb Lα,n	$YbPO_4$	0.41	0.3%	1.4%	0.4%	1.9%
								Dy $L\beta_{2,15}$	$DyPO_4$	0.24	0.3%	2.5%	0.2%	0.6%
P Kα	PET	$CePO_4$	95.4	12.42	14.23	14.98	16.47	Y Lβ	YAG	0.26	*<0.1%*	*<0.1%*	*0.2%*	*0.2%*

For xenotime only

Pb Mβ	PET	Pyromorphite	*Not determined*	Ce Lα (II)	*Not determined (insignificant in Xnt)*
Eu Lβ	LiF	EuPO4	*Not determined*	Pr $L\beta_{2,15}$, Ce $L\gamma_5$, La Lγ	*Not determined (insignificant in Xnt)*
Gd Lβ	LiF	GdPO4	*Not determined*	Nd $L\beta_{2,15}$, La $L\gamma_{2,3}$	*Not determined (insignificant in Xnt)*

Bold: important correction. Normal: minor correction. Italic red: very minor correction less than 0.5 % that can be ignored.

* Various Mnz and Xnt compositions obtained by the authors and from various sources in the literature.

4.3.4. Beam damage. The use of a focussed beam with high current and long counting time in REE-phosphate likely leads to beam damage effects and inaccurate results. In the case of Mnz or Xnt, beam damage leads to the loss of P within the first couple minutes of analysis, and the apparent gain of other elements, especially REE and Th, and even Pb-Mα (Fig. 5a). The effect is strong with a 20 nm C-coating, and can be lowered by using a dual (Al+C)-coating (Fig. 5b). Beam damage appears to be less important in Xnt, most likely due to its slightly higher thermal conductivity compared to Mnz [46, 47]. Better coating such as Au or Ag can help further, yet they are not recommended as they produce additional X-ray lines and absorption edges that complicates the analysis near Pb-Mα (Au) and near U-Mβ and Th-Mα (Ag). Also, these heavy metals have much higher absorbance compared to Al or C.

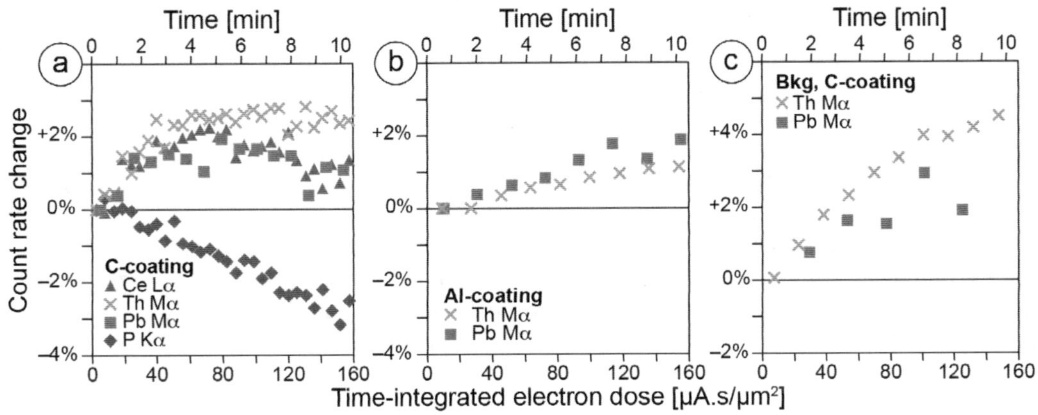

Figure 5. Count rate variation in Burnet Mnz as function of the time-integrated accumulated electron dose a) for Ce-Lα, Th-Mα, Pb-Mα, and P-Kα with 20 nm C-coating, b) for Th-Mα and Pb-Mα with a dual (Al+C)-coating, and c) on background for Th-Mα and Pb-Mα with C-coating. Analytical conditions: 15 keV, 200 nA, focussed beam.

Beam damage can be corrected on the first element analysed on each spectrometer using a time dependent intensity (TDI) or sub-counting correction by dividing the total counting time into several time intervals, and regressing these data to determine the intensity at time t = 0 s. For all other elements, it is desirable to minimize the beam damage effect, which can be done by ensuring their analysis is over before the beam damage is too strong. Two alternative solutions are either to analyse the major elements separately (e.g., at low current) and then only the trace elements at higher current, or to use the EDS to analyse P-Kα and the major REE (e.g., Ce, La, Nd, Pr, and Sm in Mnz). In this paper, a time dependent intensity correction is applied on U Mβ, Th-Mα, and Pb-Mα, and P-Kα is analysed by EDS, as this line is mostly free of any major interference in Mnz; P-Kα should still be analysed by WDS on Xnt due to the strong interference of Y-Lα on P-Kα. Work is in progress to evaluate the feasibility of major REE analysis by EDS (see below).

Another undesirable effect of beam damage is the increase in background intensity over time (e.g., Fig. 5c with C-coating). Although this effect is mostly undetectable when measuring a major or a minor element, it becomes important when measuring trace amounts. It is assumed that the densification of the matrix, an increase in space charge, and a change in the coating properties (degradation, contamination) ultimately result in a general loss of P and apparent gain in Th and REE, which is visible as a white spot surrounded by a dark rim in BSE after the analysis. This effect is clearly observed when running a blank test (see details in [14]), where the blank test results on Pb-Mα are significantly negative around -50 to -100 ppm when the background is only measured after the 10 min of counting on the peak. To counteract this effect, it is necessary to obtain a "fresh" background measurement, which is done by measuring only the background on the first point, and only the peak on the next analysis points measured right after the background analysis; the background measurement obtained on the first point is then applied on the next peak-only measurement points.

4.3.5. Matrix correction. For all analyses, the data processing and matrix correction routine was applied using the software Probe for EPMA with the most recent mass absorption coefficient table from

NIST [48]. The Phi-Rho-Z matrix correction of Armstrong [49] was used with data collect on a JEOL instrument, and the Pouchou and Pichoir [50] matrix correction was employed with data from a CAMECA instrument. Evaluation of the choice of matrix correction routine is beyond the scope of this paper. Despite the use of different matrix corrections on different instruments, the resulting U-Th-Pb_{total} age remains identical within the error, as demonstrated with several age consistency materials (Moacyr, Burnet, PR-88, RT-87, etc.). A complete evaluation of the effect of commonly applied matrix corrections on the absolute measurement of U, Th, Pb and other major elements in Mnz or Xnt remains to be done.

5. Applications

A study of an Archean aluminous quartzite from South-West Montana [51] is considered to illustrate the use of the monazite U-Th-Pb_{total} analysis in a simple case with significant Pb-content. The samples are sillimanite+garnet bearing quartzite taken a few kilometres apart. Three samples were selected for a full thin section map in order to identify all monazite. In each sample, 5 to 20 monazite grains (depending on availability) were mapped using the quantitative analysis method described above. Additional examples from the Upper Granite Gorge of the Grand Canyon [52] and from low-Pb monazite dating in low grade metasediments from upstate New York [15] are also considered.

5.1. Results

After petrological examination, samples were coated with 20 nm Al-coating and a 7 - 10 nm overlay of C-coating. The WDS full thin section element maps revealed all accessory phases, especially Mnz (e.g., Fig. 6a). Seven Mnz grains were identified in sample AA09-34B, five of which were then mapped to highlight compositional variation of the key elements Y, Si, Ca, Th and U (Figs. 6b to 6f). These element maps were quantified with standards using the programme "CALCIMAGE" from Probe Software, Inc. For accurate quantification, all elements should be mapped, but this task is unrealistic, as it would require obtaining maps for all other major and minor elements, notably REEs and P, or to know the Mnz composition of each homogeneous domain. Instead, the composition of the missing elements can be constrained with some assumptions. With the exception of Si-rich Mnz, the phosphorous content can be assumed to be constant around 28 - 32 and 34 - 38 wt% P_2O_5 for Mnz and Xnt, respectively, and the remaining non-analysed elements are chiefly REEs. In order to test the accuracy of such a quantitative mapping, five distinct fixed compositions were tested (Fig. 7): a) no specified element (only the analysed elements are considered in the calculation), b) a Gd-rich, or c) a Ce-rich Mnz (30 wt% P_2O_5; Gd_2O_3 or Ce_2O_3 by difference), d) a Ce-rich Mnz with a lower P_2O_5 content (25 wt%), or e) an average monazite composition (30 wt% P_2O_5; 13 wt% La_2O_3; 3 wt% Pr_2O_3; 11 wt% Nd_2O_3; 2.5 wt% Sm_2O_3; 2 wt% Gd_2O_3; Ce_2O_3 by difference). All maps were corrected using the MAN background correction. Results obtained in a strongly zoned monazite were identical within the error, except when the non-analysed elements were not constrained (Fig. 7).

Quantitative analysis was obtained on compositionally homogeneous domains of the unknown and on the Mnz age consistency standards Moacyr and Burnet to obtain U-Th-Pb_{total} ages (Table 8). L-type spectrometers were used for all elements except Y, Si, Sr and As (Table 4). The measurement of U, Th and Pb was performed using the multipoint background correction for improved accuracy on the background measurement, and up to 8 background positions were measured for 50 s each (Table 6). The TDI correction was applied on U, Th and Pb measurements to correct for the beam damage effect. The background was measured only on the first analysis point, followed by 4 to 8 points with a peak-only measurement. Three distinct analytical methods were evaluated to measure the REEs (Table 8): a) a shared background, or b) a MAN background correction on WDS measurement, and c) an analysis of LREE on EDS combined with a MAN background corrected WDS analysis of the other REE (combined EDS-WDS analysis). Using the element maps, several representative Mnz domains were targeted in sample AA09-34B. Lead-content in all these domains is relatively high (> 0.5 wt%), as the actinide-content is high and the age is old ($\geq$ 1.7 Ga). Whereas the cores commonly show a high

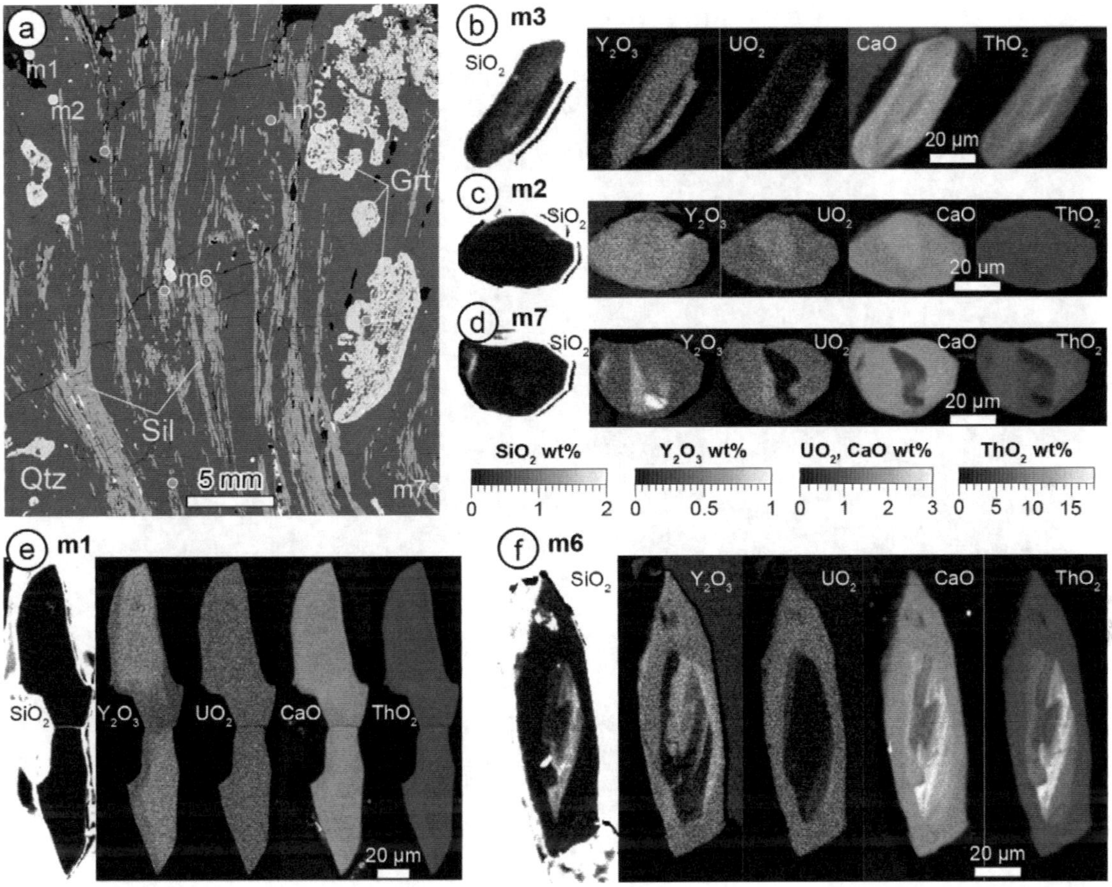

Figure 6. a) BSE image from a full thin section WDS map in sample AA09-34B with the identification of Mnz grains (yellow dots) and zircon (blue circles). b-f) Quantitative element maps of SiO_2, Y_2O_3, UO_2, CaO and ThO_2 in the Mnz identified in a).

Th-content and low U-content, the rims are typically enriched in U. A mixture of cheralite and huttonite/thorite substitutions explains the substitution of Th and U in these monazite domains, with a clear predominance of the cheralite substitution as suggested by the higher Ca content compared to Si. Work is still in progress, and several samples from this area remain to be dated. Yet, results obtained in this sample yield two distinct clusters of ages, with the Mnz cores yielding an Archean age around 2.4 Ga, and the rims yielding a Paleoproterozoic age around 1.74 Ga (Table 8; [53]).

5.2. Discussion
Tests on the quantitative element maps (Fig. 7) suggest that an average Mnz composition is sufficient to obtain accurate measurements in these low precision element maps, with minor differences in matrix correction depending on the fixed value attributed to the non-analysed elements. However, it remains necessary to constrain the composition of the non-analysed elements, especially the REEs, for instance with an average monazite composition (e.g., 30 wt% P_2O_5, some LREE, and Ce_2O_3 by difference). The quantified maps are extremely valuable for one-to-one comparison of the various compositional domains within a single sample, and even between various samples. For instance, the Y-rich cores of Mnz grains m6 and m7 (compare Figs. 6d and 6f) are similar and likely to be representative of the same event, whereas the composition of grains m1, m2 (Figs. 6c and 6e) and the rims of grains m6 and m7 is

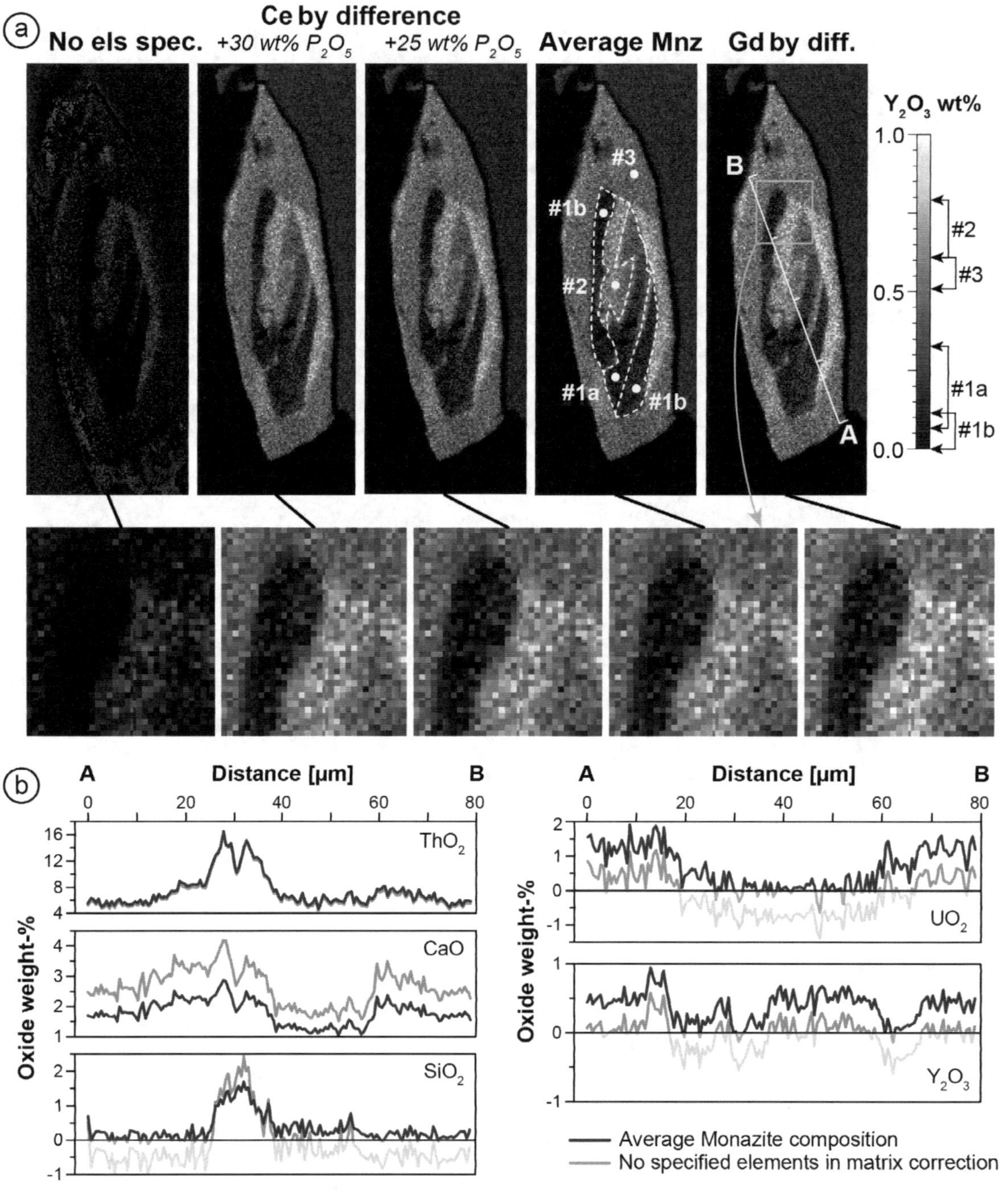

Figure 7. a) Comparison of quantitative element maps of Y_2O_3 assuming different specified composition for the non-analysed elements (see text for detail). The variation of composition in each domain is highlighted by arrows on the colour scale. b) Cross-section A-B for the calculated composition at the various fixed compositions.

Table 8. Results of quantitative analysis in sample AA09-34B and the Moacyr and Burnet age reference standards. Age in Ma (million year). Values below detection limit or negative are kept for comparison.

		Burnet	**Moacyr**	**AA09-34B**				
		Age ref. standard	*Age ref. standard*	**m6 Core #1a**	**m7 Core**	**m6 Rim #3**	**m7 Rim**	**m2 All grain**
	Age	**1085** ±14	**505** ±10	**2421** ±20	**2478** ±90	**1717** ±4	**1730** ±14	**1757** ±26
P_2O_5	EDS	27.45	28.35	28.22	30.18	30.01	29.56	29.83
SiO_2	MAN	1.38	1.07	1.28	0.11	0.18	0.36	0.09
CaO	MAN	0.85	0.46	2.43	0.86	1.74	1.89	1.84
Y_2O_3	MAN	0.50	1.22	0.11	0.94	0.38	0.26	0.45
La_2O_3	Shared	9.27	14.56	13.61	16.67	15.54	15.58	15.49
La_2O_3	MAN	9.26	14.54	13.57	16.64	15.51	15.55	15.45
La_2O_3	EDS	9.6	14.9	14.0	17.3	16.1	16.0	16.0
Ce_2O_3	Shared	29.38	29.05	23.55	29.46	26.84	26.67	26.65
Ce_2O_3	MAN	29.34	29.02	23.47	29.40	26.77	26.60	26.56
Ce_2O_3	EDS	29.7	29.1	24.2	29.3	27.5	26.5	26.8
Pr_2O_3	Shared	3.54	3.23	2.37	2.94	2.67	2.60	2.65
Pr_2O_3	MAN	3.57	3.29	2.37	2.98	2.67	2.59	2.65
Pr_2O_3	EDS	3.2	3.3	2.9	2.9	3.6	3.0	2.7
Nd_2O_3	Shared	9.44	11.14	8.19	10.66	9.40	8.99	9.35
Nd_2O_3	MAN	9.44	11.14	8.16	10.65	9.37	8.96	9.33
Nd_2O_3	EDS	9.8	10.6	8.2	10.8	9.6	9.2	9.7
Sm_2O_3	Shared	3.30	1.87	1.09	1.84	1.46	1.33	1.56
Sm_2O_3	MAN	3.33	1.90	1.14	1.88	1.56	1.39	1.61
Sm_2O_3	EDS	3.9	2.2	0.9	2.0	0.8	1.2	1.5
Eu_2O_3	Shared	-0.09	-0.03	0.09	0.18	0.24	0.22	0.23
Eu_2O_3	MAN	-0.08	0.00	0.14	0.23	0.36	0.32	0.28
Gd_2O_3	Shared	1.40	0.85	0.68	1.36	1.06	0.93	1.20
Gd_2O_3	MAN	1.48	0.85	0.66	1.41	1.02	0.85	1.18
Tb_2O_3	Shared	0.26	0.08	0.02	0.14	0.02	0.03	0.09
Tb_2O_3	MAN	0.26	0.10	0.04	0.16	0.08	0.07	0.11
Dy_2O_3	Shared	0.94	0.24	0.12	0.44	0.18	0.16	0.24
Dy_2O_3	MAN	0.93	0.23	0.09	0.43	0.17	0.13	0.23
Ho_2O_3	Shared	0.09	0.03	0.04	0.07	0.01	0.00	0.02
Ho_2O_3	MAN	0.01	-0.01	-0.07	0.00	-0.06	-0.08	-0.04
Er_2O_3	Shared	0.13	0.05	0.01	0.06	0.03	0.02	0.02
Er_2O_3	MAN	0.09	0.03	-0.04	0.03	-0.01	-0.03	-0.02
Tm_2O_3	Shared	0.01	0.00	0.01	0.01	0.00	0.00	-0.01
Tm_2O_3	MAN	0.00	0.01	-0.03	0.01	-0.01	-0.02	-0.01
Yb_2O_3	Shared	0.03	0.02	0.01	0.00	-0.01	-0.01	-0.01
Yb_2O_3	MAN	0.01	0.01	-0.05	-0.01	-0.03	-0.04	-0.03
ThO_2	MPB	9.93	6.59	15.16	2.49	6.15	7.05	5.93
UO_2	MPB	0.29	0.12	0.32	0.86	1.88	1.71	2.07
PbO	MPB	0.51	0.15	1.78	0.66	0.98	1.01	1.04

similar, as confirmed by the quantitative point analysis (Table 8). The analyst can thus select some unique compositional domains to be analysed, rather than randomly measuring domains and grains. Overall, this process increases the chance of measuring all age domains in a sample and increases the efficiency in that the number of repeated analyses of a particular compositional domain can be limited to the number needed for a statistically significant weighted mean.

Results of quantitative analysis of REE by WDS are identical within the error for all major and minor elements, independent of the background correction method. A slightly better accuracy is suggested from the shared background method for elements close to their respective detection limit, notably Yb and Ho, as suggested by the minor amount of analysis with slightly negative results compared to the results of the MAN background correction (Table 8). It should be emphasized that the MAN background correction method is suitable mostly for elements above 200 - 500 ppm, and if a better accuracy is required, the use of the shared background [14] or the MAN background correction with a "blank" correction [44] is recommended. The blank correction can be done using a standard with a precisely known trace amount of the element to be corrected, providing that this standard has a composition close to the analysed material.

Results for the major and some minor REEs by EDS are satisfactory, although the calculated compositions have a lower precision. To further evaluate this, analysis of several Mnz reference materials was done on a JEOL JSM-6390 SEM equipped with a 30 mm^2 Thermo UltraDry SDD EDS detector (Table 9) using the standard intensities (same standard as for WDS analysis) and the recommended software defaults for peak shape [54]. A one-minute analysis at 25 % deadtime (beam current ~ 2.5 nA) on this instrument yielded a counting statistics error of 4 - 7 % error for elements above 10 wt%, 10 - 15 % error at 5 - 10 wt%, and 15 - 25 % error at up to 2 wt%. Results are within the error identical with the WDS analysis. EDS analysis might thus be recommended for low precision analysis of major REE (e.g., La, Ce, Pr, Nd, Sm), and WDS should be preferred for REE below 2 wt% or for any REE for which a higher precision is desired. Unfortunately, it is currently impossible to ascertain the accuracy of the EDS versus the WDS analysis due to the lack of independent measurements. Although work is still in progress for the combined analysis of EDS and WDS, it appears feasible to measure several major REE elements, along with phosphorous by EDS, which would ultimately offer a longer analysis time on WDS to measure the minor and trace REE.

Table 9. EDS analysis of several Mnz reference materials (60 s counting).

	Burnet			Moacyr			Bananeira			Diamantina			RT-87		PR-88 hi-Th		PR-88 low-Th		
	WDS	EDS	EDS	WDS	EDS	EDS	WDS	EDS	EDS	WDS	EDS	EDS	WDS	EDS	WDS	EDS	WDS	EDS	EDS
Age	1115	-	-	508	-	-	518	-	-	694	-	-	445	-	327	-	323	-	-
P_2O_5	28.18	27.1	27.0	28.38	27.8	27.4	29.17	28.9	29.0	30.39	29.8	29.6	29.41	28.3	27.97	28.3	28.96	29.4	28.2
Y_2O_3	0.90	0.8	1.0	1.43	1.2	1.9	1.53	1.4	1.7	1.08	0.9	1.4	2.50	3.9	2.48	2.7	3.62	2.9	5.2
CaO	1.36	1.2	1.2	0.53	0.5	0.5	1.03	0.9	0.9	0.15	0.1	0.2	1.28	1.6	1.15	1.1	1.12	0.9	1.0
SiO_2	0.98	1.7	1.4	1.18	1.3	1.1	0.60	0.6	0.2	0.42	0.0	0.0	0.21	0.4	0.53	0.9	0.25	0.6	0.6
La_2O_3	8.45	8.4	8.3	14.65	14.1	14.5	11.08	10.9	11.2	15.29	15.2	15.3	12.99	12.9	11.51	11.4	11.41	12.0	11.8
Ce_2O_3	27.96	27.1	27.5	30.20	28.8	29.2	27.10	26.3	27.2	32.71	32.0	32.1	28.32	27.4	25.62	24.7	25.66	25.9	25.6
Nd_2O_3	9.34	9.0	7.7	11.69	11.0	9.7	11.79	11.0	10.1	12.82	12.2	10.4	12.02	9.9	10.93	10.6	11.21	10.8	10.0
Pr_2O_3	3.45	3.4	3.6	3.27	3.2	3.3	3.39	3.2	3.7	3.43	3.7	3.3	3.18	3.0	2.91	2.9	2.96	2.8	3.3
Sm_2O_3	4.03	3.9	4.0	1.95	1.8	1.7	5.11	4.9	4.5	2.48	2.0	2.3	2.01	1.7	2.72	2.7	2.97	2.9	2.6
Gd_2O_3	2.10	1.6	2.2	0.95	0.7	0.7	3.02	2.4	3.2	2.06	1.4	2.0	1.51	1.4	2.05	1.6	2.45	1.9	2.5
Dy_2O_3	1.50	0.6	1.7	0.31	0.1	0.3	0.72	0.3	0.8	0.69	0.3	0.9	0.79	0.9	0.93	0.4	1.23	0.4	1.4
ThO_2	11.50	12.2	12.0	6.94	7.3	7.2	6.10	6.3	6.1	0.19	0.2	0.2	5.03	5.3	8.25	8.4	5.40	5.3	5.9
UO_2	0.35	0.4	0.1	0.21	0.2	0.0	0.43	0.4	0.2	0.06	0.0	0.0	0.31	0.1	1.38	1.5	1.54	1.3	1.3
PbO	0.61	0.4	0.5	0.16	0.2	0.2	0.17	0.1	0.1	0.01	0.1	0.0	0.11	0.1	0.18	0.2	0.14	0.1	0.1
Total	100.71	97.9	98.3	101.86	98.3	97.7	101.23	97.8	99.0	101.76	97.9	97.6	99.67	97.0	98.59	97.4	98.92	97.0	99.6

5.3. Interpretation

Once a date has been accurately calculated, it should be put into a context in order to obtain an age of some specific event (i.e., "petrochronology" when linked to a petrological or a tectonic aspect; [37]). The geological context is first considered in order to compare the calculated dates with other ages obtained in the surrounding areas with different methods and minerals. For instance, based on previous works in the Northern Madison Range [51], the set of dates obtained in sample AA09-34B can be interpreted as a maximum deposition age for the quartzite defined by the youngest Archean monazite (~2.42 Ga) and a major metamorphic event affecting the entire area known as the Big Sky orogeny. The latter is dated at 1.77 to 1.72 Ga [51], and is responsible for the observed granulite facies metamorphism. Results further suggest that the Northern Madison Range in Montana represents a continuous section of the Big Sky orogenic crust [50].

The full thin section map and the individual monazite grains can help to link textural features with the growth of Mnz, for instance by looking at the preferential growth of Mnz in a pressure shadow in a strongly deformed rock [12, 55]. It is also interesting to compare the compositional variation of Mnz grains included in a phase (e.g., garnet) versus those present in the matrix (e.g., sample from the Grand Canyon; Figs. 8a and 8b). Dates obtained in Mnz included in a mineral (e.g., 1.71 ± 0.01 Ga for Mnz in garnet; Fig. 8a) can only pre-date the host mineral formation and must be viewed as an inherited age or possibly a prograde metamorphic event. The latter is preferred in this case, as the deposition age of the metasediment is older (≥ 1.74 Ga). By similarity of composition and age, some core domains in Mnz from the matrix can also be linked to this prograde event (e.g., core of m1; Fig. 8a). To further help with the geo-compositional comparison, Mnz or Xnt compositions are commonly normalised to the chondrite composition, which reflects the average composition of the Earth. Unfortunately, this normalisation usually leads to relatively similar Mnz (or Xnt) profiles due to large difference between chondrite composition (ppm range) and Mnz or Xnt (wt% to ppm range). A more sensitive way to compare the data is to normalise the Mnz or Xnt composition to a "reference" Mnz or Xnt composition, either taken from the sample itself or from one of the age consistency standards (e.g., compared to Moacyr, Fig. 8b). This normalisation will help to identify subtle enrichment or depletion in a specific REE (see also Fig. 5 in [15]). Other geo-compositional features can also be key in the age interpretation. For instance, it is well recognized that if a Mnz domain grows during the resorption of garnet, an aluminous silicate that preferentially incorporates HREE over LREE, the Mnz will show a higher content in HREE, or it can trigger the growth of Xnt (if thermodynamically stable; e.g. Fig. 8d). In several samples from the Grand Canyon, some Mnz rims and most Xnt yield a significantly younger date around 1.64 Ga (Fig. 8c). This consistency between samples and the fact that Xnt growth is associated with the resorption of garnet (Fig. 8d) strongly suggest that these ages are linked to a late re-equilibration event after the peak of metamorphism, which is dated by the majority of other Mnz around 1.70 - 1.68 Ga. Finally, a few Mnz overgrowths of distinct composition yielded an ill-defined date around 1.4 - 1.5 Ga, which could be interpreted as either an overgrowth or a (partial?) resetting of the age during a dissolution-reprecipitation event involving fluids. Ages obtained in multiple samples from a specific area are then compared, which helps to reinforce the interpretation of specific events, such as those depicted in Fig. 8c [52]. For instance, dating multiple samples was necessary in a study of low-grade Mnz from the Potsdam sandstone (see detail in [15]). In this study, actinide poor Mnz and Xnt domains with very low Pb-content (≤ 400 ppm) yielded a range of imprecise dates between ca. 500 and 200 Ma, and their interpretation in a single sample was impossible. However, when multiple samples were considered, a pattern of dates emerged and ultimately permitted the dates to be interpreted to represent fluid-events that correlate with the main orogenic events affecting the northern Appalachians between 480 and 250 Ma (see Fig. 8 in [15]).

6. Conclusion

The dating of Mnz or Xnt by electron microprobe is a very powerful technique. The microprobe can provide U-Th-Pb_{total} dates at the micron-scale, along with element maps and a complete geo-compositional characterisation of the major and minor elements that compose the Mnz or Xnt. The main limitation of the technique arises from the detectability of trace elements, especially Pb, yet, actinide-poor samples as young as 200 Ma can still be accurately dated with reasonable precision (± 10 - 20 %; see Fig. 2 in [15]). The combination of full thin section compositional mapping to reveal the location of all accessory phases, of high-(spatial)-resolution element mapping of Mnz or Xnt, and of the complete geo-compositional characterization of REE phosphate strengthens the interpretation of the geological significance of calculated dates. If a sample cannot be dated by EMP due to the low Pb-content, element mapping and geo compositional analysis of Mnz or Xnt by EPMA remains critical for establishing an essential understanding of phase compositions and for characterising the compositional zoning and potential polygenetic nature of a grain prior to laser or ion beam analysis. In any case, the "age" of a mineral is nearly meaningless if not interpreted in the context of the geological, structural, and/or reaction history.

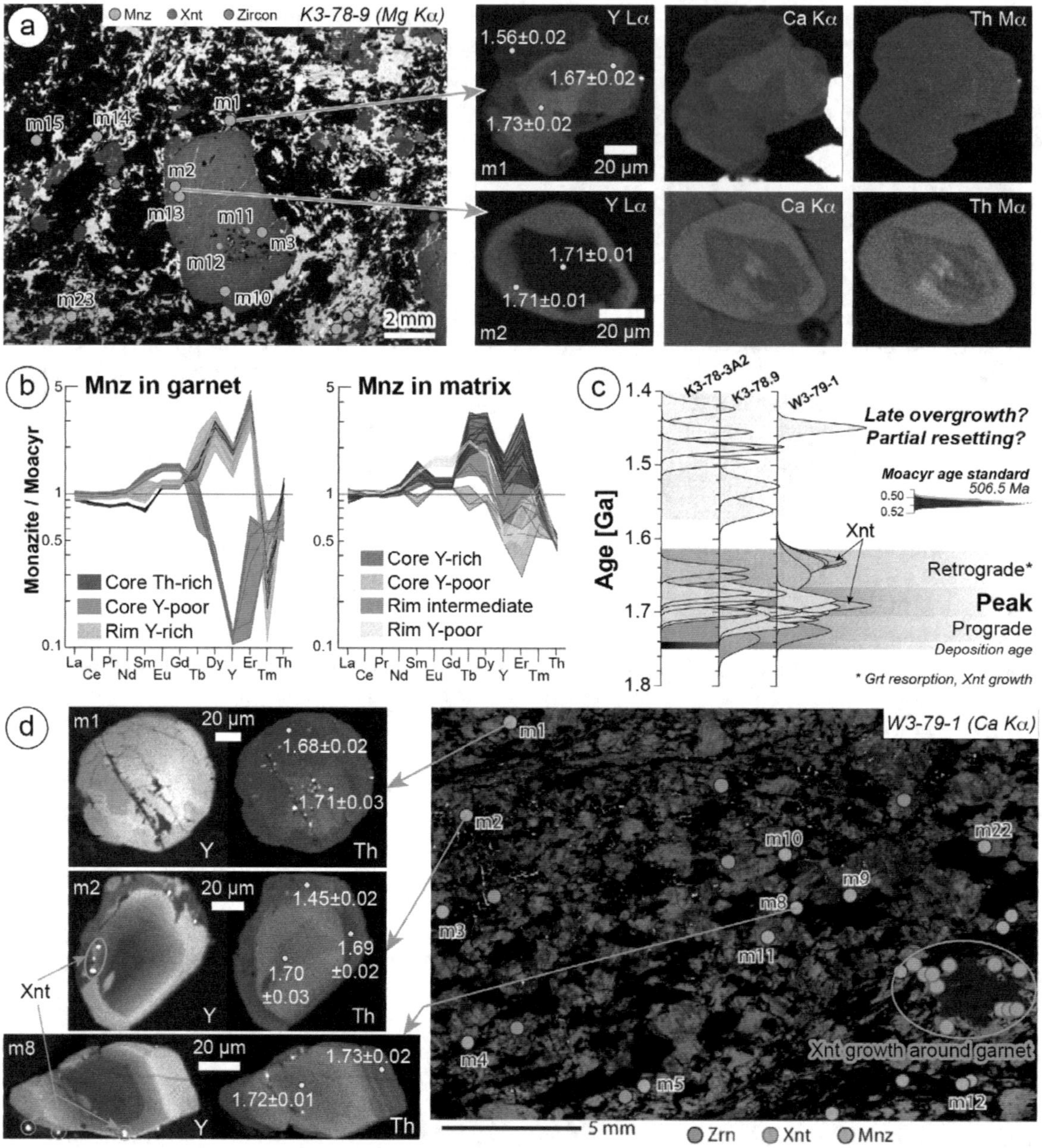

Figure 8. Example of full thin section and Mnz element maps in samples from the Upper Granite Gorge in the Grand Canyon. a) Sample K3-78-9 highlighting the difference in growth zonation of Mnz included in garnet versus Mnz in the matrix. b) Mnz composition (REE+Th) normalised to the Mnz age reference standard "Moacyr". c) Age results in three samples from the Grand Canyon revealing four distinct age groups: inherited, peak of metamorphism, late re-equilibration, and late thermal event. d) Sample W3-79-1 showing the late growth of Xnt associated to the resorption of garnet. Xnt analyses yield ages around 1.64 Ga and is interpreted as a late re-equilibration stage during retrogression.

IOP Conf. Series: Materials Science and Engineering **891** (2020) 012001 doi:10.1088/1757-899X/891/1/012001

Acknowledgements
JMA acknowledges funding from NSF EAR-1427626 and the University of Colorado Boulder for the purchase of the JEOL JXA-8230 microprobe. The authors thank K.H. Mahan for providing information and the authorisation to publish early data in samples from South-West Montana. Additional age reference materials (Diamantina, Bananeira Mnz) and REE-phosphate standards were gratefully provided by G.d.O. Gonçalves and D. Harlov, respectively. We greatly appreciate a careful, positive review by S.L. Kearns and editorial handling by B. Eske Sørensen.

References
[1] Schoene B 2013 U-Th-Pb Geochronology. in: *Treatise on Geochemistry: Second Edition.* **4** 341-378
[2] Parrish R R 1990 *Can. J. Earth Sci.* **27** 1431-1450
[3] Bowles J F W 1990 *Chem. Geol.* **83** 47-53
[4] Parslow G R, Brandstätter F, Kurat G and Thomas D J 1985 *Can. Mineralogist* **23** 543 551
[5] Montel J-M, Veschambre M and Nicollet C 1994 *Compt. Rendus Acad. Sci. Série II* **318** 1489-1495
[6] Montel J-M, Foret S, Veschambre M, Nicollet C and Provost A 1996 *Chem. Geol.* **131** 37-53
[7] Suzuki K, Adachi M and Tanaka T 1991 Sediment. Geol. **75** 141-147
[8] Montel J M, Kato T, Enami M, Cocherie A, Finger F, Williams M and Jercinovic M 2018 *Chem. Geol.* **484** 4-15
[9] Suzuki K and Adachi M 1991 *Geochem. J.* **25** 357-376
[10] Suzuki K and Kato T 2008 *Gondwana Res.* **14** 569-586
[11] Shazia J R, Harlov D E, Suzuki K, Kim S W, Girish-Kumar M, Hayasaka Y, Ishwar Kumar C, Windley B F and Sajeev K 2015 *Lithos* **236-237** 1-15
[12] Williams M L and Jercinovic M J 2002 *J. Struct. Geol.* **24** 1013-1028
[13] Dahl P S, Terry M P, Jercinovic M J, Williams M L, Hamilton M A, Foland K A, Clement S M and Friberg L V M 2005 *Amer. Mineralogist* **90** 1712-1728
[14] Allaz J M, Williams M L, Jercinovic M J, Goemann K and Donovan J 2019 *Microsc. Microanal.* **26** 1-17
[15] Allaz J, Selleck B, Williams M L and Jercinovic M J 2013 *Amer. Mineralogist* **98** 1106 1119
[16] Jercinovic M J and Williams M L 2005 *Amer. Mineralogist* **90** 526-546
[17] Cocherie A, Legendre O, Peucat J J and Kouamelan A N 1998 *Geochim. Cosmochim. Acta* **62** 2475-2497
[18] Scherrer N C, Engi M, Gnos E, Jakob V and Liechti A 2000 *Schweiz. Miner. Petrogr. Mitt.* **80** 93-105
[19] Konečný P, Kusiak M A and Dunkley D J 2018 *Chem. Geol.* **484** 22-35
[20] Overstreet W C 1967 *The geologic occurrence of monazite*. U.S. Geol. Surv. Prof. Pap. 530, 327 pp.
[21] Gramaccioli C M and Segalstad T V 1978 *Amer. Mineralogist* **63** 757-761
[22] Bowles J F W, Jobbins E A and Young B R 1980 *Mineral. Mag.* **43** 885-888
[23] Watt G R 1995 *Mineral. Mag.* **59** 735-743
[24] Van Emden B, Thornber M R, Graham J and Lincoln F J 1997 *Can. Mineralogist* **35** 95 104
[25] Seydoux-Guillaume A-M, Wirth R, Heinrich W and Montel J-M 2002 *Eur. J. Mineral.* **14** 869-878
[26] Bulakh A G, R. N A, Zaitsev A N, Pilipiuk A N, Wall F and Kirillov A S 2000 *Neues Jahrb. für Mineral. - Abhandlungen* **5** 217-233
[27] Krenn E, Putz H, Finger F and Paar W H 2011 *Mineral. Petrol.* **102** 51-62
[28] Chen W, Honghui H, Bai T and Jiang S 2017 *Resources* **6** 51
[29] Graeser S and Schwander H 1987 *Schweiz. Miner. Petrogr. Mitt.* **67** 103-113
[30] Ondrejka M, Uher P, Pršek J and Ozdín D 2007 *Lithos* **95** 116-129

[31] Jercinovic M J, Williams M L and Berman R G 2013 in: *AGU Fall Meet. V53B–2794*
[32] Steiger R H and Jäger E 1977 *Earth Planet. Sci. Lett.* **36** 359-362
[33] Suzuki K and Adachi M 1991 *J. Earth Sci. Nagoya Univ.* **38** 11-38
[34] Jercinovic M J, Williams M L, Allaz J and Donovan J J 2012 *IOP Conf. Ser. Mater. Sci. Engng.* **32** 1-22
[35] Fialin M, Rémy H, Richard C and Wagner C 1999 *Amer. Mineralogist* **84** 70-77
[36] Dumond G, McLean N, Williams M L, Jercinovic M J and Bowring S 2008 *Chem. Geol.* **254** 175-196
[37] Williams M L, Jercinovic M J, Mahan K H and Dumond G 2017 *Rev. Mineral. Geochem.* **83** 153-182
[38] Williams M L, Jercinovic M J, Goncalves P and Mahan K H 2006 *Chem. Geol.* **225** 1 15
[39] Goncalves P, Williams M L and Jercinovic M J 2005 *Amer. Mineralogist* **90** 578-585
[40] Gonçalves G O, Lana C, Scholz R, Buick I S, Gerdes A, Kamo S L, Corfu F, Marinho M M, Chaves A O, Valeriano C and Nalini H A 2016 *Chem. Geol.* **424** 30-50
[41] Jercinovic M J, Williams M L and Lane E D 2008 *Chem. Geol.* **254** 197-215
[42] Williams M L, Jercinovic M J and Terry M P 1999 *Geology* **27** 1023-1026
[43] Donovan J J and Tingle T N 1996 *J. Microsc. Microanal.* **2** 1-7
[44] Donovan J J, Singer J W and Armstrong J T 2016 Amer. Mineralogist **101** 1839-1853
[45] Donovan J J, Snyder D A and Rivers M L 1993 *Microbeam Anal.* **2** 23-28
[46] Du A, Wan C, Qu Z and Pan W 2009 J. Amer. Ceram. Soc. **92** 2687-2692
[47] Hikichi Y, Ota T, Daimon K and Hattori T 1998 *J. Amer. Ceram. Soc.* **81** 2216-2218
[48] Chantler C T, Olsen K, Dragoset R A, Chang J, Kishore A R, Kotochigova S A and Zucker D S 2005 *X-ray form factor, attenuation and scattering tables* (version 2.1). [Online] National Institute of Standards and Technology, Gaithersburg, MD http://physics.nist.gov/ffast
[49] Armstrong J T 1988 in: *Microbeam Analysis*. (Newbury D E; Ed.) (San Francisco, CA: San Francisco Press) 239-246
[50] Pouchou J L and Pichoir J F 1991 in: *Electron probe quantition*. (Heinrich K F J and Newbury D E; Eds.) (New York, NY: Plenum Press)
[51] Kellogg K S and Mogk D W 2009 *Rocky Mt. Geol.* 44 85-102
[52] Allaz J M, Williams M L and Jercinovic M J 2010 in: *AGU Fall Meet. V41D-2305*
[53] Flynn C, Mahan K H and Allaz J M 2018 in: *AGU Fall Meeting T42A-08*
[54] Seddio S M 2018 in: *M&M 2018 Pre-meeting Congr. X.61 10-1*
[55] Williams M L and Jercinovic M J 2012 *J. Metamorph. Geol.* **30** 739-752

Processing and indexing of electron backscatter patterns using open-source software

H W Ånes[1], J Hjelen[1], B E Sørensen[2], A T J van Helvoort[3] and K Marthinsen[1]

[1] Norwegian University of Science and Technology, Department of Materials Science and Engineering, 7491 Trondheim, Norway
[2] Norwegian University of Science and Technology, Department of Geoscience and Petroleum, 7491 Trondheim, Norway
[3] Norwegian University of Science and Technology, Department of Physics, 7491 Trondheim, Norway

E-mail: hakon.w.anes@ntnu.no

Abstract. A new method to increase the signal-to-noise ratio *S/N* of electron backscatter patterns (EBSPs) based upon principal component analysis (PCA) is presented. The PCA denoising method is applied to ten scans of EBSPs from the same region of interest of a recrystallised nickel sample acquired with a decreasing *S/N*, achieved by reducing the exposure time while increasing the camera gain accordingly. That PCA denoising increases *S/N* in EBSPs is demonstrated by comparing indexing success rates after both Hough and dictionary indexing (HI and DI) of the Ni patterns having undergone one of four processing routes: i) standard static and dynamic background corrections (standard corrections), ii) standard corrections and pattern averaging with the four closest neighbours, iii) standard corrections and PCA denoising, and iv) standard corrections and pattern averaging followed by PCA denoising. Both pattern averaging and PCA denoising increases the indexing success rates for both indexing approaches for the studied Ni scans, with the former processing route providing the better success rates. The best success rates are obtained after pattern averaging followed by PCA denoising. The potential of PCA denoising to reveal additional pattern details compared to standard corrections and pattern averaging is demonstrated in a pattern from an orthoclase ($KAlSi_3O_8$) grain in a geological sample. Software code, and the Ni data sets, are released alongside this article as part of KIKUCHIPY, an open-source software package dedicated to processing and analysis of EBSPs.

1. Introduction

Electron backscatter diffraction (EBSD) in the scanning electron microscope (SEM) is a technique to characterise crystallographic features [1]. Data collection and indexing is reliable and automated through commercial systems. However, in recent years, open-source software packages have emerged as alternatives that are free to use, reveal all parts of their algorithms, and can facilitate swifter implementation of new concepts. An illustrative example is contributions to the indexing of electron backscatter patterns (EBSPs), such as dictionary indexing (DI) [2] available in the EMSOFT package, and the ASTROEBSD package [3]. While the latter package and commercial systems in general rely on extraction of Kikuchi bands in patterns using Hough-transform based indexing (HI), the DI approach compares dynamically simulated patterns to the full experimental patterns. DI can therefore index patterns in which extraction of bands is difficult due to a lower signal-to-noise ratio *S/N* [4, 5], which can result from highly deformed or fine-grained materials, or high EBSD camera gain and short exposure times.

An area explored to a lesser extent in open-source software for EBSD analysis is to increase *S/N* before indexing, with pattern averaging within a kernel [5, 6] a recent example. Another possibility is to use multivariate statistical analysis (MSA), which aims to help extract the most useful information within the data. Previous applications of MSA to EBSPs have focussed on reducing the number of patterns to a set of significant component patterns [7-9], so-called dimensionality reduction, and then index only these patterns. However, one can also reconstruct the full set of EBSPs from the component patterns considered to contain useful signal while discarding the component patterns considered to contain mostly noise, effectively increasing *S/N*.

Here, we present the use of dimensionality reduction to increase *S/N* in EBSPs using principal component analysis (PCA) [10], so-called PCA denoising. PCA and other MSA methods are available via the open-source, multi-dimensional data analysis package HYPERSPY [11], which itself depends upon SCIKIT-LEARN [12]. The PCA denoising approach is outlined. Indexing results from HI and DI of patterns from ten EBSD scans from the same region of interest (ROI) of a recrystallised nickel (Ni) sample, acquired with a decreasing *S/N*, are compared. The patterns have undergone one of four processing routes: i) standard static and dynamic background corrections (standard corrections), ii) standard corrections and pattern averaging with the four closest neighbour patterns, as presented by Wright *et al.* [5], iii) standard corrections and PCA denoising, and iv) standard corrections and pattern averaging followed by PCA denoising. Also, details revealed in a pattern from a grain of orthoclase ($KAlSi_3O_8$) in a geological EBSD data set after either PCA denoising or pattern averaging are compared to details in a dynamically simulated pattern best matching the experimental pattern. The software code, and the Ni data sets, are released alongside this article as part of KIKUCHIPY [13], an open-source software package dedicated to processing and analysis of EBSPs.

2. Denoising of EBSPs with principal component analysis

Briefly, a map of EBSPs as shown in Fig. 1a is a set of n observations (here $n = n_x \times n_y$ patterns) of a number of s variables (here $s = s_x \times s_y$ detector pixel intensities). The goal of PCA is to decompose data in this format to a set of values of linearly uncorrelated, orthogonal variables called principal components (PCs), while retaining as much as possible of the variation in the data [10]. Before finding the PCs, the patterns are arranged into a matrix $\mathbf{X}$ of dimensions $(n \times s)$, as shown in Fig. 1b. Each pattern n_i in $\mathbf{X}$ is centred about their mean. A singular value decomposition (SVD) $\mathbf{X} = \mathbf{U\Sigma V}^T$ is performed where $\mathbf{U}$ and $\mathbf{V}$ are orthogonal matrices and $\mathbf{\Sigma}$ is a diagonal matrix, shown in Fig. 1c, with respective dimensions $(n \times p)$, $(p \times p)$, and $(s \times p)$, with p as the rank of $\mathbf{X}$. The PCs in $\mathbf{V}^T$ is in our case a set of component factors p sorted in order of decreasing variance, describing the underlying intensity distributions on the detector. The product $\mathbf{U\Sigma}$, the component loadings p, is then the variation of their strength from one observation point to the next.

Ideally, the first component corresponds to the crystallographic feature most prominent in the data, for example the largest grain, the next corresponds to the second largest feature, and so on, until the later components at some point contain only noise. If this is the case we can increase *S/N* in the patterns by reconstructing our data matrix $\mathbf{X}_{\text{new}} = \mathbf{U\Sigma V}^T$, keeping only the $k \leq p$ components considered to contain useful signal, as shown in Fig. 1d.

3. Methods

3.1. Materials

A recrystallised, polycrystalline sample of pure Ni was selected to demonstrate improved indexing results of patterns with a lower *S/N*. As the final preparation steps the sample was electropolished with an A2 electrolyte followed by plasma cleaning. To highlight the additional pattern details revealed by

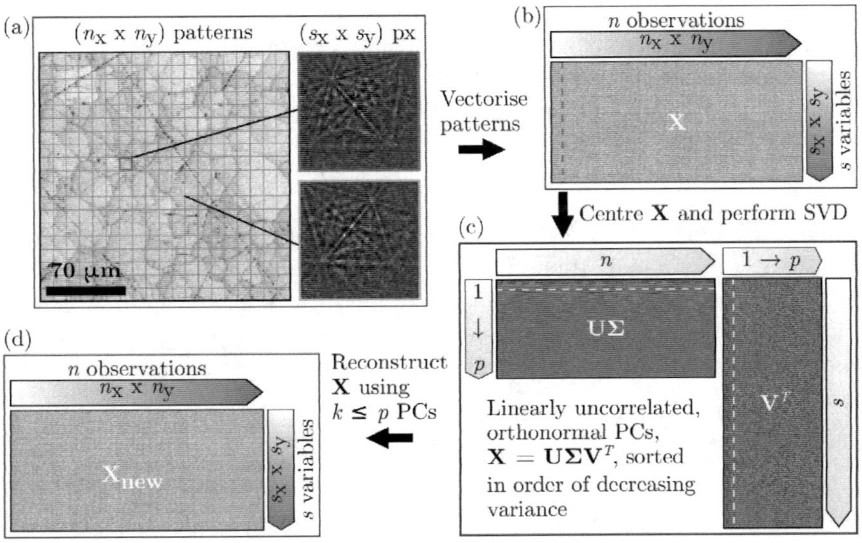

Figure 1. Processing steps to denoise (a, b) a data set of EBSPs **X** by first computing the (c) principal components p and then performing dimensionality reduction by reconstructing the data set $\mathbf{X}_{new}$ keeping only $k \times p$ components containing useful signal.

PCA denoising, a geological sample containing an orthoclase grain was studied. As the final preparation step this sample was polished with a suspension of 40 nm amorphous silica particles for 15 minutes and then rinsed in ethanol.

3.2. EBSD experiments

All EBSD data sets were obtained with a NORDIF UF-1100 camera on a Hitachi SU-6600 FEG SEM operated at 20 kV. The ten Ni data sets were collected consecutively from the same ROI with increasing camera gain and hence shorter exposure times listed in Table 1. 29,800 patterns with (60 × 60) pixels of 8 bit depth were collected for each scan in a (300 × 223.5) μm^2 square grid with a 1.5 μm step size. The geological data set comprise 94,248 patterns with (240 × 240) pixels of 8 bit depth collected in a (308 × 306) μm^2 square grid with a 5.0 μm step size and an 18.13 ms exposure time. Note that only patterns from a single grain in this data set are studied in this work.

Table 1. Scan number, camera gain, exposure time t and indices k of the principal components used in dimensionality reduction for the Ni data sets.

Scan	1	2	3	4	5	6	7	8	9	10
Gain [dB]	0	3	6	9	12	15	17	20	22	24
t [ms]	3.50	2.65	1.95	1.35	0.95	0.65	0.50	0.37	0.31	0.25
k	1-300	1-300	1-300	1-300	3-300	3-300	3-300	5-300	5,8-300	7-300

3.3. Pattern processing

All pattern processing was performed with KIKUCHIPY unless otherwise stated. The slowly varying diffuse background was removed from all patterns following standard background corrections. The correction procedure involves an initial subtraction of the raw pattern by a static background obtained by averaging multiple patterns collected over an area slightly larger than the ROI, followed by the subtraction by a dynamic (per pattern) background obtained by Gaussian blurring each pattern obtained after the first step.

Pattern averaging with the four closest neighbours was performed as indicated in Fig. 1 in Ref. [5]. Specifically, a (3 × 3) kernel, with all coefficients set to 1 except for 0 in the corners, was spatially correlated with the ($n_x \times n_y$) patterns, followed by normalisation of each pattern's intensity by dividing by the sum of kernel coefficients, which in this case was 5. Note that the implementation of neighbour pattern averaging in KIKUCHIPY allows for arbitrary 1D and 2D kernel sizes and coefficients, e.g., a Gaussian kernel.

The p first component patterns and loadings of each EBSD data set were found by performing PCA as explained above in section 2. After visual inspection, a number $k \leq p$ components, different for each data set, were chosen for dimensionality reduction. The components chosen for the Ni EBSD data are listed in Table 1, while up to 500 components were chosen for the geological EBSD data.

3.4. Indexing and indexing success rate

EDAX TSL 7.2 data collection was used for HI on the Ni patterns stretched to (80 × 80) pixels with a (9 × 9) convolution mask, a 2° θ step size, and searching for a maximum of 9 peaks, as was done in Ref. [5]. EMSOFT 4.2 was used for DI, with creation of dynamically simulated master patterns for both Ni [14] and orthoclase ($KAlSi_3O_8$) [15] with a 20 keV incident electron beam, a sample tilt of 70°, Monte Carlo energy bins from 10 - 20 kV with a 1 kV step, and to a maximum penetration depth of 100 nm with a 1 nm step. Simulated patterns were sampled from a uniform grid in orientation space with an average angular step of 1.4°, and the best matching orientation was further refined using a bound optimisation by quadratic approximation approach [16]. Pattern centres for both sets of EBSD data are given in Table 2. Note that the camera was binned from (480 × 480) px to (60 × 60) px and (240 × 240) px for the Ni and geological data sets, respectively. Before DI, both experimental and simulated patterns were processed in EMSOFT using a high-pass FFT filter with a maximum width of 0.125 and adaptive histogram equalisation with 4 and 5 regions for the Ni data and orthoclase pattern, respectively.

Table 2. Pattern centres in TSL (x^*, y^*, z^*) and EMSOFT 4.2 (x_{pc}, y_{pc}, L) conventions for the Ni and geological data sets.

Data	(x^*, y^*, z^*)	CCD pixel size [μm]	(x_{pc}, y_{pc})	L [μm]
Ni	(0.4210, 0.7794, 0.5049)	70.0	(-38.86, 132.65)	16 986.48
Orthoclase	(0.4033, 0.8739, 0.5115)	70.0	(-46.42, 179.47)	17 186.40

To assess the improvements to indexing results after PCA denoising and pattern averaging for the Ni data sets, each orientation in the scans was compared to orientations in the identical and adjacent points in a (3 × 3) kernel in scan 1, taken as the reference. The reference scans for the DI and HI data were the DI and HI results at zero camera gain and standard background corrections, respectively. If the orientation had a misorientation smaller than 5° to any of these points, it was considered to be successfully indexed. This calculation of the indexing success rate using a reference kernel (ISR^{RK}) [5], and the visualisation of orientation maps, were done with the open-source MATLAB texture toolbox MTEX [17] (note that a MATLAB license is required to run MTEX).

4. Results and discussion

EBSPs, acquired from the same ROI in Ni scan number 1, 6 and 9 after only standard background corrections (here called standard patterns) and after PCA denoising (here called denoised patterns) using the components indicated in Table 1, are shown in Fig. 2a. Also shown in the figure is a part of the orientation maps from scan 1 (left column), 6 (middle column) and 9 (right column) after Hough indexing of (b) standard patterns, (c) denoised patterns, (d) averaged patterns, and (e) averaged and denoised patterns, and (f) dictionary indexing of standard patterns, with their indexing success rates, ISR^{RK}, in the upper right corner. The signal-to-noise ratio in the original patterns decreases with increasing camera gain, however pattern averaging and PCA denoising retains some of the useful signal, as can be seen especially when comparing row (b) and (e) for scan 6 and 9. Among the scans and processing routes presented in Fig. 2, dictionary indexing of standard patterns provides the best results, except for HI of patterns in scan 9 after averaging and denoising, with respective ISR^{RK} of 69 % and 86.1 %.

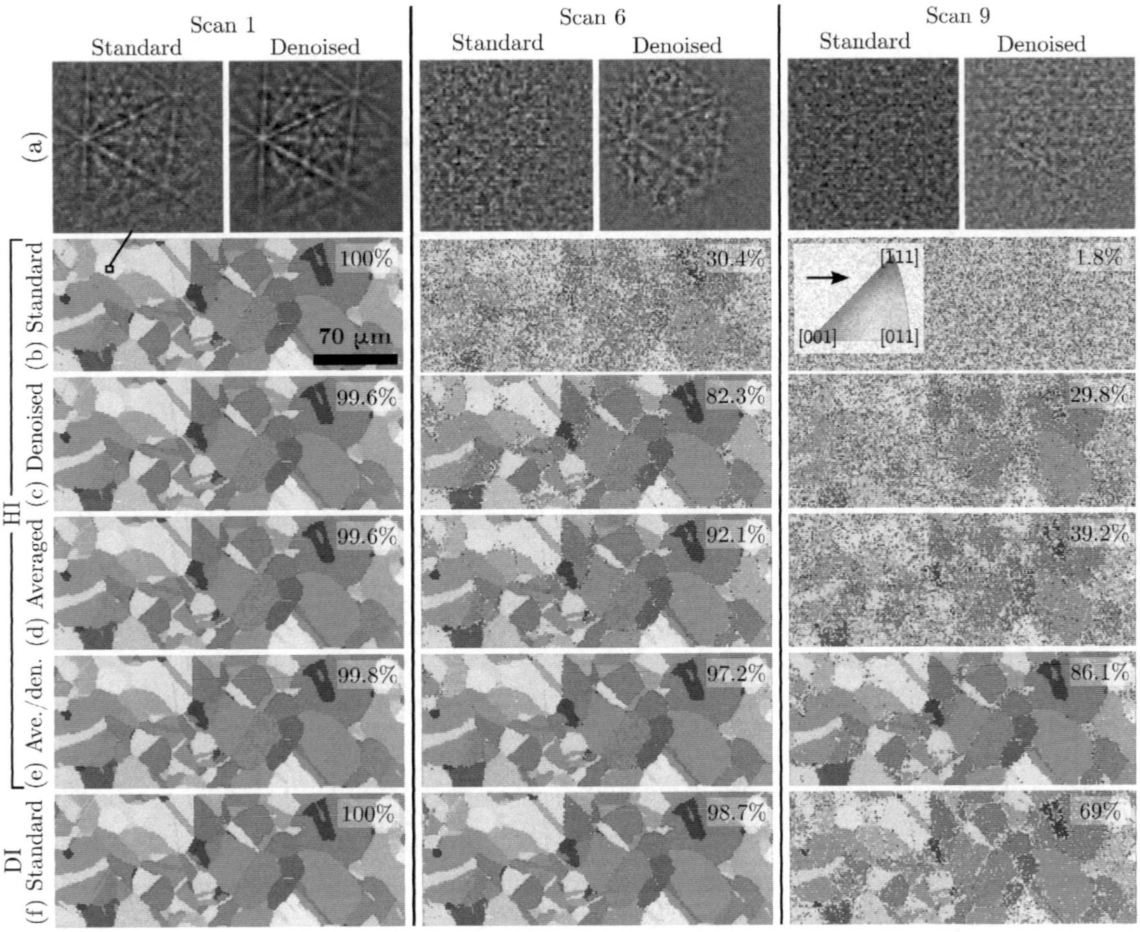

Figure 2. (a) Standard and denoised EBSPs from the same, indicated position in Ni scan number 1 (left), 6 (middle) and 9 (right) with orientation maps after HI of (b) standard patterns, (c) denoised patterns, (d) averaged patterns, and (e) averaged and denoised patterns, and (f) DI of standard patterns. The colour designates the crystal direction parallel to the right according to the colour key indicated. The ISR^{RK} is shown in the top right corner of each map.

The indexing success rates of all Ni scans after Hough and dictionary indexing of standard, averaged, denoised, and averaged and denoised patterns are shown in Fig. 3 as a function of camera gain. The DI ISR^{RK} of averaged and denoised patterns are not shown because a negligible increase in the success rates was observed compared to DI of averaged patterns. The HI and DI ISR^{RK} of standard patterns decreases from 100 % for scan 1 with zero camera gain, to 1 % and 46 % respectively for scan 10 with maximum gain. There are minor differences in the ISR^{RK} between the two indexing approaches until scan 5 where it drops to 65 % for HI, however, PCA denoising raises it to 94 %. PCA denoising also improves the DI ISR^{RK}, with the largest improvement, from 46 % to 61 %, for scan 10. Pattern averaging provides the better ISR^{RK} compared to PCA denoising, however, combining the two drastically improves the results obtained from HI. For scan 10, the HI ISR^{RK} jumps from a success rate of 17.5 % for averaged patterns to a success rate of 61.3 % for averaged patterns followed by PCA denoising. These results also show that HI might be sufficient in cases where patterns have a sufficient *S/N*. Furthermore, DI is highly robust towards noise, and should therefore be used when indexing patterns with a low *S/N* resulting in insufficient extraction of Kikuchi bands, in line with previous studies [4, 5]. This is best highlighted for scan 10, where almost all averaged patterns were successfully indexed with DI.

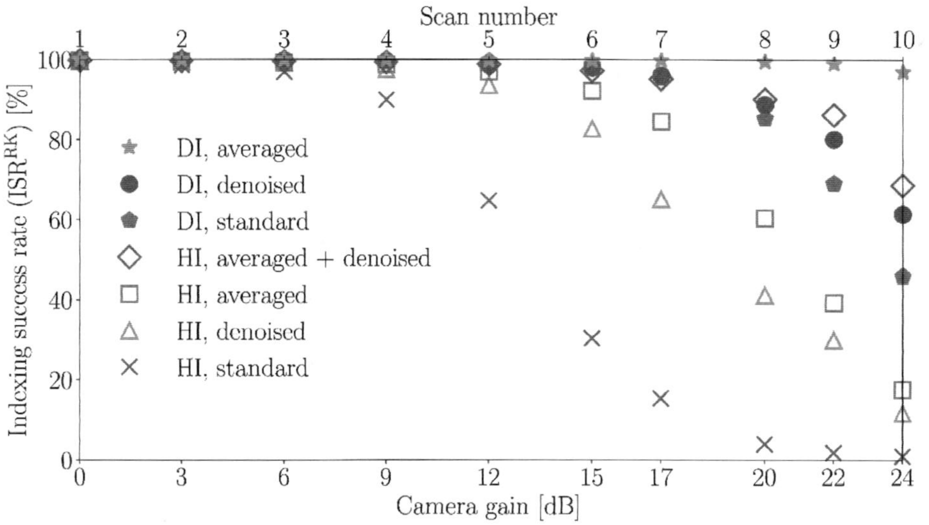

Figure 3. Comparison of the Hough and dictionary indexing success rates ISR^{RK} of standard, averaged, denoised, and averaged and denoised patterns, as a function of camera gain.

It is crucial to know how many principal components *k* to use for dimensionality reduction without loss of useful signal, as was discussed by Wilkinson *et al.* [8] and McAuliffe *et al.* [9]. The so-called scree plot of the explained variance as a function of the components sorted in order of decreasing variance is often used in multivariate statistical analysis to select this number [10]. Ideally, a significant change in the explained variance from a certain component to another can be identified, and thus the number of components to use. However, the explained variance decreases smoothly for the PCA results from the scans studied in this work, as was also observed by Wilkinson *et al.* [8]. To further complicate the selection, it is observed in this work that for the scans acquired with higher camera gain, notably scan 5 - 10, the first few components with the highest explained variance contained only noise and no discernible useful signal. These are the first few discarded components in Table 1. This demonstrates the need to inspect each principal component when selecting which components to use for dimensionality reduction for denoising of EBSPs.

To assess the number of components to use for dimensionality reduction for scan 9, the HI and DI ISR^{RK} after denoising with different number of components were calculated as shown in Fig. 4. 20 data sets were reconstructed with components in the range from 44 to 944 (from 50 to 1,000 minus the six discarded components listed in Table 1) in steps of 50. Interestingly, the trend in the ISR^{RK} is different for HI and DI. The HI ISR^{RK} is highest at 47 % when using only 44 components and decreases slowly with more components. The DI ISR^{RK} is lowest at 63 % when using 44 components and increases to 81 % with 244 components where it evens out. No significant decrease is observed with more components. An explanation for this trend in the DI ISR^{RK} is that the earlier components contain mostly signal from larger, similar crystallographic features, like grains, with the mid-range components up to about 244 containing a mix of signal from smaller grains and noise, until the later components contain only noise. An explanation for the decrease in the HI ISR^{RK} is that even though more components with signal from smaller grains are added to patterns from these grains, no indexing solution is found for these patterns. Instead, noise is added to patterns from larger grains, reducing the indexing success for these patterns. It is clear from Fig. 4 that most of the useful signal in scan 9 is found in the first 250 components, and thus the principal components up to $k = 250$, minus the first containing only noise, should be used for PCA denoising of this scan. Parts of the results from EMSOFT's DI is an indexing success rate, which is different from the ISR^{RK}, involving a computation of the smallest misorientation between the top matching orientation and the next nearest matches selected by the user. If the smallest misorientation is smaller than a threshold value the indexing is considered to be a success. This indexing success rate may be used to perform a search for the number of components to use for dimensionality reduction in cases where a reference scan is not available.

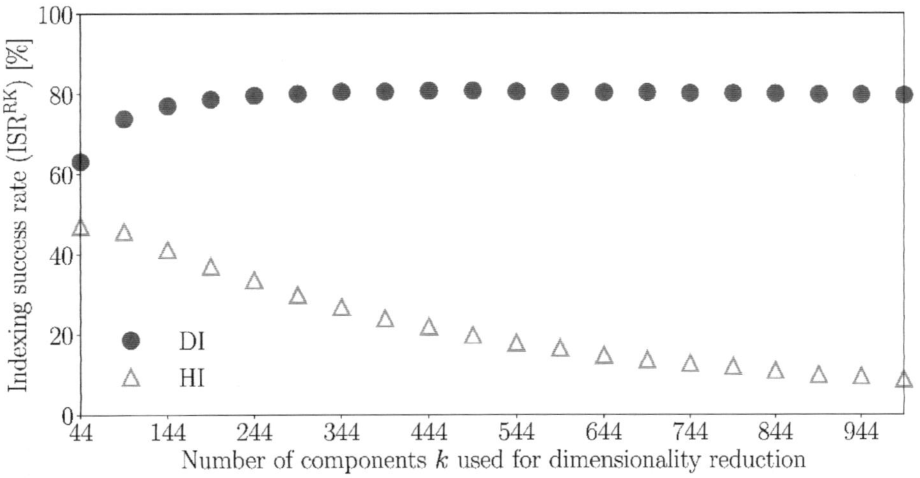

Figure 4. HI and DI ISR^{RK} as a function of number of principal components k used for dimensionality reduction in PCA denoising of scan 9.

The ability of PCA denoising to reveal fine details in EBSPs is studied in patterns from an orthoclase grain in a geological EBSD data set. A pattern from the grain after standard background corrections is shown in Fig. 5a. The average of 365 patterns containing Kikuchi bands from the grain is shown in (b), while the pattern from the same position as the one in (a) after PCA denoising is shown in (c). The best matching simulated pattern in (d) corresponds to the Euler angle triplet (261.1°, 36.0°, 272.3°). All patterns (a-d) underwent high-pass FFT filtering and adaptive histogram equalisation in EMSOFT as final processing steps. The dot product between the experimental and simulated patterns are shown in (a-c),

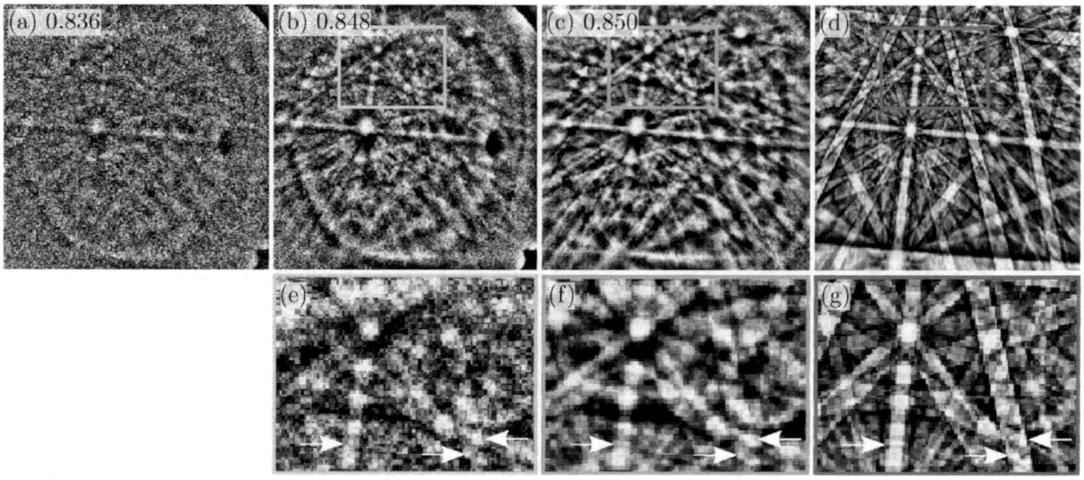

Figure 5. Comparison of an experimental pattern from an orthoclase grain after (a) standard background corrections, (b) averaging with all patterns within the grain, and (c) PCA denoising to (d) the simulated pattern with the highest dot product. The dot products are indicated, with the highest, 0.850, for the pattern in (c). Cut outs indicated in (b), (c), and (d) are magnified in (e), (f), and (g), respectively, and highlight features revealed by PCA denoising.

with the highest for the PCA denoised pattern. This and the fine details in the contrast from overlapping Kikuchi bands highlighted in the cut outs (e-g) demonstrate that PCA denoising in this example better harnesses the oversampling of the orthoclase grain than pattern averaging. Hansen *et al.* [18] found that Poisson-type noise, introduced into simulated patterns artificially and into experimental patterns by reducing the camera exposure time, can greatly inhibit the accuracy of geometrically necessary dislocation densities calculated from cross-correlation EBSD. Increasing *S/N* in patterns via PCA denoising might improve results from this type of analysis.

It should be noted that dictionary indexing is computationally intensive, mainly because each experimental pattern is compared to all patterns in the dictionary. For new users, the steps needed to create the dictionary of dynamically simulated EBSD patterns [19] might be challenging compared to performing Hough indexing via commercial software. However, it is our belief that these drawbacks are more than compensated for when considering the indexing results obtained after dictionary indexing of patterns of low *S/N* demonstrated in this work.

The patterns denoised in this work were obtained from recrystallised Ni and a large orthoclase grain. For the former, noise was introduced by reducing the exposure time and increasing the camera gain. A study of how PCA denoising can improve the quality of noisy patterns caused by high deformation and/or a fine-grained microstructure, related to the size of the interaction volume, is a natural continuation of this work. To improve upon the method of PCA denoising itself, a more robust selection of which principal components to keep for dimensionality reduction should be explored.

5. Conclusions

A method to increase the signal-to-noise ratio *S/N* in EBSPs based upon principal component analysis followed by dimensionality reduction, so-called PCA denoising, was presented. Applying PCA denoising to EBSD scans of Ni with a decreasing *S/N* demonstrated its ability to improve the indexing success rates after both Hough and dictionary indexing, effectively increasing the patterns' *S/N*. By first averaging each pattern with its four closest neighbours and subsequently applying PCA denoising, the *S/N* was increased further, and for these patterns the indexing success rates for Hough indexing was actually better than for dictionary indexing of patterns having undergone only standard background corrections. That PCA denoising is able to reveal fine details in EBSPs not visible after standard background corrections or pattern averaging was also shown by visual comparison of a pattern after these three processing routes and a simulated pattern. The choice of principal components to keep for dimensionality reduction was discussed, highlighting the need for a careful selection based upon the *S/N* in EBSPs and the number of expected crystallographic features in a scan. Pattern processing before indexing was conducted with a new open-source software package for processing and analysis of EBSPs, named KIKUCHIPY. In releasing this software package and the Ni data sets, we hope to make a useful contribution to the growing suite of open-source software for processing and analysis of EBSPs.

6. Data statement

The Ni data sets can be downloaded from Zenodo (https://doi.org/10.5281/zenodo.3265037). An up-to-date copy of KIKUCHIPY is available from the GitHub repository (https://github.com/kikuchipy/kikuchipy).

Acknowledgments

HWÅ acknowledges the Norwegian University of Science and Technology (NTNU) for financial support through the NTNU Aluminium Product Innovation Centre (NAPIC). The authors thank the staff at the Electron Microscopy Lab at NTNU for maintaining the microscopy facilities. The authors are also grateful to the reviewer for suggesting we explore neighbour pattern averaging.

References

[1] Schwartz A J, Kumar M, Adams B L and Field D P 2009 *Electron backscatter diffraction in materials science. 2nd edition.* (New York, NY: Springer)
[2] Chen Y H, Park S U, Wei D, Newstadt G, Jackson M A, Simmons J P, De Graef M and Hero A O 2015 *Microsc. Microanal.* **21** 739-752
[3] Britton T B, Tong V, Hickey J, Foden A and Wilkinson A 2018 *J. Appl. Crystallography* **51** 1-10
[4] Singh S, Guo Y, Winiarski B, Burnett T L, Withers P J and De Graef M 2018 *Sci. Rep.* **8** 1-8
[5] Wright S I, Nowell M M, Lindeman S P, Camus P P, De Graef M and Jackson M A 2015 *Ultramicroscopy* **159** 81-94
[6] Brewick P T, Wright S and Rowenhorst D J 2019 *Ultramicroscopy* **200** 50-61
[7] Brewer L N, Kotula P G and Michael J R 2008 *Ultramicroscopy* **108** 567-578
[8] Wilkinson A J, Collins D M, Zayachuk Y, Korla R and Vilalta-Clemente A 2019 *Ultramicroscopy* **196** 88-98
[9] McAuliffe T, Foden A, Bilsland C, Daskalaki-Mountanou D, Dye D and Britton T 2020 *Ultramicroscopy* **211** https://doi.org/10.1016/j.ultramic.2020.112944
[10] Jolliffe I T 2002 *Principal component analysis. 2nd edition.* (New York, NY: Springer)
[11] de la Peña F, Ostasevicius T, Fauske V T, Burdet P, Jokubauskas P, Nord M, Sarahan M, Prestat E, Johnstone D N, Taillon J, *et al.* 2017 *Microsc. Microanal.* **23** 214-215
[12] Pedregosa F, Varoquaux G, Gramfort A, Michel V, Thirion B, Grisel O, Blondel M, Prettenhofer P, Weiss R, Dubourg V, *et al.* 2011 *J. Machine Learning Res.* **12** 2825-2830
[13] Ånes H and Bergh T 2020 *KIKUCHIPY v0.1.2* https://github.com/kikuchipy/kikuchipy
[14] Villars P and Calvert L 1985 *Pearson's handbook of crystallographic data for intermetallic phases.* (Metals Park, OH: American Society of Metals)
[15] Colville A A and Ribbe P H 1968 *Amer. Mineralogist* **53** 25-37
[16] Powell M J D 2009 *The BOBYQA algorithm for bound constrained optimization without derivatives.* Report DAMTP 2009/NA06. (Cambridge, UK: University of Cambridge) 26-46
[17] Hielscher R and Schaeben H 2008 *J. Appl. Crystallography* **41** 1024-1037
[18] Hansen L T, Jackson B E, Fullwood D T, Wright S I, De Graef M, Homer E R and Wagoner R H 2017 *Microsc. Microanal.* **23** 460-471
[19] Jackson M, Pascal E and De Graef M 2019 *Integrating Mater. Manufact. Innovation* **2** 1-21

IOP Conf. Series: Materials Science and Engineering **891** (2020) 012003 doi:10.1088/1757-899X/891/1/012003

SI traceable characterisation of nanomaterials by X-ray spectrometry

B Beckhoff

Physikalisch-Technische Bundesanstalt (PTB), Abbestrasse 2-12, 10587 Berlin, Germany

E-mail: burkhard.beckhoff@ptb.de

Abstract. The development of new materials and the assessment of nanomaterials require the correlation of the materials' functionality or toxicity with their chemical and physical properties. To probe these properties, analytical methods that are both sensitive and selective at the nano- and microscales are required. The reliability of most analytical methods is based on the availability of reference materials or calibration samples, the spatial elemental compositions of which have to be as similar as possible to the matrix of the specimens of interest. However, there is a drastic lack of reference materials in particular at the nanoscale. Physikalisch-Technische Bundesanstalt (PTB) addresses this challenge by means of a bottom-up X-ray analytical method where all instrumental and experimental parameters are determined with known contributions to the uncertainty of the analytical results. This first-principle (FP) based approach does not require any reference materials but a complete characterisation of the analytical instruments' characteristics and, in addition, knowledge on the X-ray fundamental parameters related to the elements composing the sample. In order to reveal more reliable FP data in line with recent FP roadmap recommendations, PTB has been developing and using calibrated instrumentation, both energy- and wavelength-dispersive X-ray spectrometers, in conjunction with well-known synchrotron radiation (SR) of high spectral purity. Examples of recent PTB works on different FP determinations mostly of technologically relevant elements are given. SR based X-ray spectrometric methods allow for the variation of the analytical sensitivity, selectivity, and information depth needed to effectively reveal the spatial, elemental, and chemical specimen parameters of interest. Examples of particle characterisation, interfacial speciation, elemental depth profiling, as well as layer composition and thickness characterisations in advanced materials and nanostructures as well as for in-situ conditions are given.

1. Introduction

The development of new materials and the assessment of nanomaterials requires the correlation of the materials' functionality or toxicity with their chemical and physical properties. These requirements are driven by the steadily increasing needs for the integration of different technological processes in one device or spatially closely coupled devices calling for complete understandings of these processes and the related device functionalities. To probe the related chemical and physical properties reliably, analytical methods that are both sensitive and traceable at the nano- and microscales are required [1]. The reliability of most analytical methods is based on the availability of reference materials or calibration samples, the spatial elemental composition of which has to be as similar as possible to the matrix of the specimens of interest. However, there is a drastic lack of reference materials, in particular at the nanoscale. This considerably affects the quantification reliability of most analytical methods, as reference materials are used to compensate for missing knowledge on instrumental or experimental parameters of the analytical technique. In many categories (such as film thickness, nanoobjects or nanoparticles, depth profiling resolution) the number of existing nanoscaled reference materials [2] corresponds to the number of new nanomaterials introduced in the market each month. The German National Metrology Institute Physikalisch-Technische Bundesanstalt (PTB) addresses this challenge of missing nanoscaled reference materials by means of a bottom-up, International System of Units (SI)-traceable X-ray analytical method, X-ray spectrometry (XRS), where all instrumental and experimental parameters are determined with known contributions to the uncertainty of the analytical results [3]. This first principle-based approach in XRS does not require any reference materials, but a complete characterisation of the analytical instruments' characteristics and, in addition, knowledge on the X-ray fundamental parameters [4] related to the elements composing the sample. The characterisation of instrumental and experimental parameters can be reliably performed using X-ray radiometry, thus metrology of X-ray radiation, allowing for the calibration of components of X-ray instruments by comparing them to primary source or detector standards [5]. During the last decade, an international consortium of key players in academia, industry, and metrology organized a continuously updated X-ray fundamental parameters roadmap [6]. Current fundamental parameter-related research aims at improving the knowledge of fundamental parameter data by means of experimental and theoretical methods by novel instrumentation, specimens, and algorithms [4, 7, 8] allowing for a reduction in analytical uncertainties of XRS for both chemical traceability and SI traceability. Synchrotron radiation-based X-ray spectrometric methods allow for the variation of the analytical sensitivity, discrimination capability, quantification reliability, and information depth needed to effectively reveal the spatial, elemental, and chemical specimen parameters of interest. SI-traceable XRS has been successfully applied as a reference measurement technique to determine surface contamination [9] on flat substrates, perform particle characterisation with respect to elemental or species composition, and enable interfacial speciation [10, 11] or elemental depth profiling of nanolayered systems [12]. For process-oriented applications or other analytical methods, calibration samples could be qualified with respect to layer composition and thickness in advanced materials [13], surface functionalization [14, 15] or biomedical materials [16, 17]. Recent instrumental achievements provide access to liquids [18, 19], liquid-solid interfaces and even the in-situ and operando elemental analysis and chemical speciation of nanoscaled battery materials [20]. Furthermore, X-ray spectrometry under grazing incidence is capable of revealing analytical and dimensional information from layered systems, nanostructures and particles deposited on surfaces [21].

2. Reference-free X-ray fluorescence analysis

The continuing development of analytical methods based on reference-free X-ray spectrometry (XRS) and related instrumentation by the Physikalisch-Technische Bundesanstalt (PTB), Germany's national metrology institute, is enabled by the use of well-characterised X-ray sources and detectors as well as by well-known experimental arrangements allowing for the determination of atomic fundamental parameters. For radiometric purposes, PTB operates several well-characterised beamlines for monochromatised synchrotron radiation at the 1.7 GeV electron storage ring BESSY II. Two of these beamlines, a plane grating monochromator (PGM) beamline for undulator radiation and a four-crystal monochromator (FCM) for bending magnet radiation, serve as excitation sources for reference-free XRS. Employing calibrated photodiodes in conjunction with monochromatised synchrotron radiation of a pre-selectable intensity, i.e., by varying the stored electron beam current at BESSY II, the absolute counting efficiency and response behaviour of energy-dispersive X-ray detectors, such as a Si(Li)- -or SDD-detector, can be determined experimentally. With respect to the XRS beam geometry, special attention has to be paid when defining the angles of incidence and observation as well as the solid angle of detection. The methodological development of XRS at PTB is, among other issues, dedicated to high-end investigations in the R&D of semiconductor samples related to industrial applications requiring reference-free methods, in particular for new materials where not enough appropriate reference materials are available. The use of monochromatised synchrotron radiation at BESSY II has led to developments in the non-destructive investigation of wafer surface contamination, speciation and nanolayered materials by reference-free XRS.

Figure 1 shows the principal arrangements in reference-free X-ray spectrometry: total-reflection X-ray fluorescence (TXRF) at very shallow angles of incidence, grazing-incidence X-ray fluorescence (GIXRF) at intermediate angles of incidence and conventional XRF at rather high angles of incidence. In PTB's instrumentation for XRS, photodiodes are used to record the radiant power of the incident, transmitted or reflected beam.

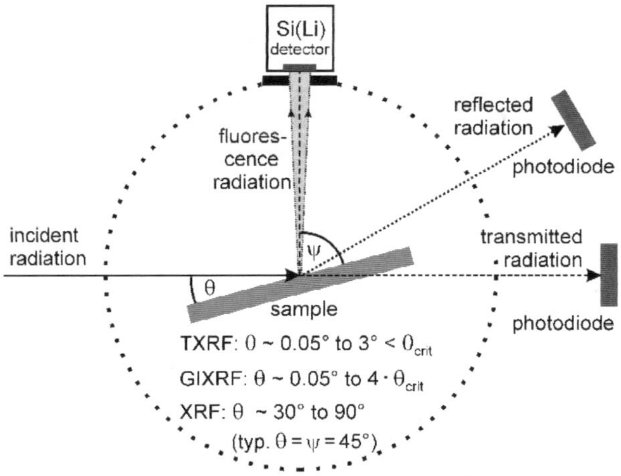

Figure 1. Typical arrangements for reference-free XRS employing monochromatised synchrotron radiation as the incident excitation radiation. The solid angle of detection is defined by a known diaphragm placed in front of the Si(Li)-detector at a given distance with respect to the centre of the sample. When detecting fluorescence radiation with one detector, the angle between the incident radiation and the fluorescence detection channel is 90°. When simultaneously detecting fluorescence radiation with two different detectors, normal incidence ($\theta = 90°$) and observation angles ψ of the detectors of 30°, 45° or 60° with respect to the sample surface, respectively, are chosen. θ_{crit} designates the critical angle of external total reflection at a flat sample surface.

If all relevant instrumental and atomic fundamental parameters are known, the concentration Ci of the element i in a homogeneous multielemental sample s of thickness T can be calculated without additional references using, for example, the following basic quantitation formula of X-ray fluorescence analysis:

$$C_i = \frac{P_i}{P_0 \tau_{i,E_0} Q \frac{\Omega_{det}}{4\pi} \frac{1}{\sin\Theta} \frac{1-\exp(-\mu_{tot,i}T)}{\mu_{tot,i}}} \tag{1}$$

where E_0 is the photon energy of the incident (excitation) radiation, $P_0 = S_0 / \sigma_{diode,E_0}$ is the radiant power of the incident radiation to be converted into the incident radiation intensity, S_0 is the signal of the calibrated photodiode measuring the incident radiation, σ_{diode,E_0} is the spectral responsitivity of the calibrated photodiode, θ is the angle of incidence with respect to the sample surface, E_i is the photon energy of the fluorescence line l of the element i, R_i is the detected count rate of the fluorescence line l of the element i, ε_{det,E_i} is the detection efficiency of the Si(Li)-detector at the photon energy E_i ,

$P_i = R_i / \varepsilon_{det,Ei}$ is the intensity of the fluorescence line l of the element i, τ_{i,E_0} is the photo electric cross section of the element i at the photon energy E_0, $\mu_{S,E}$ is the absorption cross section of the sample s at the photon energy E, $\mu_{tot,i} = \mu_{S,E_0}/\sin\theta + \mu_{S,E_i}/\sin\psi$ with ψ being the angle of observation with respect to the sample surface, Ω_{det} is the effective solid angle of detection defined by both a calibrated aperture placed at a well-known distance from the sample in front of the Si(Li)-detector and the footprint of the excitation radiation on the sample surface, ω_{Xi} is the fluorescence yield of the absorption edge X_i (of the element i), $g_{l,Xi}$ is the transition probability of the fluorescence line l belonging to the absorption edge X_i, j_{Xi} is the jump ratio at the absorption edge X_i , and $Q = \omega_{Xi}\ g_{l,Xi}\ (j_{Xi} - 1) / j_{Xi}$. In the case of flat samples excited at shallow incident angles, the modulation of the incident radiant power by the X-ray Standing Wave (XSW) intensity [10] has to be included as an additional factor. Secondary and tertiary excitation channels within a homogeneous layer or bulk sample as well as intra-layer excitation and absorption effects can be likewise included as indicated in detail in the literature [22].

When part of the experimental parameters such as the spectral distribution of the excitation radiation, e.g., in the case of an X-ray tube, or the efficiency of the detectors employed are not known well enough, one may try to compensate for this lack of a-priori knowledge by using appropriate reference materials or calibration samples. The reliability of this reference-based quantitation approach strongly depends on the use of reference materials that should be as similar as possible to the specific samples to be analysed, in particular with respect to the specimen matrix. As this requirement can often not be fulfilled easily, various techniques to correct for matrix effects have been developed and validated. In the case of inhomogeneous matrices, analytical expressions for the elemental concentrations can only be given when a complete description of the spatial distribution of the main matrix constituents can also be given. Otherwise, Monte Carlo based simulations may contribute in overcoming a lack of sufficient a-priori knowledge concerning the spatial distribution of main matrix elements or, when including multiple scattering effects for which analytical expressions are more difficult to deduce, in the spectral decomposition.

The current work describes the basic requirements and fields of applications of reference-free XRS, i.e. when all relevant experimental parameters are well known (Fig. 2). This includes, of course, sufficient a-priori knowledge of the spatial distribution of the main matrix constituents as is the case for homogeneous and layered samples. Depth gradient variations in the matrix concentrations may also be dealt with when known a-priori. In multi-elemental samples the radiation of one element can enhance the fluorescence intensity of another. This is referred to as secondary fluorescence. Taking this effect into account appropriately is crucial for producing correct results in quantitative X-Ray fluorescence analysis. This is especially relevant for thin layered samples, for which the individual mass depositions of each element in each layer are to be determined. If the respective densities are known, this knowledge then can be converted into corresponding layer thicknesses.

Many new materials, e.g., in the semiconductor industry, are essentially thin layered samples at the nanoscale. Since appropriate reference materials are often not available, a reference-free approach based upon SI traceability might be the method of choice. The theoretical tools to calculate the required fluorescence intensities for the quantification algorithms are well-known and can be found in the literature [22]. Recent works at PTB aim on investigating the validity of these higher-order algorithms for selected thin layered samples, consisting of multiple transition metals with thicknesses in the low-μm range. For these samples the contribution of secondary excitation fluorescence can easily exceed 20 %, depending on the elements involved, the spatial composition and dimensions of the layered system, the excitation energy and the angle of incidence. The calibrated instrumentation and the reference-free fundamental parameter approach [3] allow for a direct validation of the computed theoretical intensities with measured intensities. There is also a strong need in the medical device industry to decrease failure rates of biomedical devices by reducing the incidence of defect structures and contaminants during the production process. The detection and identification of defect structures and contaminants is crucial for many industrial applications. The present study exploits reference-free XRF analysis as an analytical tool for the traceable characterization of surface contaminants of medical devices, in particular N,N'-ethylene-bis(stearamide), an ubiquitous compound used in many industrial applications as a release agent or friction reduction additive. Reference-free XRF analysis as primary method has been proven to be capable of underpinning all other applied methods since it yields the absolute mass deposition of the selected N,N'-ethylene-bis (stear-amide) contaminant whilst X-ray absorption fine structure analysis determines the chemical species. Ambient vibrational spectroscopy and mass spectroscopy methodologies such as Fourier transform infrared, Raman, and secondary ion mass spectroscopy have been used in this systematic procedure providing an extensive range of complementary analyses. A calibration procedure was developed using specially designed and fabricated model systems varying in thickness and substrate material [16]. Furthermore, typical real medical devices such as both a polyethylene hip liner and a silver-coated wound dressing have been contaminated and investigated by these diverse methods, enabling testing of this developed procedure. These well-characterised samples may be used as calibration standards for bench top instrumentation from the perspective of providing traceable analysis of biomaterials and surface treatments.

3. Grazing incidence X-ray spectrometry

During the last decades total-reflection X-ray fluorescence analysis, or TXRF, became one of the most powerful and cost competitive techniques for surface chemical and trace element analyses [22]. Tuning the angle of incidence across the critical angle of total external reflection, the composition of thin layered systems of low surface and interface roughness can be probed by means of the induced X-ray Standing Wave (XSW) field [10]. The dimensions of the XSW field intensity are employed as a nanoscaled sensor in this grazing-incidence XRF, or GIXRF, technique [13]. TXRF and GIXRF have successfully demonstrated their analytical capabilities in various modern applications to bio- and nanotechnologies [12, 16, 17, 21] involving minute sample amounts, surface functionalisations, nanoparticles and nanostructures. Even information about species mass depositions and the related chemical binding states [10, 11] can be revealed when using tuneable excitation radiation.

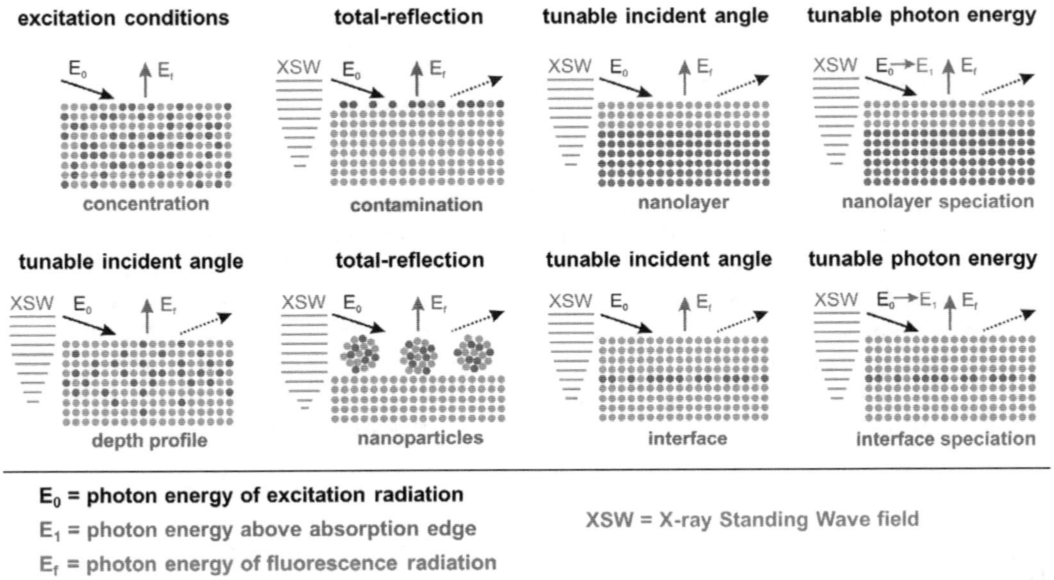

Figure 2. Different arrangements for reference-free XRS employing monochromatised synchrotron radiation as the incident excitation radiation. The analytical sensitivity and information depth can be tuned by means of appropriate operational parameters. When tuning the excitation radiation across the absorption edge of a main matrix element chemical binding state information can be revealed by X-ray absorption spectrometry.

A reliable and non-destructive metrology of in-depth elemental distributions and dimensional parameters of nanoscaled layer systems and 3-dimensional nanostructures is an essential topic in many fields of research, especially the semiconductor industry. With the decreasing minimal feature sizes and increasing complexity of the used structures, also the metrological challenges for a reliable characterization of such structures increase. As they must be characterized both with respect to their dimensional and elemental distribution parameters, a further development of the current analytical techniques is required. A combined grazing incidence X-ray fluorescence (GIXRF) and X-ray reflectometry (XRR) approach is capable of contributing to the in-depth and the dimensional characterisation of such nanoscale materials. When combining XRR with reference-free GIXRF [13], a direct access to the mass depositions (*ρd*) of the materials of interest is provided. This allows for a significant reduction of the degrees of freedom within the combined GIXRF-XRR modelling and thus improves the characterisation reliability of the methodology. The combined reference-free GIXRF-XRR approach has been applied to the in-depth analysis of thin nanolaminate stacks of Al_2O_3 and HfO_2 layers with total thicknesses in the sub-10 nm regime. For the GIXRF modelling of these nanostructures a novel approach for the calculation of the X-ray standing wave field intensity distributions [21] is necessary.

4. High resolution X-ray spectrometry

The complete understanding of the functionality of battery components requires the correlation with underlying physical and chemical properties which is the challenge for most analytical methods due to a lack of reference materials. Lithium sulphur (Li-S) batteries are promising candidates for improved batteries offering up to 5 times higher capacity than conventional lithium ion batteries. The main degradation is caused by polysulphides formed during cycling which are soluble in the electrolyte solution. For a better understanding of this process in-situ and operando characterisation techniques are required [20]. PTB used CR-2032 coin-cell formed Li-S batteries with DOL/DME (1:1 wt%) 1 mol TFSI electrolyte modified with thin cathode current collector and thin window enabling the transmission of excitation and fluorescence radiation. SI traceable X-ray spectrometric measurements were performed using radiometrically calibrated instrumentation in the PTB laboratory at BESSY II synchrotron radiation facility. By means of operando near edge X-ray absorption fine structure (NEXAFS) spectrometry at the S-K edge recorded during galvanostatic cycling with potential limitation (GCPL) measurements different sulfur species in the cell can be determined. For 8 full charge/discharge cycles (0.1 C) operando sulphur K-edge NEXAFS could be registered. After a spectral deconvolution the different sulphur species and their behaviour over time were analysed. While for the first cycles a reversible formation of polysulphides is visible further cycles show a steady increase of polysulphides due to the solubility in the electrolyte solution and, thus, the loss of cathode active material and capacity fading. The investigation of polysulphides for different states of charges (SOC) and state of health (SOH) enables a deeper understanding of the main degradation process. This information helps to improve the battery system by modifying its materials e.g., with binder components or by choosing an optimal (dis-)charge current. The simple set-up enables also investigations from the anode side. In addition, the set-up can be adapted for different types of batteries with other active materials, e.g., NCM Li-ion batteries.

Depending on the specific valence states of compounds high resolution X-ray emission spectrometry (XES) may provide valuable complementary information to XAFS. XES can equally contribute to the further development of complex nanomaterials with distinct chemical properties. Also in the field of catalysis, the identification and quantification of active sites is important for a more thorough understanding of catalytic systems. Using a physically traceable X-ray analytical approach based on calibrated instrumentation and the knowledge of atomic fundamental parameters, absolute elemental and species mass depositions or concentrations can be determined without the need for any reference material or calibration standards. For this purpose PTB build a calibrateable and compact high-resolution wavelength-dispersive spectrometer [23] for XES in the photon energy range of 2.4 keV to 19.0 keV. Using two full-cylindrical highly annealed pyrolytic graphite (HAPG) crystals as dispersive elements in modified von-Hamos geometry, a large solid angle of detection resulting in high detection efficiency is realized while achieving a moderate to high resolving power. HAPG is a synthetic type of carbon which forms mosaic crystals and can be mounted to substrates with small radii of curvature down to 50 mm. Although the peak reflectivity is smaller than for perfect crystals, the crystal mosaicity results in a rather high integrated reflectivity. The instrumental characterisation of the spectrometer included the achievable energy resolution and efficiency as well as the spectral response as a function of experimental parameters. Furthermore, the chemical speciation capability of the device has been demonstrated on the basis of different transition metal compounds. The calibration of the instrumental response of the wavelength-dispersive X-ray spectrometer enables an accurate determination of binding state related structures in X-ray spectra, thus enabling reliable identification and discrimination capabilities.

Synchrotron radiation (SR) based scanning X-ray spectroscopy (XRS) can determine both elemental concentrations and chemical binding states with high spatial resolution down to the nanoscale. At SR facilities there are currently only a few dedicated scanning X-ray microscopes that offer XRF analysis in combination with scanning transmission X-ray microscopy (STXM). A novel kind of vibration compensating XRF set-up [24] have pushed the spatial resolution towards the 100 nm regime. With the advent of UHV-compatible piezo stages with nanometre resolution, it became possible to mount optical elements in a compact manner on a single platform. Relevant parts of any external vibrations can be compensated for by such a single platform. The lay-out of its stages and related holders is to ensure amplitudes of relative vibrational movements to be smaller than the absolute ones by avoiding any resonance frequencies. PTB realised this approach in a compact set-up to align diagnostics, zone plates, order sorting apertures (OSA) and samples on a single mounting plate. This plate was mounted on a sample manipulator of an UHV XRF instrument. Its overall size offers versatile areas of operation as the compact set-up can be readily inserted in a wide range of XRF chambers through their load-locks. The optical elements of the scanning X-ray microscope consist of an Au zone plate with a 50 μm integrated central stop and an independent 50 μm OSA. All measurements were carried out with photon energy of 1,500 eV at PTB's plane grating monochromator beamline for undulator radiation at BESSY II. Two objects were used to examine the performance of the novel type of scanning nm-XRF microscope. The test object for STXM was a Siemens star on a Si_3N_4-membrane completely covered with a 133 nm thick layer of W, while for the fluorescence experiments we chose a silicon wafer with an alternating Ge-Si-wafer bar test structure. The smallest structures (approximately 100 nm) were clearly resolved in transmission while the resolution was slightly larger in XRF mode. Here, the Ge-Si-sample was measured at an angle of incidence of 60° with an energy-dispersive silicon drift detector (SDD) calibrated by PTB. The highest spatial resolution achieved in initial experiments was in the range of 130 nm to 140 nm, which is in line with the probed sample dimensions. Follow-up experiments are going to be focused at traceable XRS of appropriate and even smaller test structures.

5. Determination of atomic fundamental parameters

The quantitative analysis of X-ray fluorescence and electron induced x-ray spectroscopic methods requires good knowledge of the atomic fundamental parameters (FP) and optical constants involved. In addition, the lack of adequate reference-materials and calibration samples, in particular at the nanoscale, reliable quantification schemes in X-ray fluorescence analysis and related methods call for a high accuracy of the X-ray fundamental parameters. However, the respective uncertainties of available tabulated data are usually rather large, especially for low-Z elements or L- and M-shell fluorescence lines. In order to address this issue, different methods for experimental FP determinations have been developed, validated and applied at the PTB. Those activities on the experimental determination of X-ray fundamental parameters cover the determination of low-Z fluorescence yields, subshell photo-ionisation cross-sections and Coster-Kronig factors [25]. Many data sets widely are being used in analytical applications such as XRF where subshell or partial photo-ionisation cross-sections (PCS) are derived from the total PCS by applying constant (energy independent) jump ratios [5]. Major X-ray equipment manufacturers widely use jump ratio values in their quantification algorithms and also the International Union of Pure and Applied Chemistry proposes the use of constant ratios in a standard reference guide. The ratios are defined by the jump height of the mass absorption coefficient at the respective absorption edge and are independent of the photon energy. The ratios are then used to subdivide the total PCS into the contributions of the various subshells (fixed jump ratio approximation). As Ebel [26] indicated by comparing available data sets, this produces consistent results for the K-shell PCS and is rather questionable for the L- and M-subshells. Close to the L absorption edges both approaches deliver similar results (by definition), but they differ significantly with increasing photon energy. As the PCSs have a direct impact on measurands such as the amount of substance, both the correct approach and values of PCS data need to be experimentally confirmed (Fig. 3). Meanwhile the energy dependent slope of these coefficients could be confirmed experimentally (Fig. 4) [4]. These coefficients were calculated theoretically by Scofield, and carefully evaluated and implemented by Ebel.

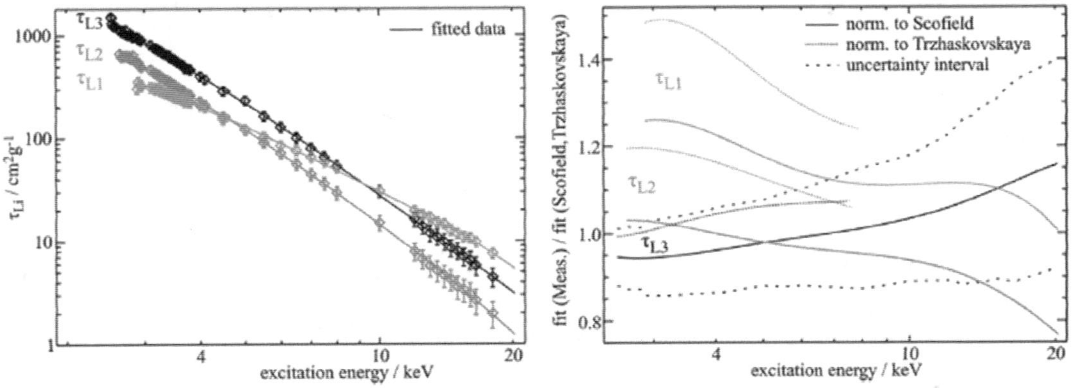

Figure 3. Experimentally determined PCS for the three Mo-L subshells (left) and the comparison to different calculated data sets (right) [4].

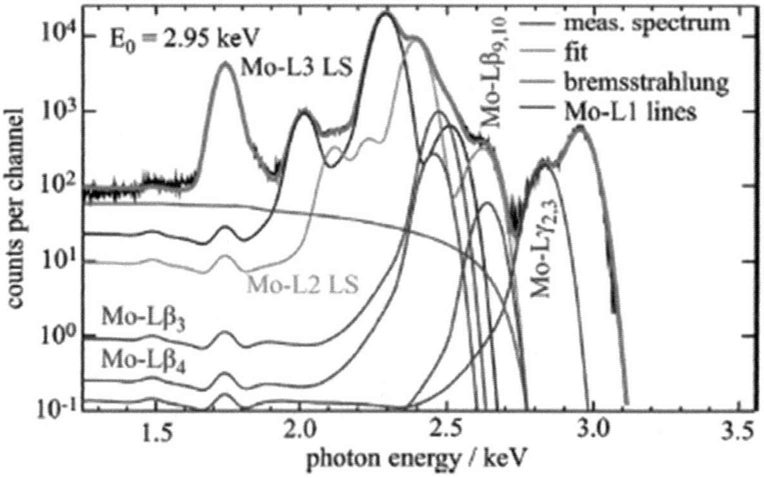

Figure 4. Response function based deconvolution of a Mo layer XRF spectrum for each L-shell used for the experimental determination of Mo-L PCS [4].

The determination of fundamental parameters (FP) for L-edges is an important part of reliable, reference-free quantitative XRF [3]. In order to reduce the uncertainties of the FP, transmission measurements of the samples have to be performed thus allowing to experimentally determine self-absorption correction factors. For reference-free XRF with lower excitation energies, the uncertainties of the fundamental parameter associated with L-edges are crucial for the total uncertainty of the quantification, especially for excitation energies above the L1-edge of the respective element. The fluorescence yields and Coster-Kronig transition probabilities for the L-edges of Ga were determined with considerably reduced uncertainties [27]. Due to the high absorption of radiation in the soft X-ray range, it is necessary to reduce the thickness of the samples for transmission measurements. Here, a nominal 300 nm thin GaSe layer was deposited on a thin silicon-nitrate window for measuring both the transmission and the fluorescence intensities simultaneously in order to perform absorption correction without having to rely on literature database values of the mass attenuation coefficients. In order to reduce the uncertainty of the spectral deconvolution, a silicon drift detector (SDD) with an optimised energy resolution was employed. The measurements were carried out at the plane grating monochromator beamline in the PTB laboratory at BESSY II using monochromatized undulator radiation and calibrated instrumentation [3].

New measurements of mass attenuation coefficients were also carried out by the French National Metrology Institute Laboratoire National Henri Becquerel (LNHB), which took the advantage of using tunable monochromatic radiation at SOLEIL synchrotron radiation facility, on the METROLOGIE beamline and using specific procedures. Within the frame of different institutional and industrial metrology research projects, the LNHB and partners studied several elements from photon energies as low as 100 eV up to 35 keV, special care was taken to ensure reliable uncertainties [8, 28]. The results of some of these collaborative studies do not agree with data previously published by other groups within their respective uncertainties.

References

[1] Proc. 2017 E-MRS Spring Meeting Symposium S ALTECH 2017 – Analytical techniques for precise characterization of nanomaterials. 2017 *Phys. status solidi C* **14** 1720017

[2] *Nanoscaled reference materials, a worldwide survey of existing reference materials at the nanoscale in different categories by the German Federal Institute for Materials Research and Testing (BAM) in cooperation with ISO/TC 229 Nanotechnologies.* see http://www.nano-refmat.bam.de/en/

[3] Beckhoff B 2008 *J. Anal. Atom. Spectrom.* **23** 845
[4] Hönicke P, Kolbe M, Müller M, Mantler M, Krämer M and Beckhoff B 2014 *Phys. Rev. Lett.* **113** 163001
[5] Beckhoff B, Gottwald A, Klein R, Krumrey M, Müller R, Richter M, Scholze F, Thornagel R and Ulm G 2009 *Phys. status solidi B* **246** 1415
[6] *Second roadmap on atomic fundamental parameters for X-ray methodologies.* 2017 International Initiative on X-ray Fundamental Parameters. see: www.EXSA.hu/fpi.php
[7] Guerra M, Sampaio J M, Madeira T I, Parente F, Indelicato P, Marques J P, Santos J P, Hoszowska J, Dousse J-C, Loperetti L, Zeeshan F, Müller M, Unterumsberger R and Beckhoff B 2015 *Phys. Rev. A* **92** 022507
[8] Ménesguen Y, Gerlach M, Pollakowski B, Unterumsberger R, Haschke M, Beckhoff B and Lépy M-C 2016 *Metrologia* **53** 7
[9] Beckhoff B, Fliegauf R, Kolbe M, Müller M, Weser J and Ulm G 2007 *Anal. Chem.* **79** 7873
[10] Pollakowski B, Hoffmann P, Kosinova M, Baake O, Trunova V, Unterumsberger R, Ensinger W and Beckhoff B 2013 *Anal. Chem.* **85** 193
[11] Pollakowski B and Beckhoff B 2015 *Anal. Chem.* **87** 7705
[12] Hönicke P, Beckhoff B, Kolbe M, Giubertoni D and van den Berg J 2010 *Anal. Bioanal. Chem.* **396** 2825
[13] Unterumsberger R, Pollakowski B, Müller M and Beckhoff B 2011 *Anal. Chem.* **83** 8623
[14] Fischer T, Dietrich P M, Streeck C, Ray S, Nutsch A, Shard A, Beckhoff B, Unger W E S and Rurack K 2015 *Anal. Chem.* **87** 2685
[15] Dietrich P M, Streeck C, Glamsch S, Ehlert C, Lippitz A, Nutsch A, Kula, N, Beckhoff B and Unger W E S 2015 *Anal. Chem.* **87** 10117
[16] Pollakowski-Herrmann B, Hornemann A, Giovannozzi A M, Green F, Gunning P, Portesi C, Rossi A, Seim C, Steven R, Tyler B and Beckhoff B 2018 *J. Pharmaceut. Biomed. Anal.* **150** 308
[17] Giovannozzi A M, Hornemann A, Pollakowski-Herrmann B, Green F M, Gunning P, Salter T L, Steven R T, Bunch J, Portesi C, Tyler B J, Beckhoff B and Rossi A M 2019 *Anal. Bioanal. Chem.* **411** 217
[18] Grötzsch D, Streeck C, Nietzold C, Malzer W, Mantouvalou I, Nutsch A, Dietrich P, Unger W, Beckhoff B and Kanngiesser B 2017 *Rev. Sci. Instrum.* **88** 123112
[19] Witte K, Streeck C, Mantouvalou I, Suchkova S A, Lokstein H, Grötzsch D, Martyanov W, Weser J, Kanngiesser B, Beckhoff B and Stiel H 2016 *J. Phys. Chem. B* **120** 11619
[20] Müller M, Choudhury S, Gruber K, Cruz V B, Fuchsbichler B, Jacob T, Koller St, Stamm M, Ionov L and Beckhoff B 2014 *Spectrochim. Acta B* **94-95** 22
[21] Soltwisch V, Hönicke P, Kayser Y, Eilbracht J, Probst J, Scholze and Beckhoff B 2018 *Nanoscale* **10** 6177
[22] Beckhoff B, Kanngiesser B, Langhoff N, Wedell R and Wolff H 2006 *Handbook of practical X-ray fluorescence analysis*. (Berlin, Germany: Springer)
[23] Wansleben M, Vinson J, Holfelder I, Kayser Y and Beckhoff B 2018 *X-ray Spectrom.* **48** 102
[24] Lubeck J, Seim C, Dehlinger A, Haidl A, Hönicke P, Kayser Y, Unterumsberger R, Fleischmann C and Beckhoff B 2018 *Microsc. Microanal.* **24** (Suppl. 2) 158
[25] Kolbe M, Hönicke P, Müller M and Beckhoff B 2012 *Phys. Rev. A* **86** 042512
[26] Ebel H, Svagera R, Ebel M F, Shaltout A and Hubbell J H 2003 *X-ray Spectrom.* **32** 442-451
[27] Unterumsberger R, Hönicke P, Colaux J, Jeynes C, Wansleben M, Müller M and Beckhoff B 2018 *J. Anal. Atom. Spectrom.* **33** 1003
[28] Ménesguen Y, Lépy M-C, Sampaio J M, Marques J P, Parente F, Guerra M, Indelicato P, Santos J P, Hönicke P and Beckhoff B 2018 *X-ray Spectrom.* **47** 341

IOP Conf. Series: Materials Science and Engineering **891** (2020) 012004 doi:10.1088/1757-899X/891/1/012004

Measurements of the quantitative analytical depth resolution at evaporated aluminium- and silver-layers with the FEG-EPMA JEOL JXA-8530F

D Berger and J Nissen

Technische Universität Berlin, Zentraleinrichtung Elektronenmikroskopie ZELMI, Strasse des 17. Juni 135, 10623 Berlin, Germany

E-mail: dirk.berger@tu-berlin.de; joerg.nissen@tu-berlin.de

Abstract. The aim of this study is to extend our general approach to the spatial resolution in the quantitative element analysis that can be achieved with a FEG-EPMA by systematic measurements of the depth resolution. The test specimens are Al and Ag surface layers with various thicknesses on top of non-fluorescent substrates. By choosing an appropriate X-ray line and by varying the primary electron energy, E_0, we determined the minimum value of E_0 for which the depth of the signal source volume is identical to the layer thickness. This value is defined as the depth resolution. With respect to the excitation energy of the characteristic X-rays, the depth resolution for the quantitative element analysis is found to be below 150 nm for Al (Kα at 3 kV) as well as for Ag (Lα at 5 kV).

The prediction of the depth resolution at low primary electron energy E_0 by means of calculations of the ionisation range of electrons or by Monte Carlo (MC) simulations is quite inaccurate because both predict better values compared to the measurements. Since we regard test specimens without disturbing secondary fluorescence it is obvious that both calculations and simulation predict too small signal source volumes for a so far unknown reason. The same discrepancy was observed for the determination of the lateral quantitative resolutions in previous studies. Therefore systematic measurement errors can be excluded. The measurements presented in this paper are very helpful for further improvements of calculations and MC-simulations at low primary electron energies E_0 because specimens without secondary fluorescence are considered only.

1. Introduction

The well-known main advantage of electron microprobes with a field emission gun (FEG-EPMA) is the high spatial resolution due to the much finer electron probe even at low primary electron energies, E_0, compared to EPMA with thermal emitters. Many papers already demonstrated the advantage of the high spatial resolution for different specimens mainly in geological applications [1-3]. In most cases, a given specimen with a given element distribution was used as a starting point and then the measuring parameter were optimised to get the optimum spatial resolution. We aim to achieve a general approach to estimate the spatial resolution from the range of the electrons, given by the mean Z-number, and the signal source volume, given additionally by the secondary fluorescence. Unfortunately, we found significant differences between calculations of the ionisation range of electrons [4] and Monte Carlo

(MC) simulations [5, 6] in comparison to the measured spatial resolutions at low primary electron energies E_0 below 7 kV [7, 8]. The reasons are not yet understood. The deviations might be examined and interpreted quite easily at specimens for which the spatial resolution is not affected by secondary fluorescence. Therefore, we prepared test specimen of thin metallic surface layers on Si wafer substrates with an intermediate layer in which nearly no secondary fluorescence is generated by the X-rays of the surface layer (see Fig. 1). The test specimen have different thicknesses of the surface layer.

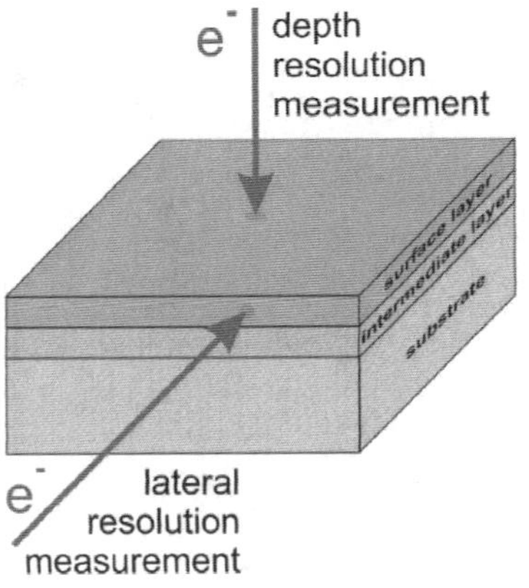

Figure 1. Measurement schemes to determine the depth and lateral resolutions from one specimen, the green intermediate layer reduces secondary fluorescence effects.

Previously, we studied the ***lateral analytical*** resolution whereby 'lateral' indicates the horizontal x- and y-directions and 'analytical' refers to the qualitative or quantitative determination of elements using characteristic X-rays [7, 8]. For determination of the quantitative lateral resolution, thin metallic layers were analysed in cross-section (Fig. 1). Hereby, quantitative lateral resolutions of 140 to 170 nm for Au layers, 150 to 210 nm for Ag layers, and 260 to 280 nm for Al layers were proved. It should be pointed out, that the "quantitative" lateral resolution is regarded, i.e., the primary electron energy E_0 is reduced until the quantification equals at least 99 % of the surface layer material. Obviously much better values of the lateral resolution are obtained if the "qualitative" lateral resolution is considered only. In accordance to the trace element detection limit, it is then enough that the layer particle or inclusion under observation emit a peak well above the background. In accordance with many other published results (see references in [9]), we measured the best value for the qualitative lateral resolution to be 26 nm at a Si-Au interface [9]. As mentioned earlier the measured values of the lateral resolution differ from calculated and MC-simulated values although the beam current- and voltage-depending probe diameter is considered.

In this study we extend the systematic consideration of the spatial resolution by measurements of the ***depth resolution*** of the quantitative element analysis. The previously prepared test specimens with different surface layer thicknesses are now excited from the top (see Fig. 1). Again the measured values are compared to calculations and MC-simulations. The depth resolution measurements offer the advantage that the results depend neither on the probe diameter nor on any beam drifts. Therefore, the depth resolution measurements are considered to be more convenient to proof the accuracy of calculations and MC-simulations. Our investigation uses test specimens with intermediate layers to reduce secondary fluorescence effects and are, therefore, an addendum to the extensive database by Bastin and Heijligers [10, 11]. Of course, our depth resolution measurements are not intended to determine the element distribution and surface layer thickness in daily use. The thin film analytic using established software like STRATAGEM or PENEPMA are more suitable for this purpose [6, 12].

We explicitly consider the spatial resolution of the analytic element quantification. Difficulties resulting from the use of low primary electron energies, low overvoltages and L-lines are discussed in several recent papers [1, 13]. The knowledge of the depth resolution is of interest for scientific research and technological developments in fields like metallic coatings on steel and glasses, hard coatings on tools, surface reactions and separation in geology and soil science, electroplating, catalysts and fuel cells.

2. Experimental and results

2.1. Preparation of the test specimens

For the experimental determination of the quantitative analytical depth resolution under different measurement conditions, aluminium and silver layers with different thicknesses were prepared by vapour deposition onto silicon wafer substrates. The layer thickness is determined from calibrated SEM images of cross-sections prepared by cleavage, Focussed ion beam milling (FIB, FEI Helios NanoLab600) or broad ion beam milling (JEOL Cross Section Polisher IB-19520CP). The thicknesses of the aluminium layers are 3850, 940, 445, 345, 245, 150 and 120 nm, the thicknesses of the silver layers 870, 270, 210, 180, 150 and 90 nm.

To avoid fluorescence effects in the Si-substrate an additional 800 nm palladium layer was deposited between Si-substrate and silver layers. Ag-Lα X-ray with an energy of 2.984 keV cannot excite Pd-Lα X-ray by fluorescence due to the higher critical ionisation energy of Pd-Lα X-ray at 3.172 keV. Figure 2 shows the layer structure of a 150 nm silver layer with additional palladium layer as an example. The Pd layer is not necessary for the specimens with Al-layers.

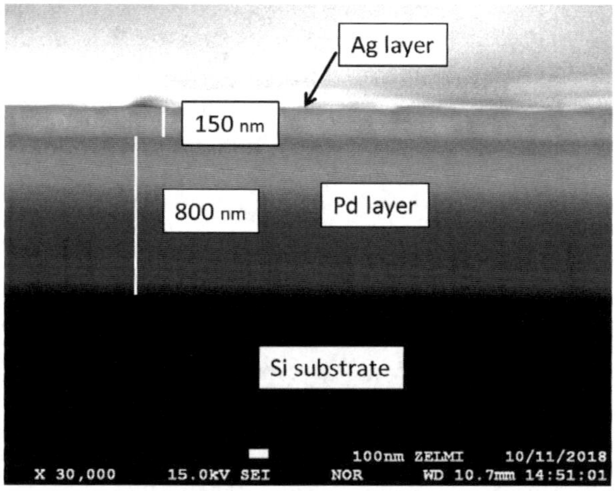

Figure 2. SEM image of a cross-section of a 150 nm silver layer with 800 nm intermediate palladium layer, cross section prepared by broad ion beam milling.

The density of both the Al and the Ag-layers were determined by the weighing method [14] to be (2.68 ± 0.02) g/cm^3 and (10.4 ± 0.1) g/cm^3 respectively. These density values are close to the bulk material densities indicating that the evaporated layers are not porous. Previously we found that sputtered layers are porous and have a different density compared to bulk samples. Finally the density values are verified by analysing quantitative measurements with the thin film software STRATAGEM.

2.2. Measured quantitative depth resolution

All test specimens are analysed at various electron energies and the Al- or Ag-content of the surface layer is determined quantitatively. Table 1 summarizes the results of the element quantification (including ZAF-correction) for the aluminium layers at the electron energies 15, 8, 7, 5, 4 and 3 keV. K_{ratio}-values are (without ZAF-correction) given in parentheses if these value are different to wt%.

IOP Conf. Series: Materials Science and Engineering **891** (2020) 012004 doi:10.1088/1757-899X/891/1/012004

Even lower energies are not useful due to too low overvoltage and low signals. To get sufficient X-ray intensity, the appropriate overvoltage to excite Al-Kα X-ray (critical ionisation energy: 1.559 keV) should not below ~ 2 [15].

Table 1. Results of the quantitative measurements of the aluminium layers in wt% Al and K_{ratio} values in parentheses if different.

Al layer thickness [nm]	15 keV	8 keV	7 keV	5 keV	4 keV	3 keV
3850 (Reference)	100.00	100.00	100.00	100.00	100.00	100.00
940	70.00 (0.6994)	100.01	99.8	99.96	100.16	99.92
445	/	/	96.08 (0.9604)	99.79	99.81	100.18
345	/	/	86.48 (0.8634)	99.59	100.19	100.18
245	/	/	/	95.25 (0.9522)	99.73	100.05
150	/	/	/	/	93.03 (0.9298)	99.72
120						98.64 (0.9863)

For each layer quantitative analyses were made using the 3,850 nm Al-layer as reference standard. Therefore, it is ensured that the reference standard has exactly the same density compared to the thin layers. In some cases, we observed a considerable difference between the densities of evaporated or sputtered films in comparison to bulk, metallurgical metals. The thickness of the reference Al-layer significantly exceeds the depth of the electron interaction volume of about 570 nm at 7 keV and 110 nm at 3 kV, respectively. Nevertheless, the continuum fluorescence in a much thicker bulk Al reference standard might originate from even higher depths compared to 3,850 nm. We assume that the difference would be small. The beam current was 10 nA and the acquisition time 20 s for the peak and 10 s for the background.

The values in Table 1 are the averages of 10 measurements in weight percent of aluminium (wt% Al). A quantification result of (100.0 ± 0.5) % indicates that the source volume of the emitted X-rays is completely inside the Al layer. The 150 nm layer is the thinnest one that can be resolved (99.72 wt% Al). Thus, a quantitative depth resolution in the range of 120 to 150 nm is achieved for aluminium at 3 keV electron energy. From our measurements it might be concluded only that the depth resolution is below 150 nm and above 120 nm. Additional test specimens would be necessary to determine the depth resolution more precisely. So far, we specify the measured depth resolution to be at least 150 nm for Al-Kα at 3kV.

The quantification results for silver are shown in Table 2. An evaporated Ag-layer is used as reference standard, whose thickness of 870 nm exceeds the depth of the electron interaction volume of about 165 nm at 7 keV and 75 nm at 5 keV, respectively. Again we assume that the continuum fluorescence nearly compares to that of a bulk reference standard. Here, the lowest practical electron energy to excite Ag-Lα X-ray (critical ionisation energy: 3.352 keV) is 5 keV at an overvoltage of 1.49. The best resulting depth resolution for silver is the range of 90 to 150 nm at 5 keV. The thinnest layer, for which an accurate analysis is still feasible is 150 nm (99.62 wt% Ag).

Table 2. Results of the quantitative measurements of the silver layers in wt% Ag and K_{ratio} values in parentheses if different.

Ag layer thickness [nm]	15 keV	8 keV	7 keV	6 keV	5 keV
870 (Reference)	100.00	100.00	100.00	100.00	100.00
270	80.92 (0.8117)	99.7	99.82	99.93	100.21
210	/	98.64 (0.9860)	99.58	99.87	100.14
180	/	/	99.36	99.82	99.62
150	/	/	/	98.49 (0.9953)	99.62
90	/	/	/	/	98.06

2.3. Monte Carlo simulation of the quantitative depth resolution

The electron scattering in aluminium and silver for electron energies between the primary energy (15, 8, 7, 6, 5, 4 and 3 keV) and the critical ionisation energy (1.559 keV for Al-K and 3.352 keV for Ag-L) is simulated with the programme CASINO (version 2.48) [5]. The simulations are performed using the experimental determined density values of the layers (see Sec. 2.1). It should be kept in mind that CASINO does not consider the secondary fluorescence of the bremsstrahlung continuum. Figures 3 and 4 shows the resulting plots of the electron paths of primary electrons (blue) and backscattered electrons (red). For the determination of the quantitative analytical resolution 99 % of the 10,000 simulated trajectories were taken into account. All electron trajectories are sorted by their penetration depth, then 1 % of the deepest are cancelled and the final depth is considered as the depth resolution, see indications in Figs. 3 and 4.

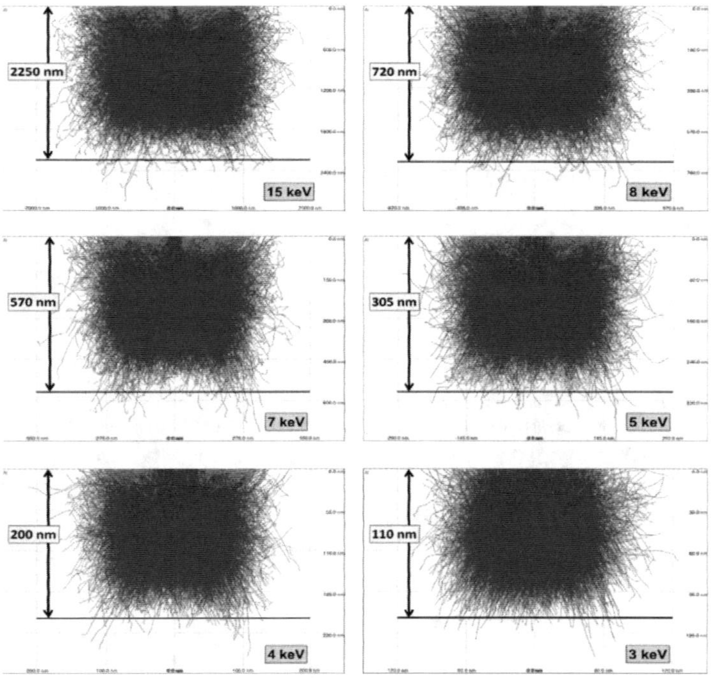

Figure 3. Monte Carlo simulations of scattered electrons in Al for different primary electron energies.

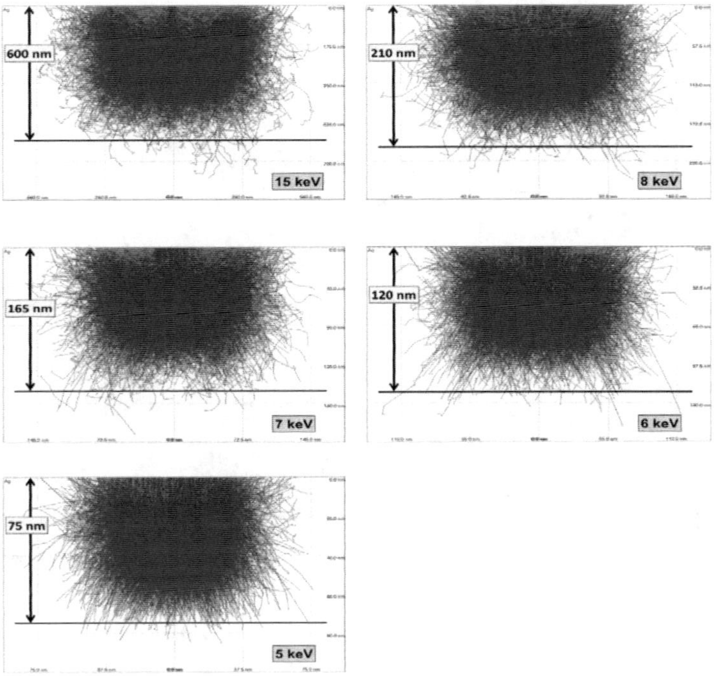

Figure 4. Monte Carlo simulations of scattered electrons in Ag for different primary electron energies.

The simulated Phi-Rho-Z curves in Figs. 5 and 6 demonstrate that, except at 15 keV, the self-absorption of Al-Kα in Al and Ag-Lα in Ag is very small, so that the range of the electron paths is a good measure for the range of emitted X-rays. The negligibility of self- and secondary fluorescence of our test specimens is considered to be a great advantage for further improvements of formulas to calculate the ionisation range of electrons or of Monte Carlo simulations at low primary electron energies.

2.4. Calculation of the quantitative depth resolution

The depth resolution is calculated by the equation of Castaing [16] as the ionisation range of electrons (X-ray range) R_x defined as the average path length that electrons travel before slowing down to an energy equal to the critical ionisation energy of the considered electron shell [4]:

$$R_X = \frac{33A}{\rho \cdot Z}\left(E_0^{1.7} - E_i^{1.7}\right) \tag{1}$$

where A is the atomic mass, ρ the measured (see Sec. 2.1) density, Z the atomic number, E_0 the energy of incident electrons and E_i the ionisation energy of the Al K-shell or Ag L-shell.

Since self-absorption of Al-Kα in Al and Ag-Lα in Ag is proven to be very small (see Phi-Rho-Z curves), absorption effects do not need to be considered.

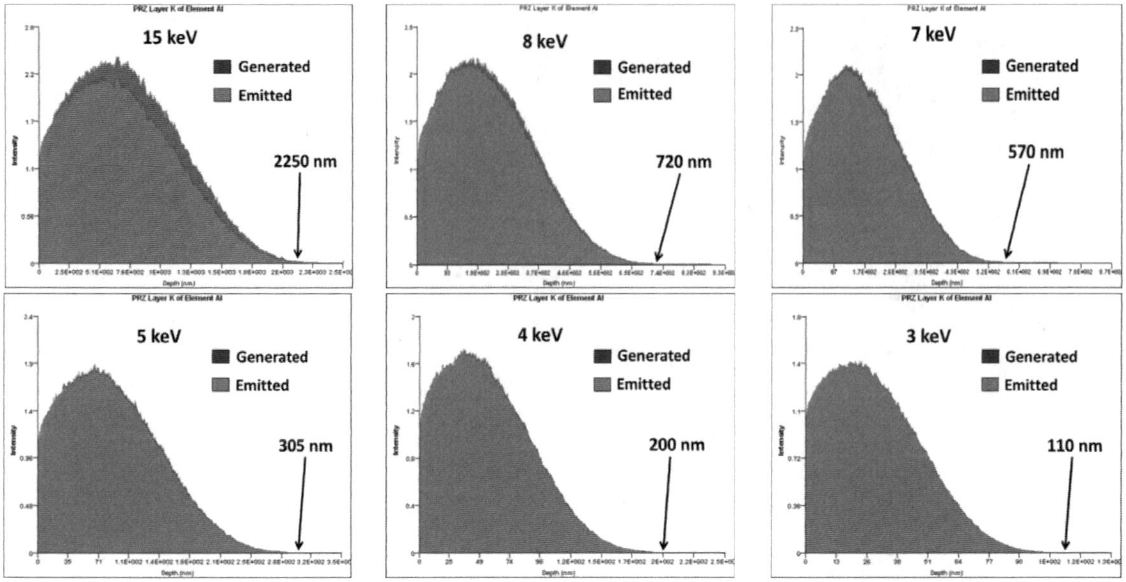

Figure 5. Simulated Phi-Rho-Z curves (intensity versus depth) of Al for different primary electron energies.

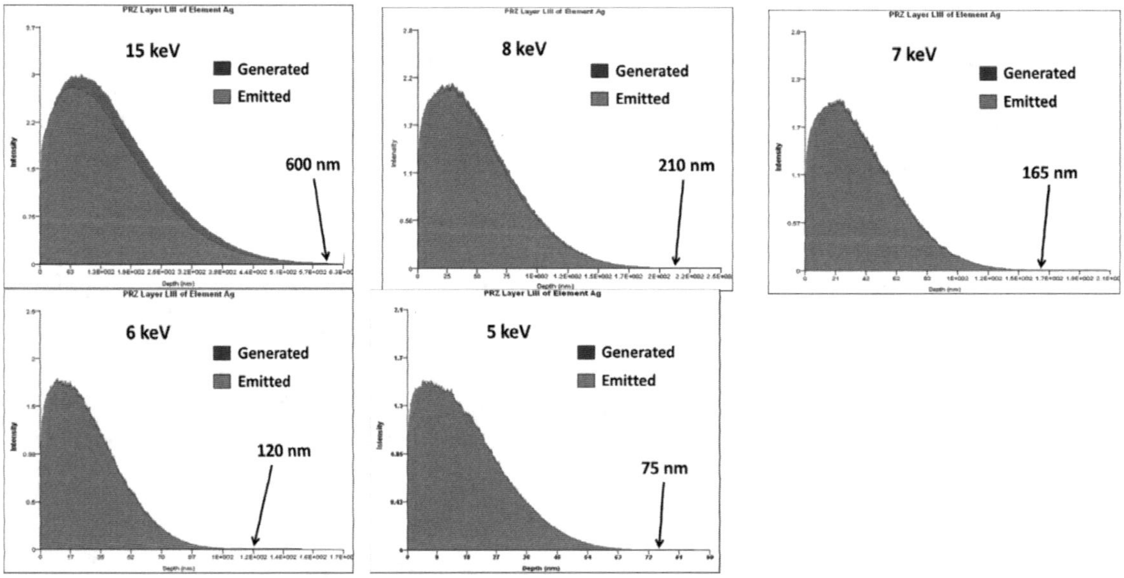

Figure 6. Simulated Phi-Rho-Z curves (intensity versus depth) of Ag for different primary electron energies.

For aluminium, Table 3 and Fig. 7 represent the comparison of the simulated, calculated and measured depth resolutions as a function of the primary electron energy E_0 of the primary electrons.

IOP Conf. Series: Materials Science and Engineering **891** (2020) 012004 doi:10.1088/1757-899X/891/1/012004

Table 3. Comparison of simulated, calculated and measured quantitative depth resolutions for Al and different primary electron energies. The ranges for the measured depth resolutions are due to the limited numbers of prepared test specimens with different layer thicknesses.

Electron energy	Calculated depth resolution [nm]	Measured depth resolution [nm]	Depth resolution from MC simulation [nm]
15 keV	2500	940-3850	2250
8 keV	823	445-940	720
7 keV	645	445-940	570
5 keV	340	245-345	305
4 keV	215	150-245	200
3 keV	111	120-150	110

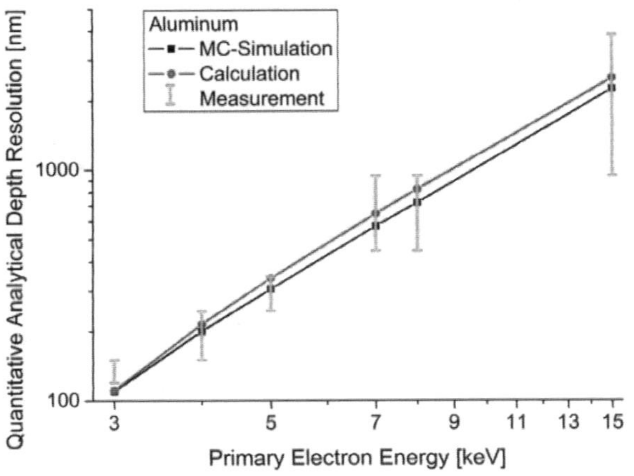

Figure 7. Plot of simulated, calculated and measured quantitative depth resolutions for Al and different primary electron energies. The error bar for the measurements are due to the limited numbers of prepared test specimens with different layer thicknesses.

The graph in Fig. 7 illustrates the improvement of the resolution with decreasing primary electron energy due to the reduction in the interaction volume. Another result is the quite good correlation between measurement and MC simulation, except at the primary electron energy of 3 keV for which the overvoltage is below 2. The reasons for this discrepancy are yet not clear. The calculated values are slightly offset to higher values than the MC simulations, but show a reasonable agreement with measured values. Nevertheless the calculated values deviate in the same manner at 3 keV. The reasons are not clear as well, but most probable both MC simulation and calculations differ from measurement for the equivalent reasons. In trying to explain the deviations it should be kept in mind that we observed the same effect for the lateral resolution as well [7, 8]. The here presented investigations of the depth resolution at least verify that the previous investigations of the lateral resolution, and in this case do not suffer from beam drift effects nor from an inappropriate consideration of the influence of the probe diameter on the lateral size of the signal source volume.

In the case of silver, Table 4 and Fig. 8 show that agreement between results from measurements and MC simulations is not as good. For all primary electron energies (5, 6 and 7 keV), for which the overvoltage is about 2, and below there is a significant deviation from the MC-simulation and the calculation for unknown reasons. One possible explanation might be that both MC-simulation and the calculation do not consider the fluorescence of Ag-L excited by the bremsstrahlung. The range of the bremsstrahlung X-rays is larger compared to the range of the electrons. Hence, if continuum fluorescence would be included, MC-simulations and the calculation would predict larger signal-source-volumes of Ag-L, which might be comparable to the measured depth resolution. The bremsstrahlung fluorescence was also used by Yuan *et al.* [17] to explain discrepancies between MC-simulations and experimental data in thin film analysis at higher acceleration voltages in the range 10 … 30 kV. Another possible explanation might be that more rigorous MC-programmes and calculation would perform better.

Table 4. Comparison of simulated, calculated and measured quantitative depth resolutions for Ag and different primary electron energies. The ranges for the measured depth resolutions are due to the limited numbers of prepared test specimens with different layer thicknesses.

Electron energy	Calculated depth resolution [nm]	Measured depth resolution [nm]	Depth resolution from MC simulation [nm]
15 keV	670	270-870	600
8 keV	193	210-270	210
7 keV	142	180-210	165
6 keV	96	150-180	120
5 keV	55	90-150	75

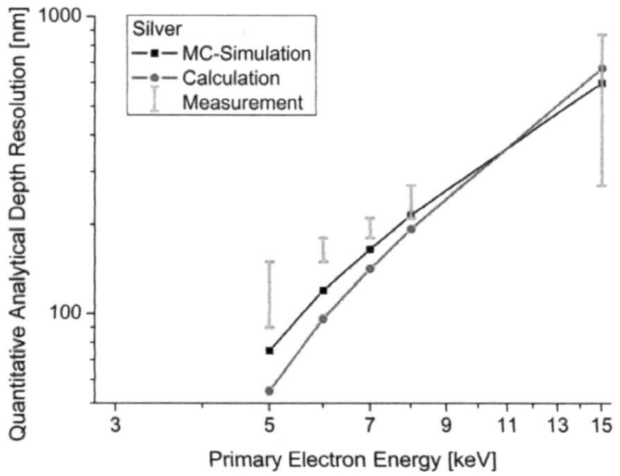

Figure 8. Plot of simulated, calculated and measured quantitative depth resolutions for Ag and different primary electron energies. The error bar for the measurements are due to the limited numbers of prepared test specimens with different layer thicknesses.

2.5. Comparison of quantitative lateral and depth resolution

The experimental results presented here together with previously published data [8] allow for a rough comparison between the quantitative lateral and depth resolution, see Table 5. As expected, the depth resolution is in all cases better compared to the lateral resolution. Obviously, the width of the electron interaction volume is larger than its depth. Even if the beam diameter (52 nm at 3 keV, 38 nm at 5 keV) is subtracted from the measured lateral resolution, the interaction volume is wide and flat. Most probably this behaviour is caused by the angular distribution of the electron scattering, which consists of a small-angle forward scattering close to the surface and a diffuse scattering in the bulk [18].

Table 5. Comparison of the quantitative lateral [8] and depth resolution for Al and Ag.

Layer material and Electron energy	Calculated *depth* resolution [nm]	Calculated *lateral* resolution [nm]	Measured *depth* resolution [nm]	Measured *lateral* resolution [nm]	*Depth* resolution from MC simulation [nm]	*Lateral* resolution from MC simulation [nm]
Al @ 3 keV	111	164	120-150	260-280	110	185
Ag @ 5 keV	55	104	90-150	150-210	75	124

A deeper insight into the shape of the interaction volume is obtained if the ratios between the depth and lateral resolution is considered, see Tables 6 and 7. Since the thickness of our test samples are optimised for the lowest electron energy, there is one number for the ratio of measured values only.

Table 6. Comparison of the ratios between analytical lateral [8] and depth resolution for Al.

Electron energy	$Ratio_C$ of depth to lateral resolution from calculation	$Ratio_M$ depth to lateral resolution from measurement	$Ratio_{MC}$ depth to lateral resolution from MC simulation
15 keV	0,64		0,74
8 keV	0,66		0,72
7 keV	0,67		0,72
5 keV	0,68		0,71
4 keV	0,69		0,70
3 keV	0,68	≈ 0,5	0,59

Table 7. Comparison of the ratios between analytical lateral [8] and depth resolution for Ag.

Electron energy	$Ratio_C$ depth to lateral resolution from calculation	$Ratio_M$ depth to lateral resolution from measurement	$Ratio_{MC}$ depth to lateral resolution from MC simulation
15 keV	0,56		0,66
8 keV	0,57		0,67
7 keV	0,57		0,68
6 keV	0,56		0,66
5 keV	0,53	≈ 0,7	0,60

The calculations and the MC simulations indicate a slimmer and deeper interaction volume for Al and a flatter interaction volume for Ag. Surprisingly, it is the other way around for the measured values. On one hand, the larger depth of the interaction volume might be due to the bremsstrahlung fluorescence as well, which is higher for larger electron energies and for Ag. The bremsstrahlung fluorescence is not considered in the calculation and in CASINO, but is present in the measurements. On the other hand, it is not yet clear why the bremsstrahlung fluorescence does not have the same influence on the lateral resolution. If this would be the case, the ratio between depth and lateral resolution in Tables 6 and 7 should not be affected by the bremsstrahlung fluorescence. Obviously, more measurements and MC simulation with other programs are necessary.

3. Conclusion

In this paper, measurements of the quantitative analytical depth resolutions are presented. The experiments are accomplished at test specimens with metallic surface layers with different thicknesses. To determine the depth resolution, the primary electron energy is reduced until the quantification of the layer material differs from 100 %. For aluminium the lowest quantitative depth resolution in the range of 120 to 150 nm is obtained for 3 keV. In comparison, the lowest quantitative lateral resolution was in the range of 260 to 280 nm [8]. For silver the lowest quantitative depth resolution is in the range of 90 to 150 nm at 5 keV. The quantitative lateral measurements resulted in the lowest resolution of 150 to 210 nm [8]. The difference between depth and lateral resolution might be explained by the beam diameter and the angular distribution of the electron scattering.

The experimental results are in accordance with Monte Carlo simulations and analytical calculations for most primary electron energies, but differ significantly for low energies for which the overvoltage is below or close to two. The cause for this discrepancy is not clear yet and has to be clarified by further investigations, in particular because the best quantitative analytical resolution is obtained at low energies for which the discrepancy is the largest. The presented measurements of the depth resolution are valuable for the first steps of further improvements and allow the testing of other MC-codes as well. These samples provide a simple test case where secondary fluorescence is negligible.

Acknowledgement

The financial support by the Europäische Fonds für Regionale Entwicklung (EFRE) within the projects "Nano-Analytiklabor" and “Nano-Werkbank” is gratefully acknowledged.

References

[1] Buse B and Kearns S 2018 *Microsc. Microanal.* **24**1-7
[2] Hughes E C, *et al.* 2018 *Amer. Mineralogist* **103** 1473-1486
[3] Fournelle J, *et al.* 2016 *IOP Conf. Ser.: Mater. Sci. Engng.* **109** 012003
[4] Merlet C and Llovet X 2012 *IOP Conf. Ser.: Mater. Sci. Engng.* **32** 012016
[5] Drouin D, *et al.* 2007 *Scanning* 29 92-101
[6] Llovet X and Savat F 2016 *IOP Conf. Ser.: Mater. Sci. Engng.* **109** 012009
[7] Berger D and Nissen J 2016 *IOP Conf. Ser.: Mater. Sci. Engng.* **109** 012001
[8] Berger D and Nissen J 2018 *IOP Conf. Ser.: Mater. Sci. Engng.* **304** 012002
[9] Berger D and Nissen J 2014 *IOP Conf. Ser.: Mater. Sci. Engng.* **55** 012002
[10] Bastin G F and Heijligers H J M 2000 *X-ray Spectrom.* **29** 212-238
[11] Bastin G F and Heijligers H J M 2000 *X-ray Spectrom.* **29** 373-397
[12] Hodoroaba V-A, *et al.* 2018 *Microsc. Microanal.* **24** 758-759
[13] Moy A, *et al.* 2019 *Microsc. Microanal.* **25** 664-674
[14] Berger D and Galbert F 2011 in: *MC2011 Conf. Proc.* (Kiel, Germany) ISBN 978-3-00-033910-3 IM7.P208
[15] Carpenter P K and Jolliff B L 2015 *Microsc. Microanal.* **21 (Suppl 3)** 1434-1444
[16] Castaing R 1960 *Adv. Electronics and Electron Phys.* **13** 317
[17] Yuan Y, *et al.* 2016 *Microsc. Microanal.* **22** 400-401
[18] Berger D 2000 PhD-thesis. (Berlin, Germany: TU Berlin) https://depositonce.tu-berlin.de/handle/11303/457

IOP Conf. Series: Materials Science and Engineering **891** (2020) 012005 doi:10.1088/1757-899X/891/1/012005

Spatial resolution limits of EPMA

B Buse and S L. Kearns

University of Bristol, School of Earth Sciences, Wills Memorial bldg., Queen's Road, Bristol BS8 1RJ, Great Britain

e-mail: ben.buse@bristol.ac.uk

Abstract. The development of field emission EPMA, has significantly improved the lateral resolution of EPMA. Two strategies are available for achieving high spatial resolution, either low overvoltage or low voltage analysis. Determining the spatial resolution for a particular analysis is complex and depends on the voltage, spot size, beam current, density of the sample, X-rays analysed and the precision and sensitivity required. Monte carlo simulations can be used to evaluate the spatial resolution for different analytical conditions and samples, provided the minimum spot size achievable at the conditions is known. Spot size is important in determining lateral resolution, which initially improves with decreasing accelerating voltages, and then increases as the minimum spot size becomes more significant than the distance electrons travel. For trace elements, the requirement for precision and sensitivity demands either higher overvoltages increasing the volume from which X-rays are generated, and/or higher beam currents increasing the spot size. At low overvoltage or low voltage conditions many additional factors must be considered: carbon contamination, coat thickness, erosion of the carbon coat, the stability of the sample and the problems surrounding the measurement of soft X-ray lines including L-lines for first row transition metals. When analysing L-lines for quantifying first row transition metals, either calibration curves must be used, or the anomalous mass absorption and variations in the fluorescence yields from partial shell occupancy must be accounted for, by using fits to experimentally determined values. By taking these factors into consideration, high quality measurements can be performed.

1. Introduction

Electron probe microanalysis (EPMA) has traditionally employed a thermionic tungsten source analysing material on the scale of several cubic micrometres (for densities of 2 - 5 g/cm^3), at accelerating voltages of 15 - 20 kV. Field emission gun (FEG) EPMA, which employs a sharpened tungsten crystal to reduce the source area of electron emission and provides a dense beam of electrons, permits higher spatial resolution to be achieved at low voltage [1]. High spatial resolution is important for analysing small phases or inclusions, diffusion profiles and particles, although geometry must be considered for the latter. Monte Carlo simulation is often used to image the scattering and deceleration of the electron beam on entering a material. At high accelerating voltages (15 - 20 kV) the electron range (the distance electrons travel into the sample) is several microns (Fig. 1) whilst the beam diameter (spot size) is comparatively small (10's - 100's nm) even for a conventional tungsten source. At low accelerating voltages (< 10 kV) the electron range is 100's nm (Fig. 1), making the spot size (10's - 100's nm) an important component of the lateral resolution (Fig. 2), exacerbated by the increase in beam diameter at lower accelerating voltages [1].

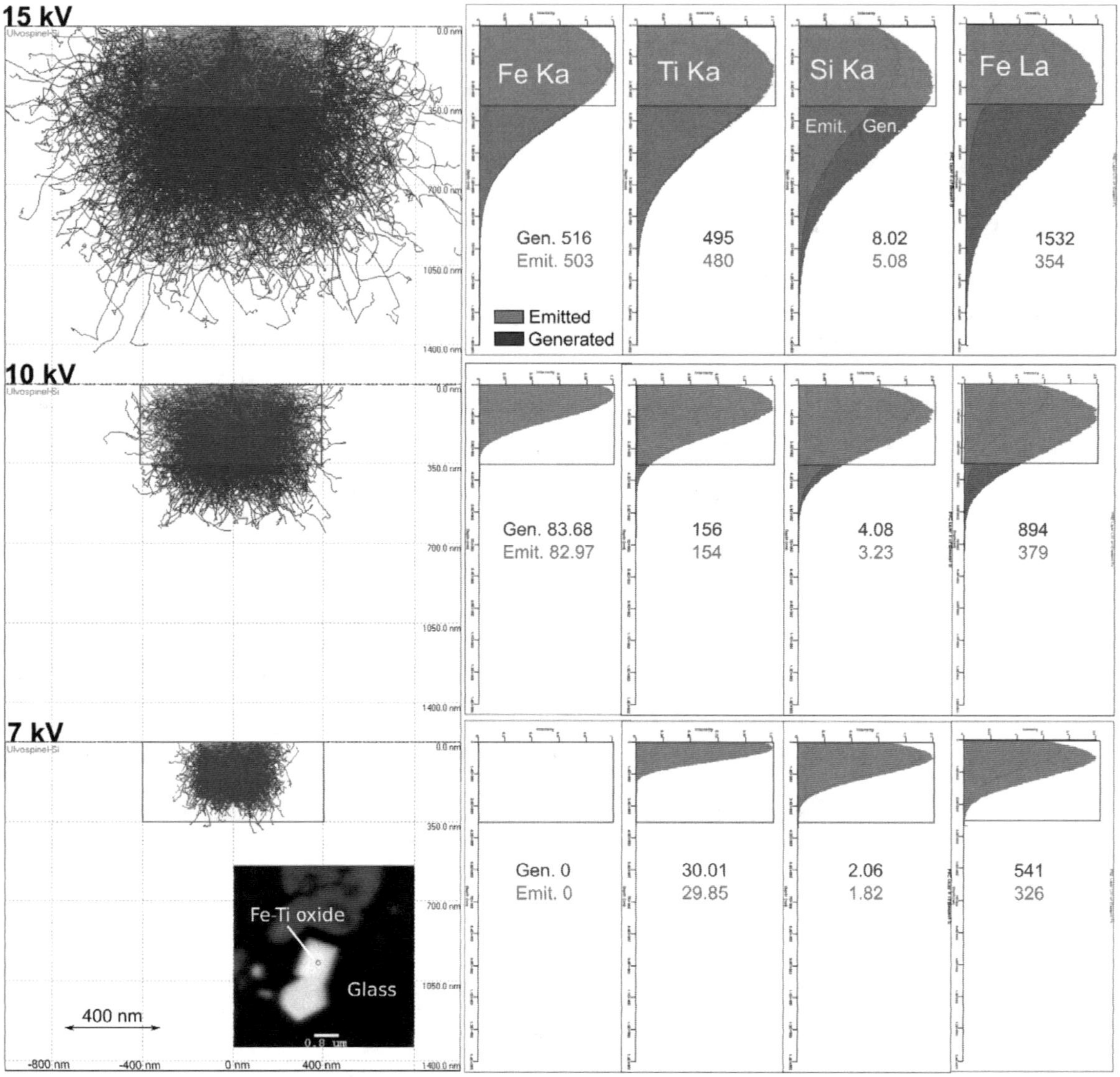

Figure 1. Monte Carlo simulations (using CASINO [2]) for bulk ulvospinel (Fe_2TiO_4) with trace Si and a density of 4.78 g/cm^3 for a beam diameter of 10 nm. On the left is the electron distribution, with backscattered electrons coloured red; on the right are the phi-rho-z curves for different elements. The blue and red values refer to the number of X-rays generated (Gen.) and emitted (Emit.) respectively. The yellow box represents the size of a 800 x 350 nm crystal, a typical size of minerals formed during high pressure experiments; see BSE image insert. Table 1 gives analyses for this crystal.

Pinard and Richter [3] and McSwiggen [4] have well characterised the interaction volume for a focussed beam as the accelerating voltage is reduced for FEG-EPMA at a constant beam current. The interaction volume initially reduces and then increases reflecting the interplay of electron range and spot size, with reduced accelerating voltage resulting in the electron range decreasing whilst spot size is increasing. Minimum interaction volumes were found to be typically in the 5 - 8 kV range. This may be calculated using Monte Carlo simulations for a range of beam currents, if the change in spot size with beam current and accelerating voltage is known, allowing the detection limit and precision to be considered and therefore the ability for trace elements to be measured at a given spatial resolution.

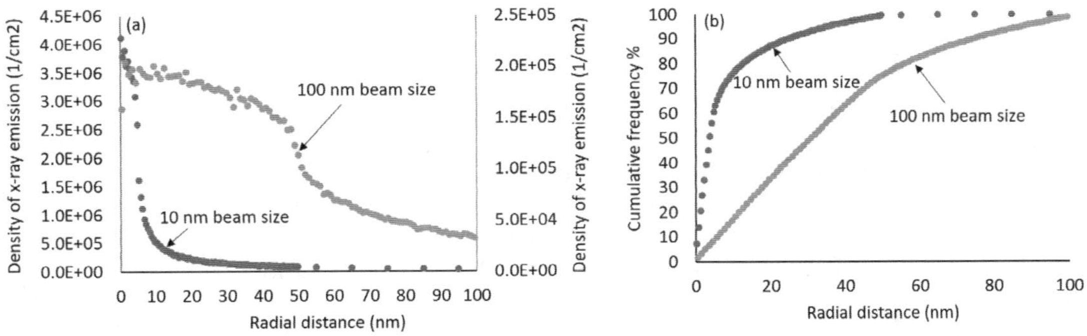

Figure 2. Fe-Lα radial distribution of X-rays calculated using PENEPMA Monte Carlo simulation [5] for ulvospinel at 7 kV and spot sizes of 10 nm and 100 nm.

Pinard and Richter [3] and McSwiggen [4] instead of considering the electron range and electron interaction volume, consider the volume from which X-rays are emitted. This determines the spatial resolution of chemical analysis and is termed 'analytical volume' by McSwiggen [4]. Given that different X-rays have different critical excitation energies at a single accelerating voltage, different X-rays will have different analytical volumes. If the overvoltage (the ratio of accelerating voltage to the critical excitation energy of the X-ray) is small, X-rays are generated close to the point at which the electron beam enters the sample (e.g., Fe Kα at 10 kV, Fig. 3). The low overvoltage means that only a small loss of energy is required before the beam is unable to excite the X-ray of interest. Whereas if the overvoltage is large, the analytical volume will closely correspond to the electron interaction volume (e.g., Si Kα at 10 kV, Fig. 3). The depth of the analytical volume is also dependent on the amount of absorption of the X-ray by the sample, determined by the mass absorption coefficient (MAC); in the case of oxygen Kα or Fe Lα which are soft X-rays (< 1 keV) and have high MACs only those X-rays generated near the surface are emitted and detected (e.g., Fe Lα at 20 kV, Fig. 3). This however has little effect on the lateral width of the analytical volume.

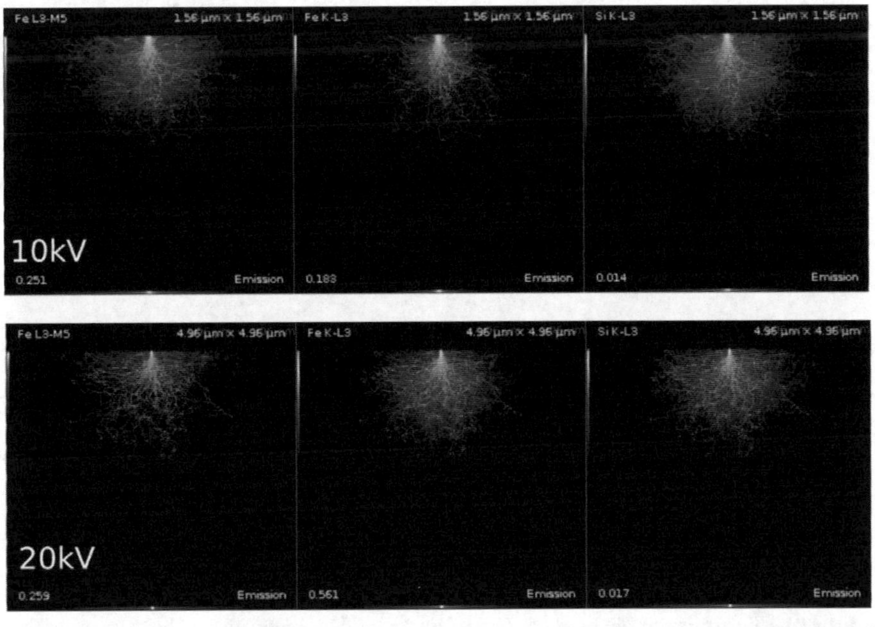

Figure 3. Integrated 2D X-Z slices displaying the source of emitted x-rays generated from Monte Carlo simulations using DTSA2 [6]. Red dots represent sites at which emitted X-rays are generated, green lines trace the electron trajectories.

1.1. Strategies for obtaining high spatial resolution

Two strategies are available for obtaining high spatial resolution in bulk samples (e.g., [3, 4, 7]; see [8] for very thin samples), either a low voltage resulting in a small electron interaction volume or a low overvoltage, which may be at relatively high accelerating voltage for a high energy X-ray, resulting in a small analytical volume for the X-ray of interest. Using low accelerating voltages allows multiple X-rays to be measured at a single accelerating voltage, with spatial resolution equal to or less than that of the interaction volume. In the example of the 350 nm thick ulvospinel in Fig. 1, all the elements could be analysed at 7 kV. Low voltage, however, limits the X-rays that are excited (e.g., Fe-L rather than Fe-K at 7 kV), often restricting the analyst to soft X-rays, which may be problematic. Problems include the close spacing of X-rays with the potential for peak overlaps, X-ray peaks may be subject to shifts and peak shape changes for different compounds (e.g., K-lines of B, C ,N and O [9]) or axial orientation (e.g., B-Kα [9]) and in the cases of Lα X-rays of first row transition metals and Mα X-rays of rare earth elements anomalous self absorption occurs (e.g., [10-12]). Surface layers also become significant, reducing the beam energy entering the sample and absorbing soft X-rays. The following must be considered: (1) thickness of conductive coat [13], including erosion of a carbon coat by the electron beam, where high current density is used (small beam size and high beam current) particularly at low accelerating voltages [14]; (2) presence of oxide layers on metal samples [15]; and (3) carbon contamination, commonly deposited in a ring around the electron beam (see Fig. 4) affecting closely spaced analyses [16, 17].

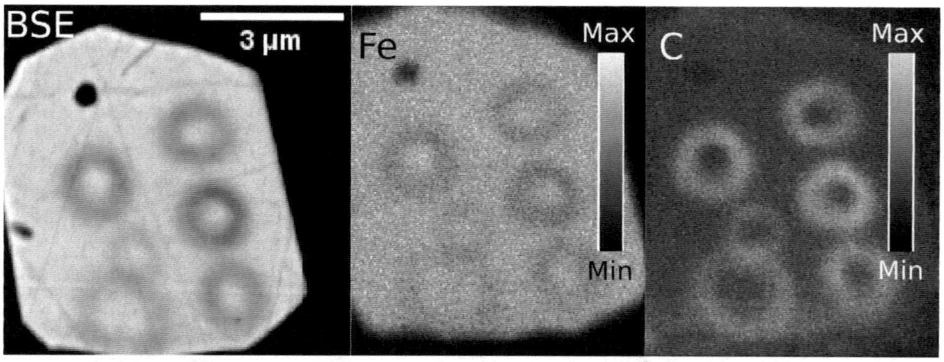

Figure 4. Backscatter electron (BSE) image and C and Fe X-ray intensity maps of analysis spots on a Fe-Ti spinel. Carbon contamination forms rings around the point of electron beam impact. At low overvoltage (accelerating voltage: 8.5 kV, critical excitation energy of Fe-Kα: 7.1 keV) carbon contamination strongly reduces Fe-Kα intensity.

Analysis at low overvoltage conditions at high voltages, avoids the necessity to use soft X-rays but low overvoltage conditions will only be satisfied for a few elements at a single voltage requiring multiple voltage analysis. In the example of the 350 nm Ulvospinel of Fig. 1, Fe Kα can be analysed at 10 kV, whilst Ti and Si must be analysed at 7 kV, to keep the analytical volume within the ulvospinel (see Table 1 for results). At low overvoltages surface layers (contamination, conductive coats, and oxide layers) become very significant (see figure 4, also [14, 17]), with a small reduction in beam energy entering the sample resulting in a large drop in X-rays generated.

Methods to mitigate the problems of surface layers include reducing carbon contamination by using liquid nitrogen or Peltier-cooled cold fingers (e.g., [17-19]), empirical correction [17], oxygen air jets for samples without a conductive carbon coat [18] and sample preparation by focussed ion beam (FIB) milling or cross-section polishing by Ar-ions to remove surface contaminations [20]. Recently for metal

Table 1. Quantification of a 800 nm wide ulvospinel in a Si-rich, Fe-poor glass at multiple voltages, using the Kα X-ray line for Fe. To achieve a reasonable quantification, Fe must be analysed at 10 kV, whilst the other elements analysed at 7 kV (see Fig. 1).

	MgO	SiO_2	CaO	FeO	Na_2O	Al_2O_3	K_2O	TiO_2	Total
15kV	1.24	26.37	0.71	59.15	1.13	6.47	0.72	11.39	107.15
10kV	1.29	4.73	0.18	76.48	0.04	1.64	0.09	15.47	99.90
7kV	1.32	1.36	0.24	76.48[a]	0.01	0.98	0.05	17.52	97.96

[a] 10 kV measurement used.

samples, Yamashita *et al.* [21] demonstrated the effectiveness of in-chamber plasma cleaning, sample heating together with a liquid nitrogen cold trap. Problems of the electron beam erosion of the carbon coat can be mitigated by: (1) reducing the current density, which may compromise either resolution or precision, (2) switching to a different conductive coat material, or (3) correcting for X-ray intensity change using time-dependent-correction (TDI) [14].

1.2. First row transition metal l-line analysis

For the analysis of many materials, first row transition metals are important. For low voltage analysis it may not be possible to excite the K-line X-rays, requiring analysis of the L-line X-rays. There has been considerable research exploring the complications in quantifying these lines. In 1985, Pouchou and Pichoir [22] demonstrated for Ni alloys, that the Lα X-rays suffered from anomalous self-absorption and partial fluorescence yields, with absorption varying not just with element concentration but the structure of the valance band, which is affected by the chemical environment (chemical bonding of atoms) requiring different MACs for different compositions. Similarly, the partial fluorescence yield varies with composition. Pouchou and Pouchoir [22] showed that for a binary series the MAC and partial fluorescence yield can be calculated using regression, allowing quantification. Fialin [23] noted the problem in MACs changing with electronic structure for Zn minerals and, therefore, standards and unknowns may have different MACs. More recently, Llovet *et al.* [24] applied this approach to Ni silicides, and Buse and Kearns [25] to olivines and showed that if solid-solution MACs and partial fluorescence yields are incorporated in matrix corrections, quantification is possible to an accuracy of ≤ 4 % relative. In examining olivines, Buse and Kearns [25] illustrated some of the difficulties in this approach; it requires measuring off-peak for some compositions and measuring the MAC across the absorption edge, where it is highly sensitive to the position of measurement. Moy and Fournelle [26] have instead of using conventional WDS peak and background measurements, have derived peak intensities through spectral deconvolution, from which MACs and partial fluorescence yields have been derived.

Remond *et al.* [27] tried a different approach to deal with anomalous absorption of Lα and Lβ, by acquiring spectra, from which absorption spectra were calculated and used to correct the spectra. Using this method, they successfully analysed Fe_2O_3 with a FeO standard.

Other approaches to the analysis of first row transition metals have been tried. Gopon *et al.* [28] analysed Fe silicides using the Fe-Lℓ noting that unlike the Fe-Lα and -Lβ X-ray lines it was not affected by bonding. This method has yielded improved quantification compared to using the Fe Lα-line without correcting for bonding effects. Statham and Holland [7] tested the Lℓ-line for measurement of steels, again finding improved quantification compared to Lα, although Cr-Lℓ was 9 % relative below the expected value. The Lℓ typically has low intensity compared to the Lα-, Lβ-lines limiting sensitivity and precision [7, 28]. Another solution is the use of calibration curves, allowing the use of the higher intensity Lα- and Lβ-lines whilst avoiding a knowledge of the variation in MACs and fluorescence yields, such as demonstrated by Buse and Kearns [29] for olivine. Moy and Fournelle [30] have investigated the use of non-machine specific calibration curves, by using the integrated intensity across the Lα- and Lβ-emission lines. The calibration curve is material specific, in this case Fe-silicides. Errors using a machine specific calibration curve are < 3.6 % relative, excluding two samples which are believed to be tilted.

2. Achieving high spatial resolution for Fe-bearing minerals

Fe-bearing minerals demonstrate the problems in analysing first row transition metals at high resolution. In this section, we analyse Fe-bearing minerals using the two strategies outlined above. To address the problems of L-line analysis several approaches are used: (1) Fe-Lℓ analysis, (2) calibration curve for Fe-Lα analysis when re-peaking on each sample, (3) calibration curve for Fe-Lα analysis with a fixed position, and (4) extending the method of Buse and Kearns [25] using a regression-fit to experimentally measured MACs and calculated partial fluorescence corrections to cover silicate minerals with complex substitutions and varying oxidation states. Measurements were made on a JEOL 8530F FEG EPMA at the University of Bristol. A Peltier cold finger [19] was used during analysis to minimise contamination and analyses were widely spaced to avoid carbon contamination from previous points. The measurements were made using a 10 μm beam size to avoid the possibility of eroding the carbon coat, thereby allowing the ability of quantification procedures to be assessed whilst removing other affects.

2.1. Fe-Kα low overvoltage measurements

Analysing the Fe Kα-line with an accelerating voltage of 10 kV, gives an overvoltage of 1.4 and yields low ionisation with net counts ca. 100 cps/nA on Fe metal using LIF compared to 1,200 - 1,600 cps/nA at 20 kV for the same crystals. Comparable precision therefore requires longer count times and/or higher beam currents. High precision was obtained for the measurements e.g., 0.8 % rel. (2 σ) for St John's Island Olivine with 7.5 wt% Fe, using a high beam current (100 nA) and long count times (60 seconds on peak) and two spectrometers combined. The results are shown in Fig. 5, comparing measured values to reference values. The results are accurate to within 5 % relative and generally insensitive to the choice of standard.

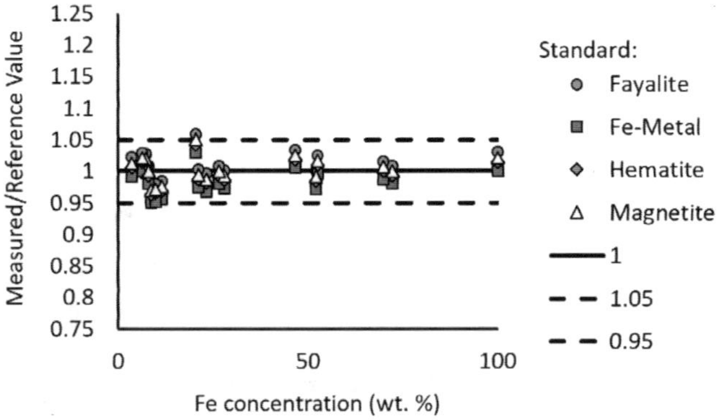

Figure 5. Fe-Kα measurements at 10 kV compared to reference values, using 4 different materials as a standard.

2.2. Fe-Lℓ measurements

The Fe Lℓ-line was measured at 7 kV, 200 nA for 120 seconds on peak using a TAPH crystal. Careful background selection is required to avoid interference with Fe-Lη and Mn-Lα peaks; in this case only the upper background measurement was used. Fe-Lℓ has low net intensities of 4.37 cps/nA on Fe metal. This is primarily a function of the refracting crystal, for despite the low fluorescence yield of L3-shell vacancies, this is compensated by the large number of ionisations. L3-shell vacancies are either filled by electrons from the M5-shell to yield Lα X-rays, or the M1-shell to yield Lℓ X-rays. Lα X-rays dominate, although the relative intensity depends on the occupancy of the 3d electron shell, which varies with element [31, 32] and bonding.

The low intensities are reflected in the larger errors at the low concentrations (Fig. 6a). The accuracy of the results varies considerably between samples, with some samples differing from reference values by > 5 % relative. The calculated concentrations are sensitive to the standard used (Fig. 6b), with magnetite standard giving the best results for the dataset, whilst Fe metal performing poorly. A look at the peak positions for fayalite and Fe metal shows that this poor performance is not due to peak shift or shape.

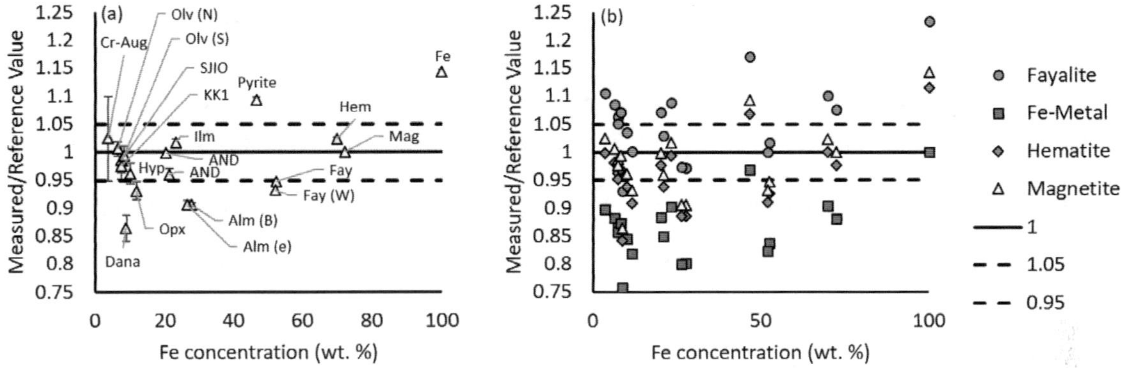

Figure 6. Fe-Lℓ measurements at 7 kV compared to reference values using a) magnetite, and b) 4 different materials as the standard. 1 sigma error bars shown for (a).

2.3. Fe-Lα measurements using calibration curves

Fe Lα was measured at 7 kV, 200 nA, on a TAP crystal, counting for 60 seconds on peak. Two methods of measuring the Fe Lα-line were tested, one using a fixed peak position (that of Fe metal), the other peaking up on each sample. To understand the changes in peak position, Fig. 7 shows the shift in peak positions (see [33]). The Lα-peak position shifts to high wavelengths with increasing Fe content as the low wavelength side is strongly absorbed by the Fe-L3 absorption edge. As Hofer and Brey [34] for garnets and Fialin *et al.* [33] for a range of minerals have shown Fe self-absorption is greatest for Fe^{2+}, with Fe^{3+} resulting in a shift in the absorption edge away from the Lα-peak maximum. Therefore, as Fialin *et al.* [33] showed, on Fig. 7 the samples with Fe^{3+} have peak positions at lower wavelengths consistent with less self-absorption.

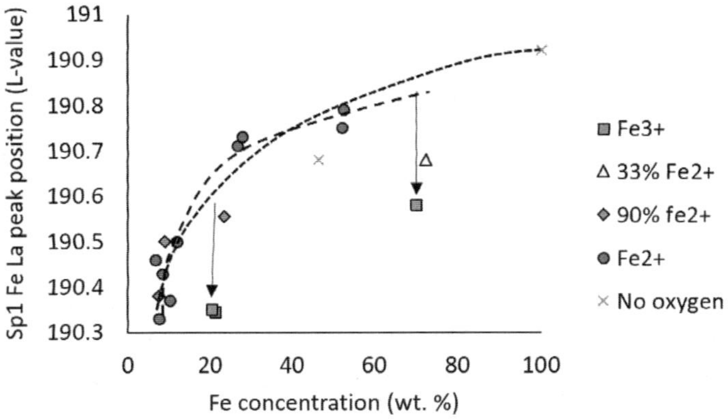

Figure 7. Fe Lα-peak, colour coded in terms of their Fe oxidation state. There is uncertainty where to plot the trend line, and whether it is continuous with Fe metal.

Figures 8 and 9 give the results for Lα using re-peaking on each sample and a fixed position. To interpret the results we must consider the measurement position relative to the peak maximum, the true emission peak position (the peak maximum without self-absorption), and the position of absorption maximum. In Fig. 8 each measurement is at peak maximum, but the peak maximum moves away from the true emission peak (and the absorption maximum) as absorption erodes the low wavelength side of the peak. In Fig. 9 the measurements are at the Fe metal peak, which is easy to accurately define, and reflects the emission peak for Fe metal. However, whilst this is close to the peak maximum for the minerals with high Fe content (see Fig. 7) as a result of absorption eroding the high-energy side of the peak, it is a long way from the true emission peak, approximated by the low Fe content samples which have suffered less absorption (see Fig. 7, also [25, 33]). The Fe peak position, has lower energy than the absorption edge for both Fe^{2+}- and Fe^{3+}-minerals, making absorption similar at this position. Both datasets show a tendency for calculated *k*-ratios to overestimate, particularly at high concentrations where self-absorption is higher. This corresponds to elevated MACs for minerals containing oxygen compared to Fe metal, consistent with the findings of Buse and Kearns [25] and Moy and Fournelle [35] for olivine. The opposite is seen for pyrite (FeS_2), where the calculated *k*-ratio underestimates, this is consistent with pyrite having a lower MAC than Fe metal, similar to that observed by Llovet *et al.* [24] for Ni-silicides.

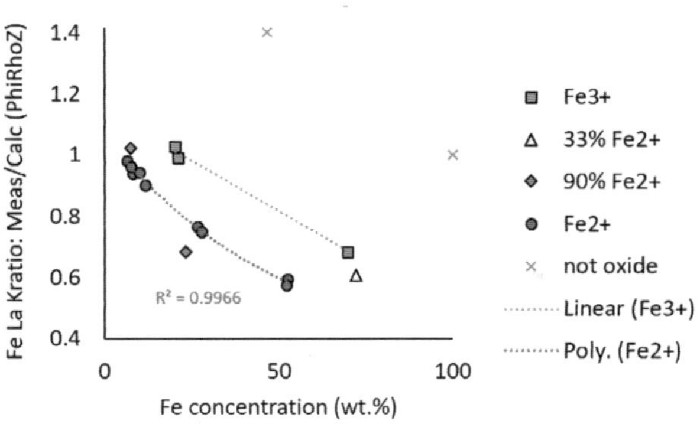

Figure 8. Fe-Lα measurements, on peak maximum, with peak search for each sample. Colour coded for Fe chemical state.

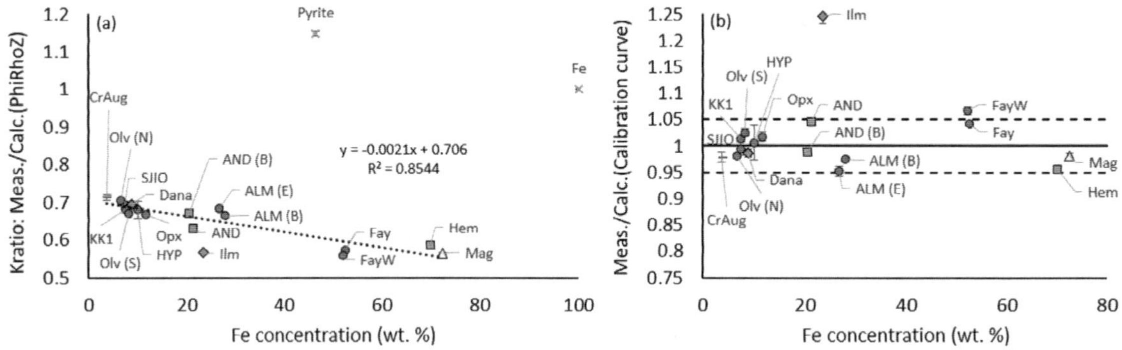

Figure 9. Fe-Lα measurements using the Fe metal peak position. a) Measured *k*-ratios compared to calculated *k*-ratios using phi-rho-Z models. Regression does not include pyrite and Fe metal; b) Measured *k*-ratios compared to calculated *k*-ratios using a calibration curve, here pyrite and Fe are off-scale. 1 sigma error bars are plotted. Colours and symbols are the same as for Figs. 8 and 7 and correspond to Fe chemical state. Cr-Augite has 80 % Fe^{2+}.

If we compare the two sets of measurements, it can be seen that by re-peaking on each sample, the discrepancy between measured and calculated *k*-ratios is largely removed for minerals with low Fe-content, where measurement is now close to the true emission peak and absorption is minimal. At high Fe contents, re-peaking has little effect, as the peak maxima are close to Fe metal. Crucially whilst measurement using peak maximum differentiates between Fe^{2+}- and Fe^{3+}-minerals and is, therefore, sensitive to the differences in self-absorption with Fe valence state; measurement at the Fe-metal peak position is not sensitive to this. This appears to result from the measurement being acquired below the absorption edge for both Fe^{2+}- and Fe^{3+}-minerals, resulting in similar absorption. This approach was tested on an extended range of minerals shown in Fig. 10, in this case to achieve similar precision correction for matrix effects from the other elements was required.

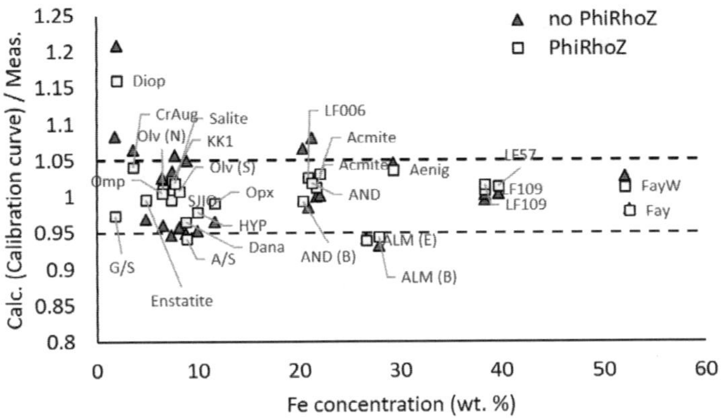

Figure 10. Measurements at the Fe metal peak (as in Fig. 9) for an extended range of materials. Agreement between measured values and calculated values using a calibration curve is improved when correcting for the matrix effects on Fe from the other elements in the material. Minerals include: pyroxences (CrAug, enstatite, diop, hyp, salite, acmite), olivines (Olv(N), Olv(S), LF57, LF109, LF006, SJIO), garnets (AND, ALM, G/S, A/S, Dana), aenigmatite (Aenig) and amphiboles (Omp, KK1).

2.4. *Fe Lα measurements: experimental MACs and partial fluorescence yields*

To investigate the application of the method of Buse and Kearns [25] to a range of silicate minerals with different valence states, MACs were measured for a range of materials (Fig. 11a). MACs were measured using the procedure outlined in Buse and Kearns [25], based on the method of Pouchou and Pouchoir [22], whereby the MAC is varied until a good agreement is reached between measured and calculated values for a range of voltages. It can be seen from Fig. 11a, that the MAC for Fe-Lα by Fe is a function of valence state, consistent with the observations above. Coordination does not appear to have a strong effect. Therefore, if the valence state is known, the MAC can be calculated. To determine the valence state, the ratio of Lα to Lβ at fixed positions was used as a proxy, as shown in Fig. 11b. This allows the MAC to be calculated from the Lα/Lβ-ratio and the Fe concentration.

Figure 12a shows the calculated partial fluorescence yield correction. Partial fluorescence yields were calculated using the method of Pouchou and Pouchoir [22] and Llovet et al. [24]. The calculated partial fluorescence yields appear to vary as a function of Fe concentration, with oxidation state having minimal influence. By using the experimentally measured MACs and the calculated partial fluorescence yield corrections, concentrations were calculated to within ± 10 % relative of the expected values (Fig. 12b).

3. Discussion

The measurements made for Fe using the Kα X-ray at low overvoltage and L-line X-rays at low voltage provide us with a good position for examining the various options for high spatial resolution and their merits. The Fe-Kα measurements gave the most reliable results both accurate and precise for all the materials measured including both pyrite and Fe metal. The measurements were also insensitive to the standard used. The limitation being that whilst the spatial resolution for Fe-Kα measurments at 10 kV

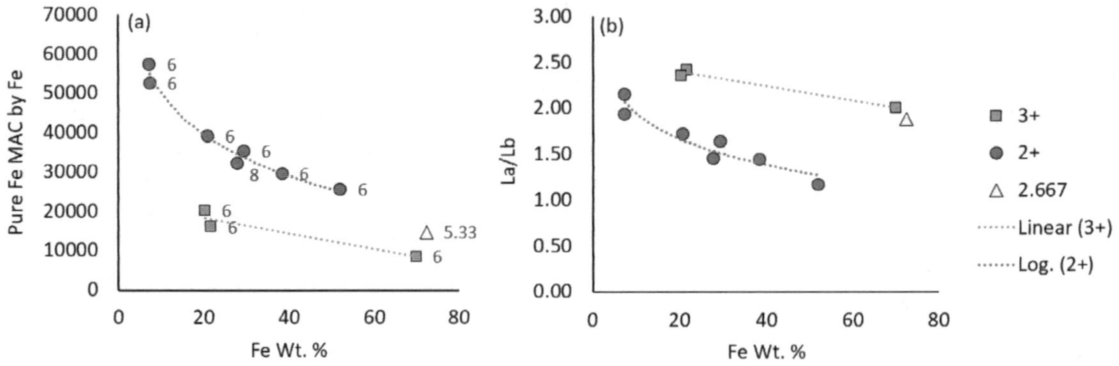

Figure 11. a) Experimentally measured MACs for Fe-Lα by Fe, and b) measurements of Lα/Lβ-ratio for materials with varying Fe oxidation state (colours - see legend), and coordination number (labels on a).

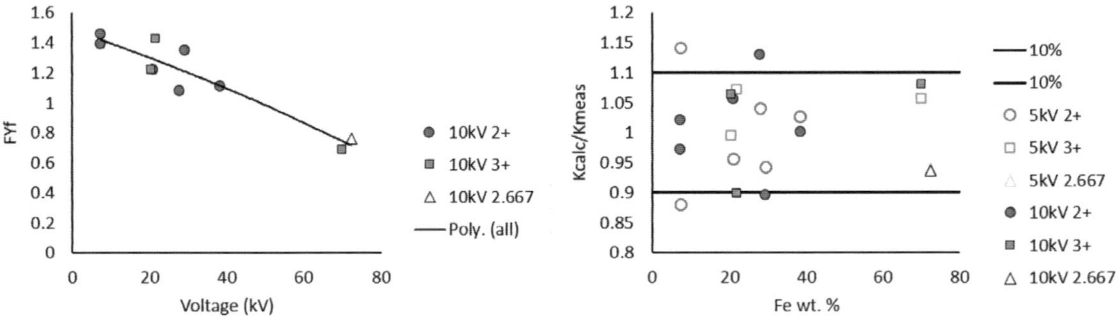

Figure 12. a) Variations in partial fluorescence yield, and b) agreement between calculated and measured *k*-ratios using calculated partial fluoresecence yields and mass absorption coefficients for the range of samples.

is very good (depth < 350 nm for ulvospinel, Fig. 1) to achieve a similar resolution for the other elements such as Si or more usefully Mg, Al and Ti in the case of ulvospinel, measurement at several voltages is required (as given in Table 1). To achieve high lateral resolution a small spot size is required. At 10 kV, 100 nA beam current, a spot size of 100 nm can be achieved [3]. This degrades the lateral resolution by ca. 100 nm (see Fig. 2) making it significantly worse than the depth resolution. Poorer lateral resolution is acceptable for polished cubic phases where the depth is typically substantially less than the width. At small spot sizes and high power, there would be the potential for carbon erosion - this should be assessed and if necessary, a correction applied using TDI. Alternatively, a different coating material may be used.

The measurement of Fe using L-line X-rays allows measurement at low voltage providing high spatial resolution (< 350 nm depth at 7 kV for ulvospinel, Fig. 1) for all the X-rays which can be excited. The results show, in keeping with previous findings (e.g., [7, 28]), that the Lℓ-line provides reasonable results (accuracy within ≤ 10 % relative), which is a substantial improvement from conventional Lα-line measurements for first row transition metals. The Lℓ measurements clearly require careful selection of the standard, with this having a significant effect on accuracy. Errors are large at low concentrations due to the weak intensity of this line. The intensity of the Lℓ-line relative to Lα varies with element, becoming more favourable as the atomic number is reduced towards Ti and the occupancy of the 3d shell declines reducing L3-M5 transitions [31, 36]. The results show that whilst conventional Lα measurements are highly inaccurate (with discrepancies up to 45 % relative) the use of a calibration curve to calculate Fe concentration can yield good results with accuracy approximating ≤ 5 % for a range of minerals (excluding Fe metal, pyrite and ilmenite) and are rendered insensitive to Fe oxidation state when a fixed position is used. Whilst this accuracy is less than achieved in standard high voltage analysis (typically 1 - 2 % relative) and is less robust than the low overvoltage measurements made using the Fe-Kα X-ray line, it is a significant result. The accuracy is comparable to measurements made of olivine where the solid-solution MACs and partial fluorescence yields were determined [25], achieving accuracies of ≤ 4 %. The main advantage of this calibration curve approach is the ability to measure a range of minerals (olivine, pyroxene, garnet, and some oxide minerals) of varying oxidation state. Simple solid-solution MACs are unable to account for the variations in MAC with Fe valence changes. Initial tests in calculating MACs and partial fluorescence yields for a range of silicate minerals with varying oxidation states, are encouraging but require further investigation to see if similar precision can be achieved. If we consider the lateral resolution for the Lα calibration curve method; at 200 nA and 7 kV the spot size would exceed 150 nm [3], making the lateral resolution worse than the depth resolution. At 7 kV carbon coat erosion is expected to be worse [14], although the effect on X-ray intensities may be similar or less given the larger overvoltages. Again, TDI or other coating materials could be considered. Improvements to the Fe-L depth resolution could be made by dropping the accelerating voltage, but below 4.97 kV Ti-K cannot be excited, and with reduction in voltage there is an associated increase in spot size, particularly at high beam currents [3].

In the above cases, use of high beam currents (100 nA - 200 nA) are necessary to provide high precision. To achieve higher lateral resolution by reducing spot sizes lower beam currents could be used. For Fe Kα at low overvoltage the higher intensities would permit a reduction in beam current whilst still maintaining reasonable precision (e.g., 1.6 % relative 2σ for St John's Island olivine with 7.5 wt% Fe at 25 nA using 2 spectrometers), although trace element sensitivity would be reduced. In both the case of Lα X-rays and even more so using Lℓ X-rays, the lower X-ray intensities using a TAP crystal limit the potential for reducing the beam current whilst maintaining reasonable precision. While many mineral species are stable under the electron beam (e.g., olivine, pyroxene, garnet and spinel), it is not true for others minerals such as carbonates, phosphates, micas and natural glass. In the case of plagioclase feldspar the resolution achievable is significantly affected by the stability of the sample requiring a defocussed beam. Saunders *et al.* [37] found that while a spatial resolution of ≤ 350 nm could be achieved for orthopyroxene with a 30 nm beam at 5 kV, plagioclase required a 500 nm defocussed beam yielding a spatial resolution of ca. 750 nm to maintain sample stability.

4. Conclusions

The optimum analytical conditions for a given analysis are complex to determine and the maximum spatial resolution achievable is dependent on the elements and material analysed. Low overvoltage and low voltage provide two mechanisms of achieving high resolution, with the resolution a function of the electron range, the spot size and the overvoltage and absorption of a particular element (e.g., [3, 4]). The spot size depends on the precision and sensitivity required, together with the intensity of an X-ray line on a particular diffraction crystal. Maximum spatial resolution will be achieved for major element analysis of X-ray lines > 1 keV, where low beam currents can be used in combination with low overvoltage or low voltage settings, for dense materials (e.g., [1]).

The measurements demonstrate that low overvoltage measurements can give robust results, regardless of the material being analysed, but are limited to one or a few elements at a given voltage. Low overvoltage analysis is very sensitive to carbon erosion which occurs at high current density, and carbon contamination, which is especially problematic for tightly spaced analyses. Robust results require mitigation or avoidance of these affects. Mitigation methods include TDI correction and the use of other coating materials for carbon erosion [14] and the use of an anticontamination cold-finger and empirical corrections for contamination [17]. Depending on the sample, the use of an oxygen air-jet, stage heating, chamber plasma cleaning or sample preparation using FIB milling or argon cross-section polishing may be suitable to reduce contamination [17-21].

Low voltage analysis provides high spatial resolution for all the X-rays excited, equal or less than the interaction volume. It does however restrict the X-rays which can be used. The L-line X-rays of transition metals are particularly problematic. The Fe-Lα measurements show that despite anomalous absorption and partial fluorescence yields (see [24]), reasonably accurate results can be obtained using either a calibration curve for a range of minerals, or experimentally determined solid-solution MACs and partial fluorescence yields for a single mineral solid solution [25]. Initial application of experimentally determined MACs and partial fluorescence yields for a range of silicate minerals are encouraging, but currently yield larger errors. The calibration curve using the Fe metal peak position provides a method of measurement which removes sensitivity to Fe valence state. A consequence of using the TAP crystal for L-line measurements is the low intensity, requiring high beam currents for high precision particularly at low concentrations, thereby limiting the lateral resolution through larger spot sizes. The choice of low overvoltage or low voltage L-line method will therefore depend on the precision and accuracy required and the material analysed, with L-lines requiring a range of standards for calibration curves, or pre calculated solid solution MACs and partial fluorescence yields.

Acknowledgments
We would like to thank the Smithsonian Institute for supplying the Rockport, Fayalite standard (NMNH 85276).

References

[1] Kimura T, Nishida K and Tanuma S 2006 *Microchim. Acta* **155** 175-178
[2] Hovington P, Drouin D and Gauvin R 1997 *Scanning* **19** 1-14
[3] Pinard P T and Richter S 2014 *IOP Conf. Ser.: Mater. Sci. Engng.* **55** 012016
[4] McSwiggen P 2014 *IOP Conf. Ser.: Mater. Sci. Engng.* **55** 012009
[5] Llovet X and Salvat F 2016 *IOP Conf. Ser.: Mater. Sci. Engng.* **109** 012009
[6] Ritchie N W 2009 *Microsc. Microanal.* **15** 454-468
[7] Statham P and Holland J 2014 *IOP Conf. Ser.: Mater. Sci. Engng.* **55** 012017
[8] Kubo Y, Hamada K and Urano A 2003 *Ultramicroscopy* **135** 64-70
[9] Bastin G F and Heijigers H J M 1990 *Scanning* **12** 225-236
[10] Chopra D 1970 *Phys. Rev. A* **1** 230-235
[11] Fischer D W and Baun W L 1967 *J Appl. Phys.* **38** 4830-4836
[12] Lábár J L and Salter C 1991 in: *Electron probe quantitation.* (Heinrich K F J and Newbury D E; Eds.). (New York, NY: Plenum Press) 223
[13] Matthews M B, Kearns S L and Buse B 2018 *Microsc. Microanal.* **24** 83-92
[14] Matthews M B, Kearns S L and Buse B 2018 *Microsc. Microanal.* **24** 612-622
[15] Matthews M B, Kearns S L and Buse B 2019 *Microsc. Microanal.* **25** 1112-1129
[16] Pinard P T, Schwedt A, Ramazani A, Prahl U and Richter S 2013 *Microsc. Microanal.* **19** 996-1006
[17] Buse B and Kearns S 2015 *Microsc. Microanal.* **21** 594-605
[18] Bastin G F and Heijligers H J M 1988 in: *Microbeam analysis.* (Newbury D E; Ed.). (San Francisco, CA: San Franciso Press) 325
[19] Buse B, Kearns S, Clapham C and Hawley D 2016 *Microsc. Microanal.* **22** 981-986
[20] Augustyn E, Hallstedt B, Wietbrock B, Mayer J, Schwedt A and Richter S 2012 *IOP Conf. Ser.: Mater. Sci. Engng.* **32** 012001
[21] Yamashita T, Tanaka Y, Yagoshi M and Ishida K 2016 *Sci. Rep-UK* **6** 29825
[22] Pouchou J L and Pichoir F 1985 *J. Microsc. Spectrosc. Electron.* **10** 291-294
[23] Fialin M 1990 *X-ray Spectrom.* **19** 169-172

IOP Conf. Series: Materials Science and Engineering **891** (2020) 012005 doi:10.1088/1757-899X/891/1/012005

[24] Llovet X, Pinard P T, Heikinheimo E and Louhenkilpi S 2016 *Microsc. Microanal.* **22** 1233-1243
[25] Buse B and Kearns S L 2018 *Microsc. Microanal.* **24** 1-7
[26] Moy A and Fournelle J 2018 *Microsc. Microanal.* **24** (Suppl. 1)
[27] Rémond G, Myklebust R, Fialin M, Nockolds C, Phillips M and Roques-Carmes C 2002 *J. Res. Natl. Inst. Stand. Technol.* **107** 509-529
[28] Gopon P, Fournelle J, Sobol P E and Llovet X 2013 *Microsc. Microanal.* **19** 1698-1708
[29] Buse B and Kearns S L 2011 in: *Book of Abstracts of the European Microbeam Analysis Society 12th European Workshop*. (Angers, France; 15-19 May)
[30] Moy A and Fournelle J 2019 *Microsc. Microanal.* **25** 664-674
[31] Wyckoff R W G and Davidson F D 1965 *J. Appl. Phys.* **36** 1883-1885
[32] Okada M 1981 *J. Radioanal. Chem.* **63** 201-204
[33] Fialin M, Wagner C, Métrich N, Humler E, Galoisy L and Bézos A 2001 *Amer. Mineralogist* **86** 456-465
[34] Hofer H E and Brey G P 2007 *Amer. Mineralogist* **92** 873-885
[35] Moy A and Fournelle J 2017 *Microsc. Microanal.* **23** (Suppl. 1)
[36] Holliday J E 1962 *J. Appl. Phys.* **33** 3259-3265
[37] Saunders K, Buse B, Kilburn M R, Kearns S and Blundy J 2014 *Chem. Geol.* **364** 20-32

IOP Conf. Series: Materials Science and Engineering **891** (2020) 012006 doi:10.1088/1757-899X/891/1/012006

Upscaling of 2D mineralogical information to 3D volumes for geoscience applications using a multi-scale, multi-modal and multi-dimensional approach

A R Butcher

Geological Survey of Finland (GTK), PO Box 96, Vuorimiehentie 2, 02151 Espoo, Finland

E-mail: alan.butcher@gtk.fi

Abstract. Bridging the scales of geological observation is an on-going challenge to earth scientists, but recent advances in the development of 2D and 3D imaging and analysis technology has allowed a workflow to be developed that starts to make this possible. Using concepts borrowed from allied industries (petroleum geoscience and materials science), this study shows how a multi-scale, multi-modal and multi-dimensional approach can lead to a better understanding of rock's mineralogy and texture, especially ore-bearing drill cores, which in turn can be useful for elucidating the meso- and macro-features of ore bodies, leading ultimately to improved mineral exploration success.

1. Introduction

There have been significant recent advances in the last 20 years or so in the way geologists examine rocks both mineralogically and petrographically. Traditional 2D methods, such as light microscopy and electron beam-based techniques, were for many years largely analogue, and have now thankfully been fully digitised, allowing for huge improvements in the efficiency of imaging, analysis, interpretation, and reporting of polished surfaces and/or thin slices of rock samples (Fig. 1).

With the emergence of new cutting edge scanning XRF and hyperspectral imaging techniques, geoscientists are now able to scan geomaterials (such as drill cores) at increasingly larger scales than that covered by a single petrographic thin section. This has allowed us to start to bridge the different scales of geological observation, from the mega- to the nanoscale, that previously were not easily accessible to most earth scientists (Fig. 2).

2. The emergence of cross-disciplinary activity

We are living in an age of digital communication, whereby scientists and engineers can freely communicate and share their ideas. Cross-disciplinary collaboration often leads to extraordinary developments and breakthroughs.

As geologists, we are taught to think in 3D, and it has always been a limitation that we have been mostly restricted to examining rocks in the laboratory using 2D surface-based technologies (SEM, EPMA, LA-ICP-MS, etc.). But this has all changed in the last decade. By using so-called workflows, borrowed and adapted from allied disciplines (including materials and biological sciences, and

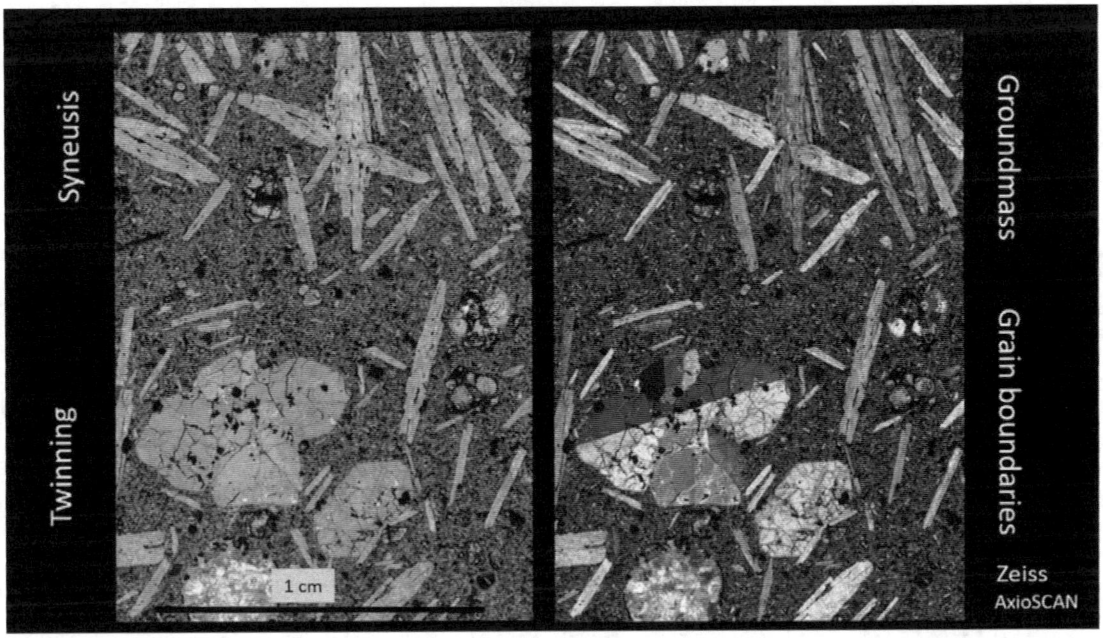

Figure 1. Recently developed technology allows for the automated collection of high quality, full colour, digital images of rock thin sections, both in plane and polarised light, allowing textural features such as zoning, twinning, and crystal and groundmass behaviour to be observed and recorded.

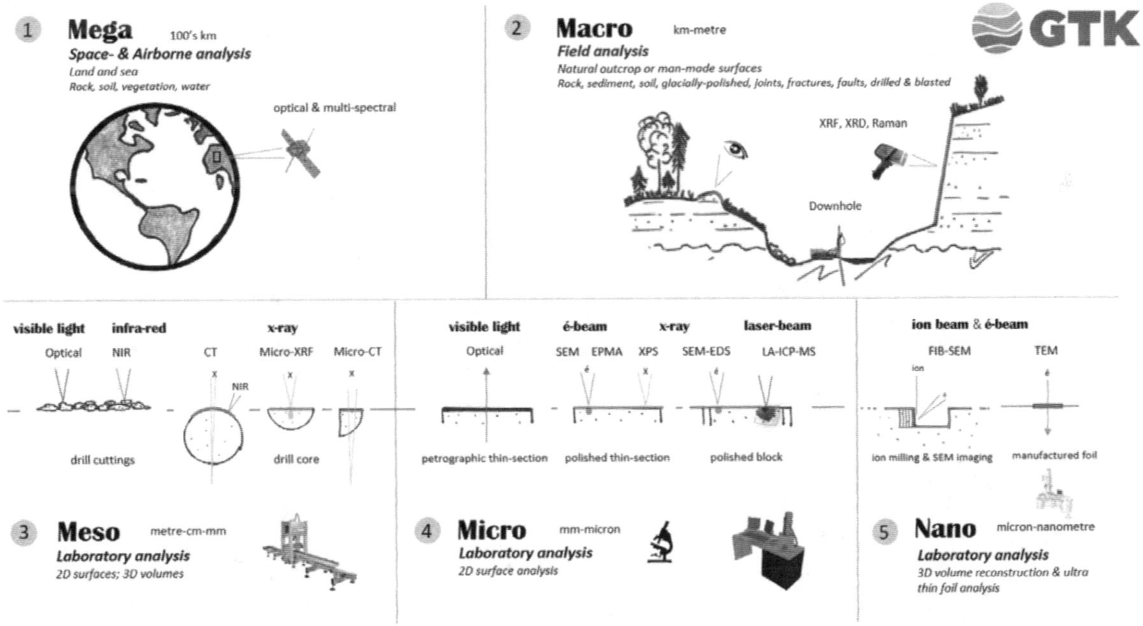

Figure 2. Generalised view of the study of geology in terms of scales of observations and technologies used to examine different features.

petroleum science and engineering), geologists can also now image rocks using powerful X-ray computed tomography-based techniques (X-CT), and focussed ion beam-SEM technologies (FIB-SEM).

3. Lessons from the oil industry

By way of example, cuttings and cores from a hydrocarbon reservoir drilling programme are the only way to know what is actually beneath the ground. There is no chance of digging a trench, or opening up a trial pit, as in mineral exploration and mining. Making the most of every centimetre of core, or gram of ditch cutting sample, is a key driver for innovation and development in the oil industry, especially when it comes to rock and mineral analysis. Nothing is left to chance, the risks are too big, and aspects such as correct stratigraphic correlation and rock sequence prediction must be as accurate as possible in order to reduce the chances of placing a gas or oil well in the wrong place.

As a result, just about every possible geo-analytical technique is used to characterise the physical properties of the overburden, cap rock, reservoir, source, and even the basement. It all starts with geophysics, moves to petrophysics, then reservoir quality, and finally petroleum engineering. Every physical attribute of a rock or mineral is exploited to further understand what is present and how it will behave at reservoir conditions – these include: mineralogy, texture, 3-D grain network, pore structure, rock density, magnetic susceptibility, radioactivity, conductivity, density, brittleness, and response to NIR, X-ray and electron beams – to name just a few parameters that are typically measured.

From experience, the key to success in the petroleum industry seems to be a multi-scale, multi-dimensional, and multi-modal approach to rock imaging, analysis and interpretation. There is no one technique that can be considered all encompassing (Fig. 3). Only by the integration of technologies can a complex problem of, say, multi-phase flow in porous geomaterials, be fully understood. Even then this apparently rigorous approach is sometimes not enough. Mistakes are made. Unforeseen problems occur because of the essentially 1D nature of drilling into what are actually complex 3D sedimentary basins, often undertaken precariously on offshore rigs, typically to great depths, and where sedimentary facies can change rapidly over short lateral distances, without notice, and unconformities or faults can unexpectedly change even the most robust and cherished depositional models of the reservoir geologist.

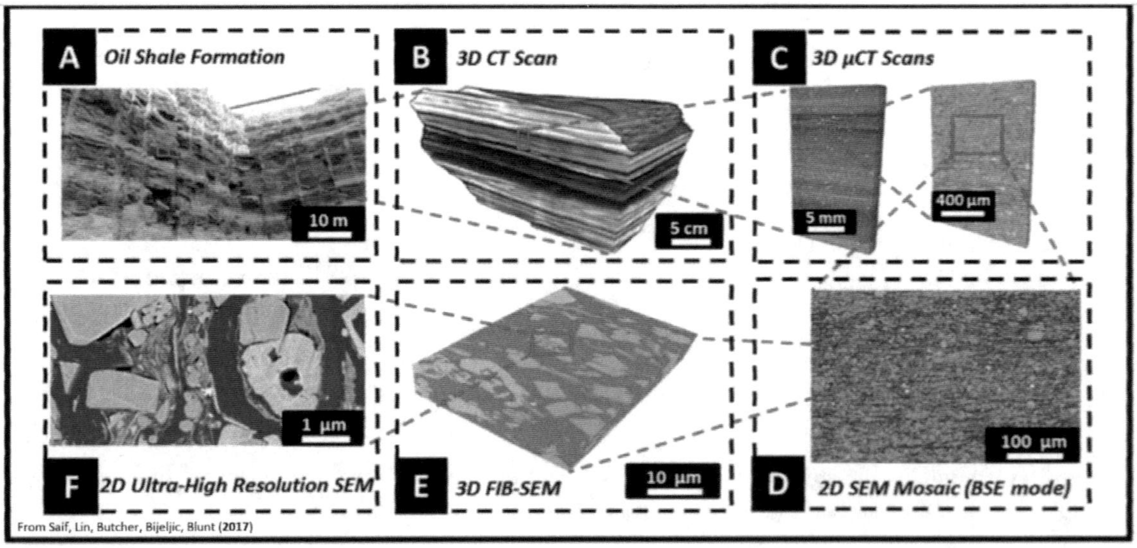

Figure 3. Multi-scale, multi-modal, imaging and analysis study carried on oil shale to allow for upscaling of features of commercial interest such as pores, cements, fractures, and bedding structures (from [1]).

IOP Conf. Series: Materials Science and Engineering **891** (2020) 012006 doi:10.1088/1757-899X/891/1/012006

4. Applications to mineral exploration

In the mining world, we clearly face similar but also different challenges. Having undertaken airborne or ground geophysical surveys, observations and interpretations to select a target, everything in mineral exploration still has to be ultimately ground-truthed. Potential targets need to be drilled to prove the presence (or not) of an ore-rich interval, as perhaps indicated by the physics. But once we have hit the zone of interest, do we make the most of our core? Is every length of core that is pulled from the ground fully digitally scanned before it is cut, split or sawn in half and quartered? Definitely not. Why not? Well, one reason is that the budgets and risk factors in mining are not on the same scale as in the oil industry. But accepting this, we can still learn much from each other.

The concept of digital rock analysis, whereby a digital rendition of a real sample is used to visualise and model rock properties, is an excellent potential cross-over concept for mining, now widely used in the petroleum world. By creating a 3D version of a core, based on reality and not a model, it is now possible to perform virtual experiments and observations without damaging the original.

With this mind, consider the case of a high-grade ore-rich exploration drill core sequence (Fig. 4). Imagine being able to visualise the 3D structure of the core, along with the minerals of commercial interest within it, and calculate their grain size, see their associations, and provide a degree of prediction of processing performance, initially without cutting open the rock. Once regions-of-interest have been found, they can then be further investigated by destructive techniques such as slabbing, crushing, or thin section preparation. Automated 2D imaging and analysis technologies can then be used to calibrate the original 3D volumes, as well as digitally capture what the grains of gold, platinum or sulphides (or whatever) actually look like in extreme detail, down to and beyond the micron-scale, and provide quantitative micro-compositional information on them as well.

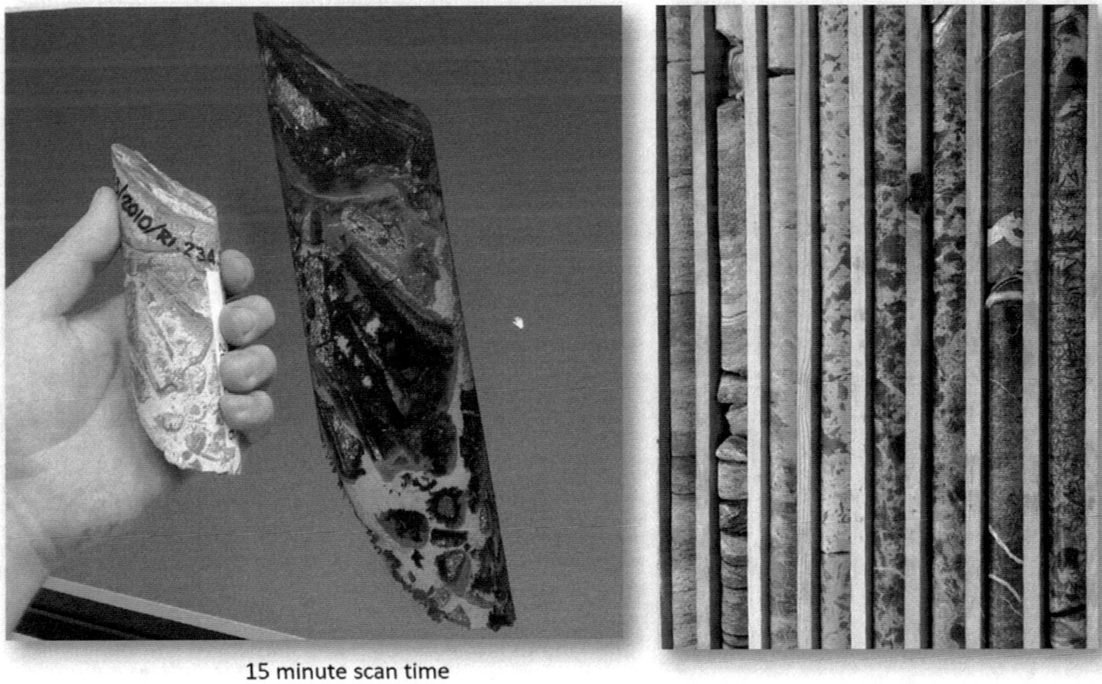

Figure 4. Example of a mineralised drill core taken from an exploration core tray and imaged rapidly using X-ray computed tomography technology (X-CT), to reveal 3D textural and mineralogical information without any damage to the sample.

Even the most highly mineralised ore, containing massive sulphides and precious metals, can now be imaged and rendered into a useful 3D digital volume, which can be later interrogated, quantified, archived and retrieved, using advanced image processing and data storage know-how (Fig. 5).

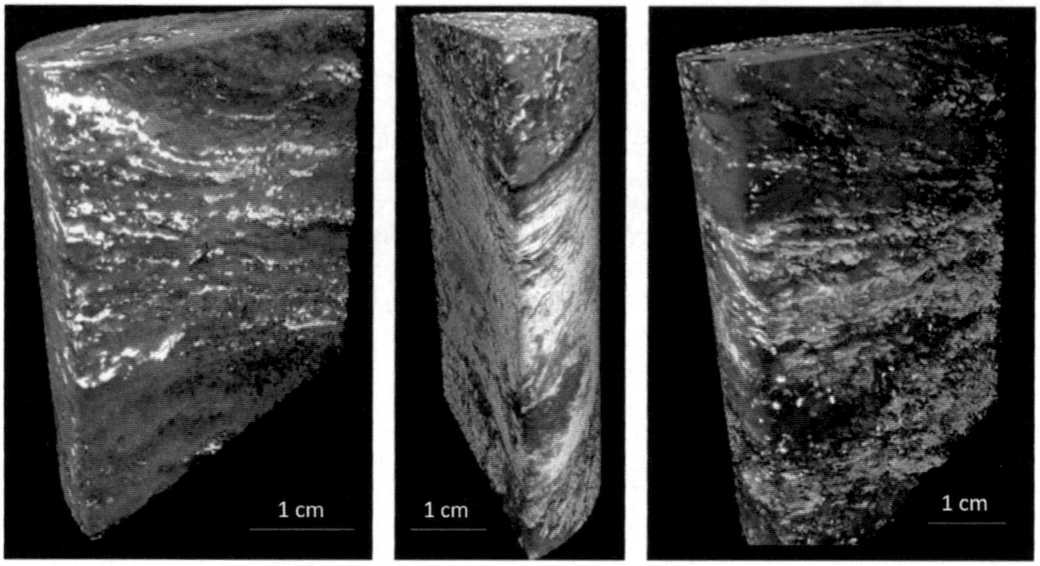

Figure 5. Three exploration drill cores – all containing multi-deformed pyrrhotite layers (bright phases) – within a silicate matrix (greys). Such images allow areas of special interest to be highlighted and can guide thin section area selection. Samples courtesy of Mawson Resources.

5. Summary

Multi-scale, multi-modal and multi-dimensional workflows have opened up a whole new world of analytical possibilities to us as a community. For the exploration geologist, it means that their valuable drill core intersections can now be examined and archived before it goes off for destruction (slicing with a diamond saw and crushing of half core for routine chemical assaying), thus preserving all the contained characteristics (bedding, layering, folding, mineralisation), see Fig. 6.

For the mineral processing engineers, 3D liberation analysis is now a real possibility, something that has been requested by them for many years, and there is encouraging work starting to be published [2, 3]. And to the ore deposit geologist, it allows for a superior textural and structural understanding of mineralised rocks, all the way from the field (mega- and macro-scale) to the laboratory (micro- and nanoscale).

This workflow approach is currently being implemented at the Geological Survey of Finland in order to reduce the risks in mineral exploration; gain insights into the commercial mineralogy of ores; contribute to smarter and greener mineral processing behaviours; and to overall improve mineral deposit knowledge and understanding via an integrated 2D - 4D approach.

Once these workflows have been properly and securely set up, it is then possible to generate databases that can be used to create realistic ore body models that are based not only on grade, or mineralogy alone, but also on, for example, more enlightened aspects such as ore paragenesis. This in turn can lead to better predictive metallurgical models, because the geologist and metallurgist will together now have a full understanding of the nature of minerals of commercial importance, from grind size estimate, to liberation behaviour, and finally elemental deportment. This is key information that exploration companies ideally need upfront, at the time they are investing heavily in costly drilling

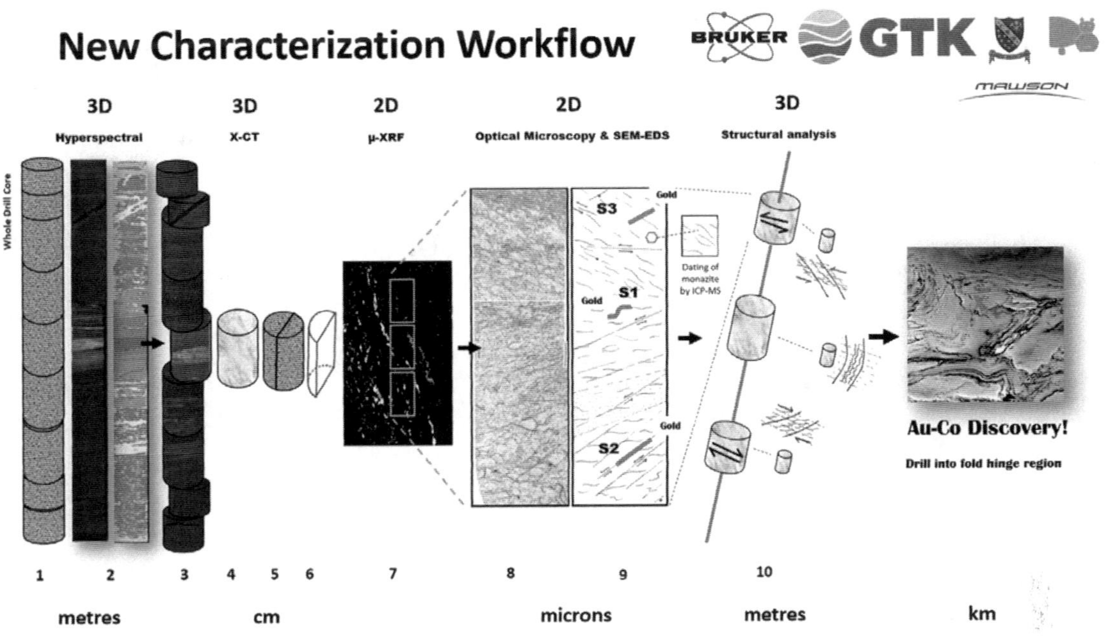

Figure 6. Multi-scale, multi-modal, multi-dimensional workflow used to characterise exploration drill core in order to optimise the discovery of gold and cobalt mineralisation. Example courtesy of Mawson Resources, and developed in collaboration with Camborne School of Mines (University of Exeter), Bruker Nano Analytics (Berlin), and Hippo Geoscience.

campaigns, as it reduces their risk and uncertainty, and improves understanding of the yet-to-be-fully-discovered ore body (Fig. 6). Menace minerals, either from an environmental view point, or purely a practical processing aspect (poor separation potential), can be flagged up early on in the exploration life cycle, and therefore save much pain later on down the track. This information is also invaluable if it can be made available prior to blasting activities in operational mines.

6. The future

Finally, the world of material science is another much underutilised area of expertise and knowledge used by geologists. It typically involves bringing together the doers (technologists) with the knowers (scientists), which is not commonly practiced for some reason. There is much literature and experience that can be readily borrowed from technologists and brought to bear on the study of earth materials – such as fracture analysis, crystallisation and annealing processes, and micro-chemical reactions – all of which can lead to a better understanding of minerals and rocks, their discovery, extraction, and recovery potential.

Acknowledgements

The author would like to acknowledge contributions from the following in providing content for this paper: Nick Cook (Mawson Resources); Gavyn Rollinson (Camborne School of Mines, University of Exeter); Sayab Muhammad and Jukka Kuva (GTK); Pieter Botha (Hippo Geoscience); Shaun Graham and Ben Tordoff (Carl Zeiss); Juhani Ojala (AB Scandinavian GeoPool Ltd); and Sam Scheller and Andrew Menzies (Bruker Nano Analytics).

IOP Conf. Series: Materials Science and Engineering **891** (2020) 012006 doi:10.1088/1757-899X/891/1/012006

References

[1] Saif T, Lin Q, Butcher A R, Bijeljic B and Blunt M J 2017 Multi-scale multi-dimensional microstructure imaging of oil shale pyrolysis using X-ray micro-tomography, automated ultra-high resolution SEM, MAPS Mineralogy and FIB-SEM. *Appl. Energy* **202** 628-647

[2] Reyes F, Lin Q, Cilliers J J and Neethling S J 2018 Quantifying mineral liberation by particle grade and surface exposure using X-ray microCT. *Minerals Engng.* **125** 75-82

[3] Guntoro P I, Tiu G, Ghorbani Y and Lund C 2019 Application of machine learning techniques in mineral phase segmentation for X-ray microcomputed tomography (μCT) data. *Minerals Engng.* **142** 105882

Quantifying phase transformation during the manufacturing process of AISI 430 ferritic stainless steel

I Collado[1], A Núñez Galindo[2], A Ruiz[2], J F Almagro Bello[2] and F J Botana[1]

[1] Universidad de Cádiz, Escuela Superior Ingeniería, Departamento de Ciencia de los Materiales e Ingeniería Metalúrgica y Química Inorgánica, LABCYP Group, Avda. Universidad de Cádiz 10, 11519 Puerto Real, Spain
[2] Acerinox Europa S.A.U, Departamento Técnico / Laboratorios e Investigación, Pol. Industrial Palmones s/n, 11379 Los Barrios, Spain

E-mail: irene.collado@uca.es

Abstract. The effect of ferrite to austenite transformation phenomenon on microstructure and annealing performance of AISI 430 (EN 1.4016) ferritic stainless steel 16%Cr-0.04%C and 17%Cr-0.02%C was studied by electron backscatter diffraction and X-ray diffraction. Hot-rolled and annealed specimens of each ferritic stainless steel were collected from manufacturing route, where different states were analysed. Chemical composition shift of AISI 430 is correlated to microstructural behaviour of each material, while annealing temperature impact over grain growth phenomena is discussed.

1. Introduction

AISI 430 ferritic stainless steel is becoming highly demanded on cold-forming processes [1], such as household appliances and, in some applications, as a replacement of austenitic stainless steel [2]. Compered to austenitic stainless steel, which provides an outstanding metallurgical property, the absence or low content of nickel in AISI 430 results in an attractive economic solution for many applications. Therefore, the aim of metallurgical manufacturers involves the enhancement of surface quality and appearance of ferritic stainless steel finishing [3]. Surface defects related to AISI 430 ferritic stainless steel, are caused in many occasions by their structural complexity. This feature structure is related to the ferrite/austenite transformation that takes place within the ferritic matrix of this stainless steel, mostly, during hot rolling [4]. The critical temperature at which phase transformation takes place is described as A_{c1}, and depends on many factors, such as alloy chemical composition and annealing rate.

For that reason, annealing is a crucial production step. During this process, stainless steel is exposed to high temperatures below A_{c1} long enough for the whole austenite to ferrite + carbides transformation to take place. If this process is performed badly, ferritic matrix would retain austenite. Consequently, the remaining austenitic phase will turn into martensite when the steel reaches a certain temperature during cooling where martensitic phase starts to nucleate, also known as Ms temperature. During annealing, AISI 430 theoretically recover its ferritic mono-phase before ending this heat treatment. Moreover, after being the austenite completely transformed into ferrite, annealing process must ensure the re-crystallisation of the ferritic structure and, therefore, recuperate its mechanical property according to its composition.

Overall, this research attempt to measure the remaining martensite as a clear indicator of the previous phase transformation by using AZTEC RECLASSIFY PHASE from electron backscatter diffraction (EBSD) and X-ray diffraction (XRD). At the same time, the microstructure of these materials is characterised during hot rolling and annealing. The study has been carried out forward comparative analysis of two sorts of AISI 430 ferritic stainless steel. One of these materials represents AISI 430 standard composition whereas compositional and manufacturing changes are assumed in the other.

2. Materials and experimental procedures

This work used two AISI 430 ferritic stainless steel compositions, whose samples were collected from hot rolling, annealing and pickling and bright annealing. Specifications of each specimen are listed in the Table 1, where stainless steel composition was analysed by X-ray fluorescence (XRF).

Table 1. Composition and characteristic of the samples collected from manufacturing process of flat stainless steel product. Eq_{Cr} was calculated by the Schaefler equation [5] (Eq_{Cr} = %Cr + %Mo + 1.5·%Si + 0.5·%Nb + 2·%Ti).

AISI 430	Composition, %	Eq_{Cr}	Measurement point (manufacturing)	Specimen thickness, mm
1A	16Cr 0.3Si 0.4Mn 0.045C 0.040N_2	11.3	Hot rolling	26
			Annealing and pickling	4
			Bright annealing	0.7
2A	17Cr 0.4Si 0.3Mn 0.020C 0.033N_2	13.2	Hot rolling	26
			Annealing and pickling	3.5
			Bright annealing	0.4

AISI 430 composition has slightly changed from 1A to 2A, where chromium, carbon and nitrogen shifts have a great impact on the ferrite - austenite transformation phenomenon, as shown in Fig 1.

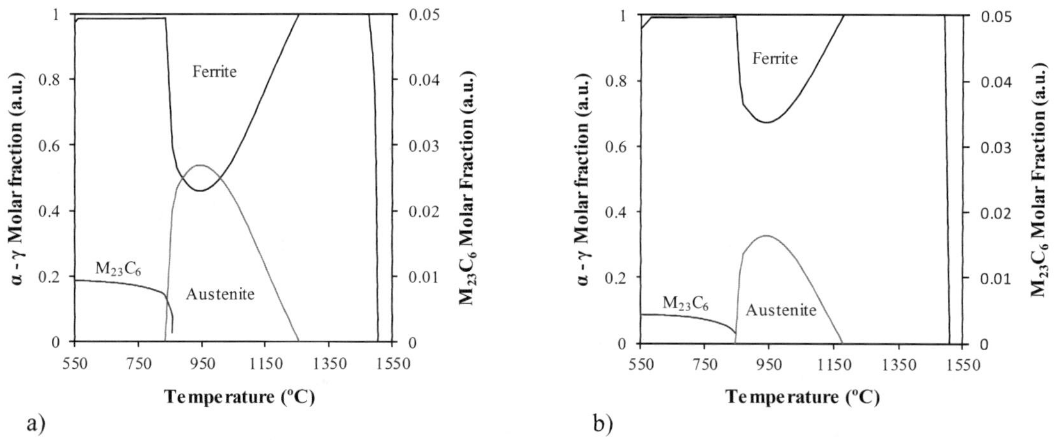

Figure 1. Molar fraction of phase transformation in equilibrium of AISI 430; a) material 1A, and b) material 2A, by THERMOCALC (TCW5) [6].

Moreover, Fig. 1 shows $M_{23}C_6$ precipitation, which is formed in the course of $\alpha \rightarrow \gamma + (Fe, Cr)_{23}C_6$ transformation during cooling. The more austenite nucleates before A_{c1}, the more carbides precipitate due to the carbon solubility decrease from austenite to ferrite phase. The precipitation of this carbide occurs in α-α grain boundary and grow into α ferrite [7]. Grain coarsening, sensitization from grain boundaries carbides precipitation and martensite formation from the remaining austenite within the ferritic matrix, are the main causes of failure and surface defects among AISI 430 stainless steel [8].

The experimental setup of this research is based on the analysis of the samples listed above (Table 1) by XRD and EBSD. X-ray diffraction (XRD) measurement of each sample kept the rolling direction, where the surface of each specimen was analysed. However, hot-rolled samples were polished with colloidal silica in order to remove roughness from the area of analysis. The 2θ scanning range, step size and count time were 40 - 130°, 0.03° and 2 sec/step, respectively, and Co-Kα radiation (1.79 Å) was used in a Bruker D2Phaser device.

The sample preparation for characterisation by electron backscatter diffraction (EBSD) was carried out with cloth disc and diamond paste auto-polished and a final polishing of colloidal silica suspension. Rolling direction were kept for all the analysis. The EBSD results were acquired by Zeiss Ultra 55 FEG-SEM. Working conditions for the EBSD analysis were: 20 kV voltage and 16.5 mm working distance.

3. Results and discussion

3.1. X-ray diffraction

The phases and their crystalline structures of 1A and 2A ferritic stainless steel were analysed by XRD analysis (Fig. 2). Two manufacturing states were studied, where hot-rolled 1A and 2A correspond to the samples recollected during hot rolling at 1,033 and 920 °C, respectively; besides, annealed 2A specimen was taken after hot rolling and annealing.

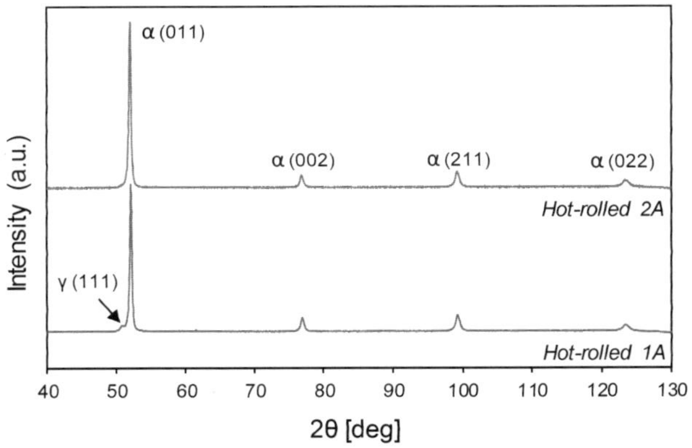

Figure 2. XRD patterns for hot-rolled 1A and 2A ferritic stainless steel.

Results were obtained by Bruker TOPAS based on Rietveld analysis [9]. Diffraction peaks of α (011), (002), (211) and (022) correspond to a ferritic phase. The XRD profile of a hot-rolled 1A shows a ferritic structure, though, a low intensity γ (111) peak is detected near α (011) peak. Besides, 7 % of remaining austenite was measured within the ferritic structure by TOPAS. However, no martensitic could be detected since this phase cannot be distinguished from ferrite via diffraction analysis [10].

Hot-rolled 2A shows an XRD pattern which is in good agreement with ferritic structure. In this case, no visible diffraction for γ (111) peak was detected by TOPAS. The higher ferritic structure of hot-rolled 2A led to increase α (011) peak intensity.

Generally, austenite turns into martensite after quenching when reached Ms temperature. In accordance with *Sung et al.* [7], Ms decreases with increasing Cr_{Eq} and N_2 content in the material composition. Therefore, a non-transformed austenite can be retained within ferrite after fast cooling. This is caused by an exchange of carbon by nitrogen, which promotes atomic ordering of chromium atoms and stabilizes the austenite [10].

This effect of N_2 and Cr_{Eq} on Ms temperature of both specimens, was established using empirical equation for Ms determination. The $\gamma \rightarrow \alpha'$ transformation by cooling is influenced by numerous factors, such as stainless steel composition, grain size, structural defects, etc. Among these, numerous authors [11] considered the stainless steel composition to be the most important.

Eichelman and Hull [9] have developed an empirical equation to calculate Ms temperature. Ms was acquired for both studied materials: $Ms_{(1A)}$: 466 °C; $Ms_{(2A)}$: 487 °C. These results are in good agreement with the previously established around the effect of ferritic stainless steel composition on retention of non-transformed austenite.

Figure 3 shows wide-angle XRD pattern of hot-rolled and annealed 2A ferritic stainless steel series. Annealed 2A diffraction α peaks intensity increased as a result of increasing the ferritic phase of 2A from hot-rolled state to annealed. This points out that the complete transformation of austenite took place during annealing process, and mono-phase ferritic phase was reached after this heat treatment.

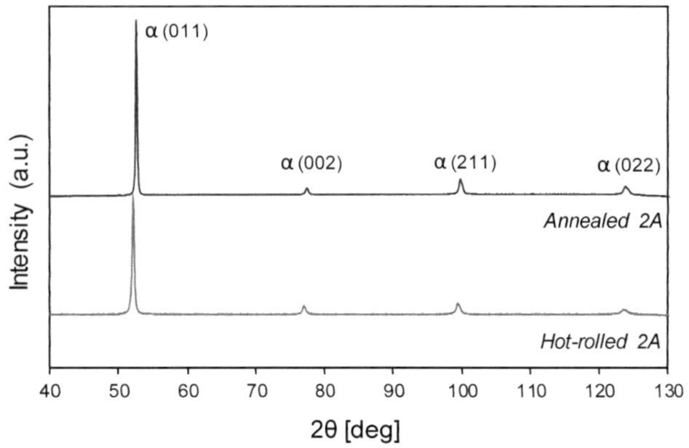

Figure 3. XRD patterns for 2A ferritic stainless steel during hot rolling and after annealing.

3.2. Electron backscatter diffraction (EBSD)

Hot rolling involves not only plastic deformation, but also re-crystallisation happens during this stage, as well as the nucleation of austenite takes place, as it was provided by THERMOCALC in Fig. 1. During the entire hot rolling, both ferritic stainless steels, are in the ferrite - austenite transformation region. From the hot rolling input temperature (Tin_{HR}) to the highest austenite transformation temperature ($T\gamma_{max}$), austenite grows around ferrite grain boundaries, increasing dislocation density, which can prevent ferrite grains from growing [12]. After reaching $T\gamma_{max}$, the reaction is reversed to $\alpha \rightarrow \gamma + M_{23}C_6$ and, therefore, the mechanism around grains growth behaves differently, as will be exposed below.

Figure 4a shows EBSD band contrast (BC) , coincidence site lattice (CSL) and grain boundaries (GB) map of 1A after 87 % of hot rolling reduction at 1,033 °C, where the average aspect ratio is 2.15. CSL and GB allowed identifying sub-nucleation phenomenon of ferrite grains as a result of hot rolling deformation along the normal direction (ND), and re-crystallisation of non-transformed ferrite due to the high temperature of this process. Moreover, BC was also part of this map because it provides a significant contrast difference between ferrite and martensite based on the variation of dislocations and local stress showed by both phases. Equiaxial grains can be observed, while some of them are elongated.

In this specimen, austenite nucleated along the ferrite-ferrite boundaries, which promote the formation of a thin layer of austenite on either side of these boundaries [13]. After quenching, martensite preserved this ordering. EBSD results provided that 70.4 % of the structure was re-crystallised at this point of hot rolling.

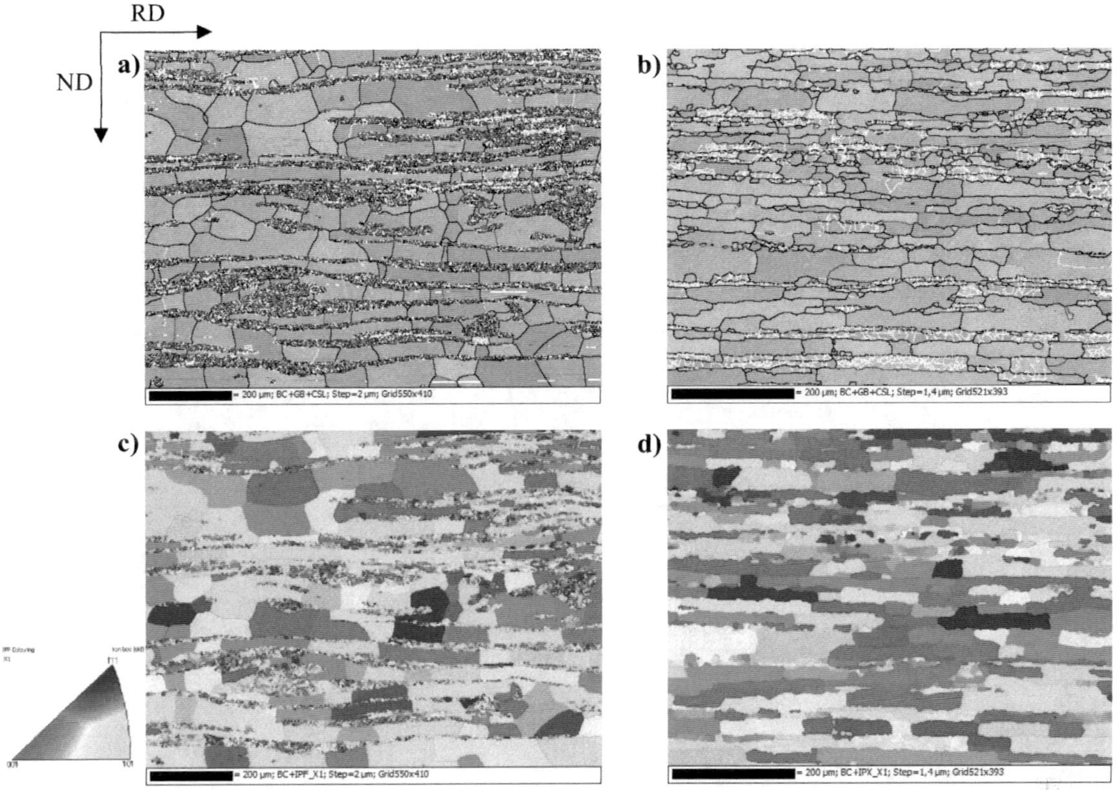

Figure 4. a) Band contrast, grain boundaries and coincidence site lattice (BC+GB+CSL) map; and c) inverse polar figure (IPF) maps of hot-rolled 1A. b) BC+GB+CSL, and d) IPF maps of annealed 1A.

Figure 4b provides annealed 1A BC+CSL+GB map after 98 % of total thickness reduction and initial annealing temperature of 843 °C. It can be observed that coarse grains are mostly elongated along the rolling direction, with an average aspect ratio of 3.27. Furthermore, fine grains are located at coursed grain boundaries; therefore, a wide grain size distribution is reached by 1A. The remained austenite from the previous hot rolling process has completely turned into ferrite and carbides during annealing. Moreover, according to EBSD results, 88.8 % of the structure was recovered after this stage.

Material 2A showed a higher ferritic behaviour during hot rolling than 1A. This performance was a consequence of a change in chemical composition of this alloy, which was based on a higher chromium and lower carbon content, see Fig. 1, leading to a higher Cr_{Eq} of 13.2, compared to 1A specimen. Moreover, as it is shown in Fig. 1, this change in alloy composition led the ferrite - austenite transformation of 2A to be lower than 1A under hot rolling conditions.

The BC+CSL+GB map, displayed in Fig. 5a, shows the 2A structure after 87.1 % reduction at 920 °C. This specimen consisted in both equiaxial and elongated grains, where the average aspect ratio is 2.43. On the other hand, martensite formed within the ferrite structure is thinner and lower in 2A than in the previous hot-rolled 1A sample. These martensite chains were not formed along RD direction since during hot rolling the force is made in the ND plane and some turbulence may occur in the nucleation of austenite. Therefore, austenite transformation depends not only on the plastic deformation direction during hot rolling, but also on the grain size distribution of the material. In accordance with EBSD analysis, the structure is 80.3 % re-crystallised in this point. This leads to consider that austenitic transformation is the driving force of the re-crystallisation performance of ferritic stainless steel during hot rolling.

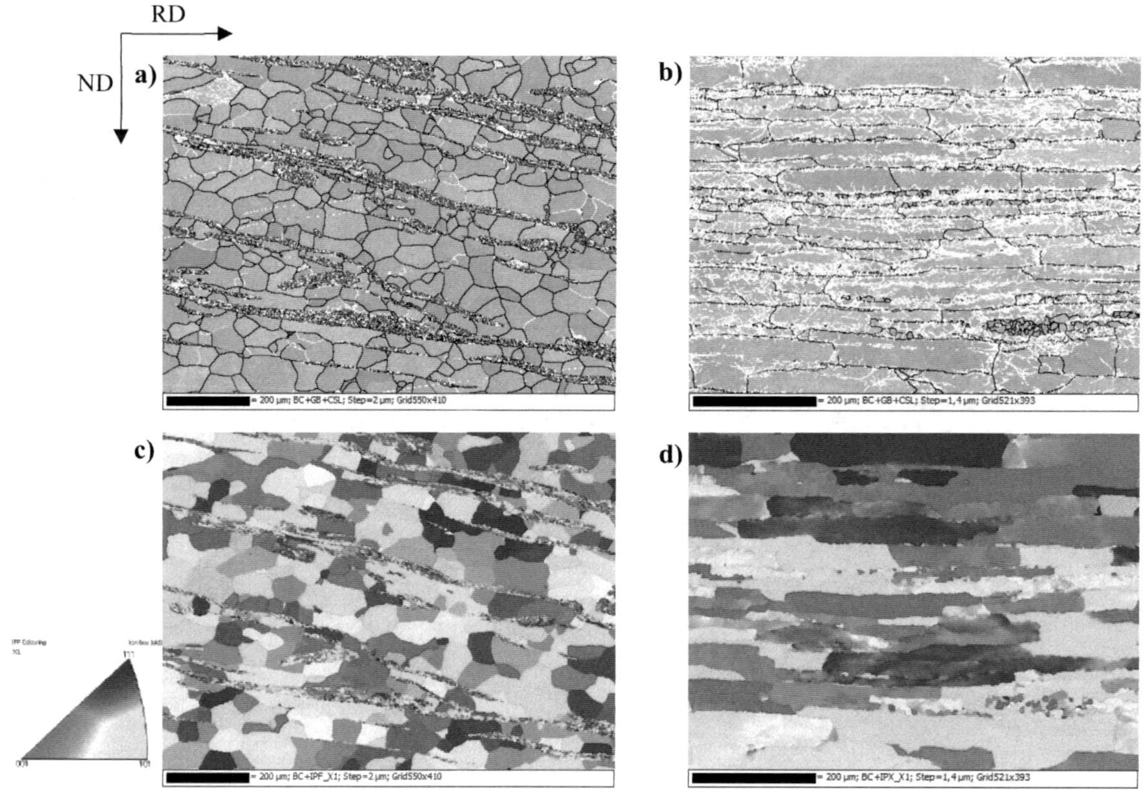

Figure 5. a) Band contrast, grain boundaries and coincidence site lattice (BC+GB+CSL) map; and c) inverse polar figure (IPF) maps of hot-rolled 1A. b) BC+GB+CSL and d) IPF maps of annealed 1A.

Figure 5b shows 2A annealed structure after 98.3 % of total thickness reduction and initial annealing temperature of 826 °C. Coursed grains are highly elongated and showed local stress, see Fig. 5d, which suggests that two considerations must be taken into account. Firstly, in 2A, austenite was rapidly dissolved sufficiently before reaching annealing process. This was due to the amount of austenite transformed during hot rolling was low enough to be completely transformed into ferrite in this stage. Secondly, since the entire annealing energy was invested in the re-crystallisation of mono-phase ferritic structure of 2A, there was an excess of annealing time resulting in excessive ferrite grain growth. This phenomenon during annealing is a common behaviour of ferritic stainless steel type AISI 441 stabilised with Nb and Ti additions to the alloy [14], which prevents the ferrite - austenite transformation at high temperatures, and, therefore, a complete ferritic behaviour is promoted.

EBSD showed that only 3.7 % of the structure was re-crystallised, whereas 56.1 % corresponded to substructure as a result of sub-grains nucleation within coarse ferrite grains, with an average aspect ratio of 2.42.

It is crucial to measure the amount of martensite retained within the ferrite matrix after quenching. This martensite corresponds to the previous austenite transformed at certain temperature during hot rolling. However, martensite and ferrite cannot be discriminated by EBSD so far since their crystal structure are fundamentally equal [15, 16].

For this purpose, a dedicated software (AZTEC RECLASSIFY PHASE [12]) was used in order to quantify that martensite considering EBSP pattern quality parameters. Martensite shows a distorted crystal lattice due to a higher density of crystalline defects, which makes the martensite get a poorer quality, as shown in Fig. 6b, than ferrite phase, shown in Fig. 6a.

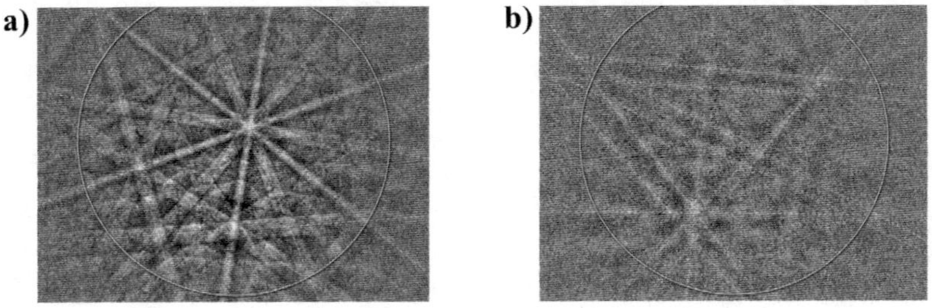

Figure 6. EBSD Kikuchi patterns of a) ferrite and b) martensite obtanined by AZtec Oxford Instruments.

This special feature can be identified by "Band Contrast" (BC) and "Band Slope" (BS) EBSP quality parameters [15], and this is used by the software to identify the martensite areas. Furthermore, these results were checked by non-quantitative conventional light-optical microscopy.

Figure 7 shows BS maps of both AISI 430 studied in this paper during hot rolling, where the martensite (marked in red) is analysed based on its lower pattern quality represented by the darkest areas.

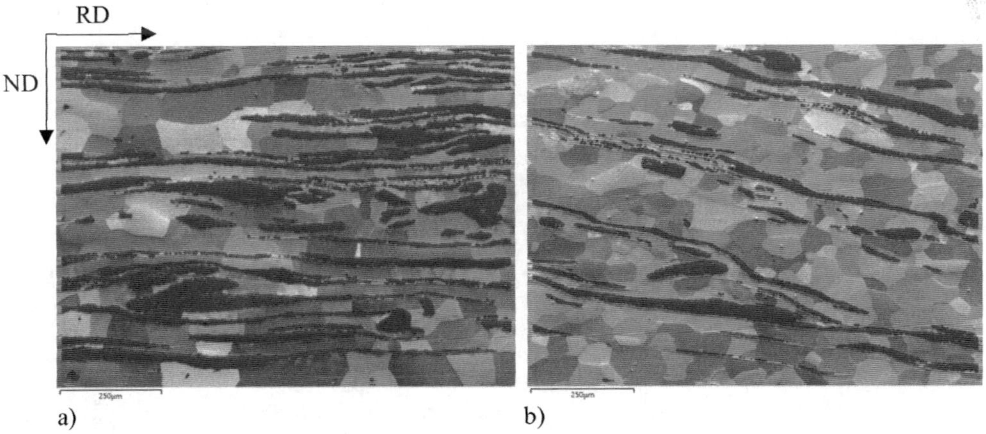

Figure 7. Band slope maps of hot-rolled a) material 1A, and b) material 2A, with martensite reclassified by AZTEC RECLASSIFY PHASE (red areas).

Table 2 lists the remained martensite, quantified by AZTEC RECLASSIFY PHASE, and total deformation of both materials during hot rolling and after annealing, measured by "Recrystallised Fraction" (DefRex) component in Channel 5 HKL. Annealed state of the specimens does not exhibit any retained martensite in the ferrite matrix.

Table 2. Martensite quantified by AZTEC RECLASSIFY PHASE, and deformed fraction of both hot-rolled (HR) and annealed (A) 1A and 2A specimens from EBSD analysis.

ID	State	Martensite, %	Deformation, %
1A	HR	30.8	25.4
	A	0.0	6.2
2A	HR	16.4	15.3
	A	0.0	40.1

3.3. *Effect of AISI 430 composition and temperature on microstructure*

According to Liu *et al.* [13], the heterogeneous grain-size distribution of both studied materials, during hot rolling and after annealing process, is a consequence of the austenite-ferrite transformation mechanism. A two-stage transformation takes place when the temperature is dropping during hot rolling and the austenite maximum temperature is reached. Therefore, the transformation is inverse from γ to α. The first stage is an interface-controlled reaction, characterised for a fast austenite transformation, which results in large ferrite grains; whereas the second stage leads to small ferrite grains due to slow diffusion-controlled transformation of the austenite [17].

Figure 8a displayed average grain size of both studied materials, where 2A achieved courser average grain size than 1A after the first-stage of hot rolling. Consequently, the austenite transformation in 2A is delayed and slowed down as a result of coarse initial grains and, therefore, the heating rate has a high impact on the phase transformation [18].

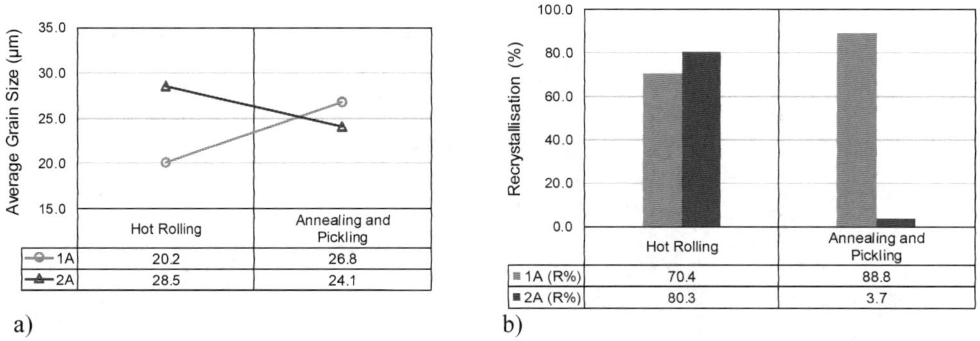

Figure 8. a) Average grain size of 1A and 2A during hot rolling and after annealing (EBSD BC+CSL+GB map). b) Re-crystallisation of 1A and 2A during hot rolling and after annealing (EBSD BC+DefRex).

Thickness reduction of 2A during the first-stage of hot rolling was essentially the same as 1A, unlike hot rolling temperature, which was kept lower in 2A. Despite this lower hot rolling temperature, it is observed in Fig. 8 that 2A structure does not seem to suffer from plastic deformation in terms of grain morphology and re-crystallisation. Material 2A shows an 80.3 % re-crystallised structure (Fig. 8b) at the hot rolling measurement point, versus 70.4 % for material 1A.

Nonetheless, the re-crystallised fraction of 2A dropped after annealing. This leads to point out the importance of cooling rate after hot rolling and annealing time. Despite having retained less austenite than 1A, the cooling time of 2A after hot rolling not only ensured the complete transformation of austenite into ferrite, but also the growth of ferrite grains started to take place. Thus, annealing time was exceeding for this material and, therefore, ferrite grains were elongated and showed a substructure, carried out by sub-grains nucleation.

3.4. Impact of martensite on AISI 430 manufacturing performance

Manufacturing performance of both AISI 430 were evaluated as a function of remaining martensite within ferritic structure. For this purpose, AZTEC RECLASSIFY PHASE was used to measure that martensite fraction.

This investigation is addressed to study microstructural behaviour of flat ferritic stainless steel products. In many occasions, most defects are located at the coil edges; therefore, this work has evaluated microstructural shifts between sheet middle and edges, in order to establish annealing performance in recovering ferritic structure.

Specimens from 1A and 2A sheet middle and edge were taken after both annealing process (annealing and pickling and bright annealing). These samples were analysed by EBSD, where grain size and aspect ratio of the specimens are shown in Fig. 9.

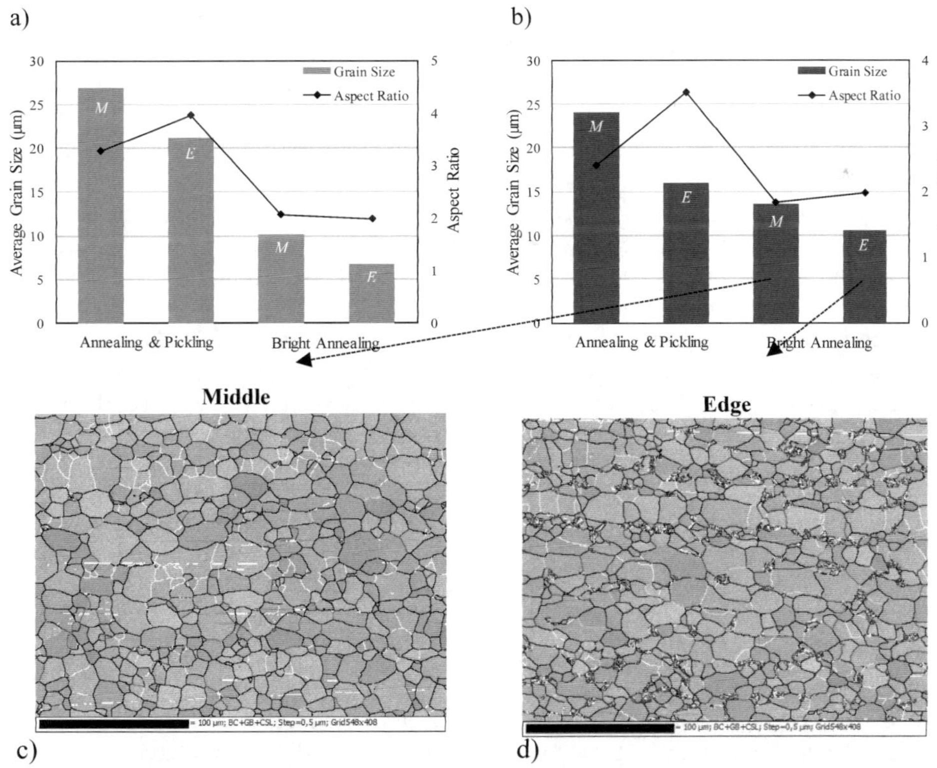

Figure 9. Average grain size and aspect ratio development of annealed a) material 1A, and b) material 2A after annealing and pickling and bright annealing. c-d) EBSD BC+CSL+GB maps of annealed 2A specimens from coil middle and edge after bright annealing (from Fig. 9b).

Re-crystallisation from EBSD analysis and martensite measured by AZTEC RECLASSIFY PHASE for the specimens above are listed in the Table 3.

Table 3. Re-crystallisation and martensite fraction measured by AZTEC RECLASSIFY PHASE of both 1A and 2A specimens collected from sheet middle and edge after annealing processes (annealing and pickling and bright annealing).

AISI 430	Annealing stage	Re-crystallisation %		Martensite %	
		Middle	Edge	Middle	Edge
1A	A&P	88.8	30.6	0	0
	BA	93.3	62.5	5.1	26.3
2A	A&P	3.7	0.7	0	0
	BA	98.2	89.9	0	8.1

These results suggest that the microstructure of 1A and 2A ferritic stainless steel has significantly changed from middle to both edges of each coil because of temperature drop between both points. This different grade of re-crystallisation between the edge and middle site of the coil is also influenced by the geometry of the strip (width) and the energy balance inside the annealing furnace. According to [19], radiation heat transfer is in charge of the over-annealing phenomenon on strip edges.

4. Conclusions

After the last rolling pass, 1A was kept enough time to cool down slowly before reaching the annealing (A&P) temperature. Controlled cooling ensures the $\gamma \rightarrow \alpha + M_{23}C_6$ reaction to partially occur before approaching the annealing stage. During annealing, this material was initially based on ferrite and remaining austenite, and annealing heat led to recover and re-crystallise the ferritic structure (88.8 % R) during cooling from A_{c1} to annealing output temperature.

The aspect ratio of hot-rolled 1A increase after being annealed from 2.15 to 3.27. The coarsening grain was promoted by large cooling time before annealing. Considering that initial microstructure of 1A was based on ferrite and re-crystallisation already started, soak time during annealing caused the coarsening of ferrite grains of annealed 1A.

The 2A cooling range, after the last rolling pass, is shorter than 1A. This means that 2A had less time to complete $\gamma \rightarrow \alpha + M_{23}C_6$ reaction than 1A. Nevertheless, the austenite fraction in 1A is higher than 2A, which can explain that 1A needs slower cooling time than 2A after hot rolling. Despite faster cooling after hot rolling, results have shown that 2A approached annealing process with most of the ferritic structure recovered. Therefore, annealing time should be lower in 2A than 1A, in order to ensure there is not an excess of annealing and subsequent over-growth of grains. According to the aspect ratio of hot-rolled and annealed 2A (2.43 and 2.42, respectively), there is no evidence of grain coarsening phenomenon.

Both 1A and 2A annealed specimens showed a microstructure based on large and small ferrite grains. This heterogeneous grain size distribution depends on austenite to ferrite transformation before A_{c1}, which was promoted by a two-stage reaction controlled by interface and diffusion of austenite transformation, respectively.

Coil middle and edges microstructure of annealed specimens (A&P and BA) were analysed by EBSD and it proved to be an important technique to characterise surface defects related to martensite formation as a result of a temperature shift through the sheet width during annealing controlled by radiative heat transfer inside bright annealing furnace. The site of strip placed by the edges is heated sooner than any site in the middle due to strip geometry and point distance to the furnace interior wall. Furthermore, the refinement of ferrite grains took place as well as the aspect ratio slightly increase from sheet middle to edges. This indicates that the temperature drop through sheet width generates most surface defects and mechanical failures located in coil edges of AISI 430 ferritic stainless steel.

Acknowledgements
The present work is included in FERRINOP project. Co-funded by CDTI.

References

[1] Núñez A, Llovet X and Almagro J F 2013 Analysis of chemical changes and microstructure characterization during deformation in ferritic stainless steel. *Microsc. Microanal.* **19** 959-968
[2] Ghosh N, Pal P K and Nandi G 2017 GMAW dissimilar welding of AISI 409 ferritic stainless steel to AISI 316L austenitic stainless steel by using AISI 308 filler wire. *Engng. Sci. Tech. Int. J.* **20** 1334-1341
[3] Mehtonen S V, Karjalainen L P and Porter D A 2013 Hot deformation behavior and microstructure evolution of a stabilized high-Cr ferritic stainless steel. *Mater. Sci. Engng. A* **571** 1-12
[4] Vafaeian S, Fattah-alhosseini A, Mazaheri Y and Keshavarz M K 2016 On the study of tensile and strain hardening behavior of a thermomechanically treated ferritic stainless steel. *Mater. Sci. Engng. A* **669** 480-489
[5] Guiraldenq P and Hardouin Duparc O 2017 The genesis of the Schaeffler diagram in the history of stainless steel. *Metall. Res. Technol.* **114** 1-9
[6] Andersson J O, Helander T, Höglund L, Shi P and Sundman B 2002 Thermo-Calc & DICTRA, computational tools for materials science. *Calphad Comput. Coupling Phase Diagrams Thermochem.* **26** 273-312
[7] Ben Rhouma A, Amadou T, Sidhom H and Braham C 2017 Correlation between microstructure and intergranular corrosion behavior of low delta-ferrite content AISI 316L aged in the range 550–700 °C. *J. Alloys Comps* **708** 871-886
[8] Sarkari Khorrami M, Mostafaei M A, Pouraliakbar H and Kokabi A H 2014 Study on microstructure and mechanical characteristics of low-carbon steel and ferritic stainless steel joints. *Mater. Sci. Engng. A* **608** 35-45
[9] Coelho A A (2018) TOPAS and TOPAS-Academic: an optimization program integrating computer algebra and crystallographic objects written in C++. *J. Appl. Crystallogr.* **51** 210-218
[10] Sung J H, Kong J H, Yoo D K, On H Y, Lee D J and Lee H W 2008 Phase changes of the AISI 430 ferritic stainless steels after high-temperature gas nitriding and tempering heat treatment. *Mater. Sci. Engng. A* **489** 38-43
[11] Lo K. H, Shek C H and Lai J K L 2009 Recent developments in stainless steels. *Mater. Sci. Engng. R: Reports* **65** 39-104
[12] Lu H H, Li W Q, Du L Y, Guo H K, Liang W, Zhang W G and Liu Z G 2019 The effects of martensitic transformation and (Fe, Cr) 23 C 6 precipitation on the properties of transformable ferritic stainless steel. *Mater. Sci. Engng. A* **754** 502-511
[13] Haghdadi N, Cizek P, Hodgson P D, Tari V, Rohrer G S and Beladi H 2018 Effect of ferrite-to-austenite phase transformation path on the interface crystallographic character distributions in a duplex stainless steel. *Acta Mater.* **145** 196-209
[14] Sello M P and Stumpf W E 2010 Laves phase embrittlement of the ferritic stainless steel type AISI 441. *Mater. Sci. Engng. A* **527** 5194-5202
[15] Kang J Y, Park S J and Moon M B 2013 Phase analysis on dual-phase steel using band slope of electron backscatter diffraction pattern. *Microsc. Microanal.* **19** 13-16
[16] Nowell M M, Wright S I and Carpenter J O 2009 Differentiating ferrite and martensite in steel microstructures using electron backscatter diffraction. in: *2009 Mater. Sci. Technol. Conf. Exhib. (MS T'09)* **2** 933-943

[17] Liu Y, Wang D, Sommer F and Mittemeijer E J 2008 Isothermal austenite-ferrite transformation of Fe-0.04 at.% C alloy: Dilatometric measurement and kinetic analysis. *Acta Materialia* **56** 3833-3842
[18] Caballero F G, Capdevila C and García de Andrés C 2001 Modelling of Kinetics of Austenite Formation in Steels With different initial microstructures. *ISIJ Int.* **41** 1093-1102
[19] Niederer M, Steinboeck A, Strommer S and Kugi A 2013 Analysis of radiative heat transfer in an indirect-fired strip annealing furnace based on integral equations. *IFAC Proc. Vols.* **46** 403-408

IOP Conf. Series: Materials Science and Engineering **891** (2020) 012008 doi:10.1088/1757-899X/891/1/012008

Exploring Mars at the nanoscale: Applications of transmission electron microscopy and atom probe tomography in planetary exploration

L Daly[1,4,5], M R Lee[1], P Bagot[2], J Halpin[3], W Smith[3], S McFadzean[3], A C O'Brien[1], S Griffin[1], L J Hallis[1] and B E Cohen[1]

[1] University of Glasgow, School of Geographical and Earth Sciences, Gregory Building, Glasgow G12 8QQ, Great Britain
[2] University of Oxford, Department of Materials, Oxford OX1 3PH, Great Britain
[3] University of Glasgow, School of Physics and Astronomy, Materials and Condensed Matter Physics, Glasgow G12 8QQ, Great Britain
[4] Curtin University, Space Science and Technology Centre, School of Earth and Planetary Sciences, GPO Box U1987, Perth, WA 6845, Australia
[5] University of Sydney, Australian Centre for Microscopy and Microanalysis, Sydney, 2006, NSW, Australia

E-mail: luke.daly@glasgow.ac.uk

Abstract. The upcoming Mars Sample Return (MSR) mission aims to deliver small quantities of Martian rocks to the Earth. Investigating these precious samples requires the development and application of techniques that can extract the greatest amount of high quality data from the minimum sample volume, thereby maximising science return from MSR. Atom probe tomography (APT) and transmission electron microscopy (TEM) are two complementary techniques that can obtain nanoscale structural, geochemical and, in the case of atom probe, isotopic information from small sample volumes. Here we describe how both techniques operate, as well as review recent developments in sample preparation protocols. We also outline how APT has been successfully applied to extraterrestrial materials in the recent past. Finally, we describe how we have studied Martian meteorites using TEM and APT in close coordination in order to characterise the products of water/rock interactions in the crust of Mars – a key science goal of MSR. Our results provide new insights into the Martian hydrosphere and the mechanisms of anhydrous-hydrous mineral replacement. In light of the unique results provided by these tools, APT and TEM should form a crucial part at the culmination of a correlative analytical pipeline for MSR mission materials.

1. Introduction

Rocks from Mars are among the rarest meteorites and are the only physical samples we currently have of the red planet. They represent a total of 277 individual stones, with a combined weight of 162.5031 kg [1] split between five main lithologies (four igneous and one breccia): the shergottites, nakhlites, chassignites, black beauty paired stone, and the anomalous Alan Hills (ALH) 84001 [1-4]. Cosmogenic exposure ages indicate that they sample 4 - 8 ejection sites on Mars [3]. A Mars sample return mission is being planned to recover samples that had been previously collected and stored by the Mars 2020 rover; these samples will add a further ~ 0.5 kg of fresh oriented rocks to our collection - potentially including sedimentary lithologies [5, 6].

Martian meteorites have proven invaluable in providing insights into volcanic activity, the action of water and impact processes on Mars [2, 4, 7-11]. However, the rarity of Martian samples means that as analytical scientists we are duty bound to maximise the amount of data collected from the smallest sample volume to provide the greatest insight into Martian planetary processes. Here we describe our recent application of transmission electron microscopy (TEM) and atom probe tomography (APT) techniques to Martian meteorites. TEM and APT are two nanoscale techniques that provide the high-resolution culmination of a correlative analytical pipeline (Fig. 1) that maximises data output from the minimum sample volumes to address research questions pertinent to the red planet.

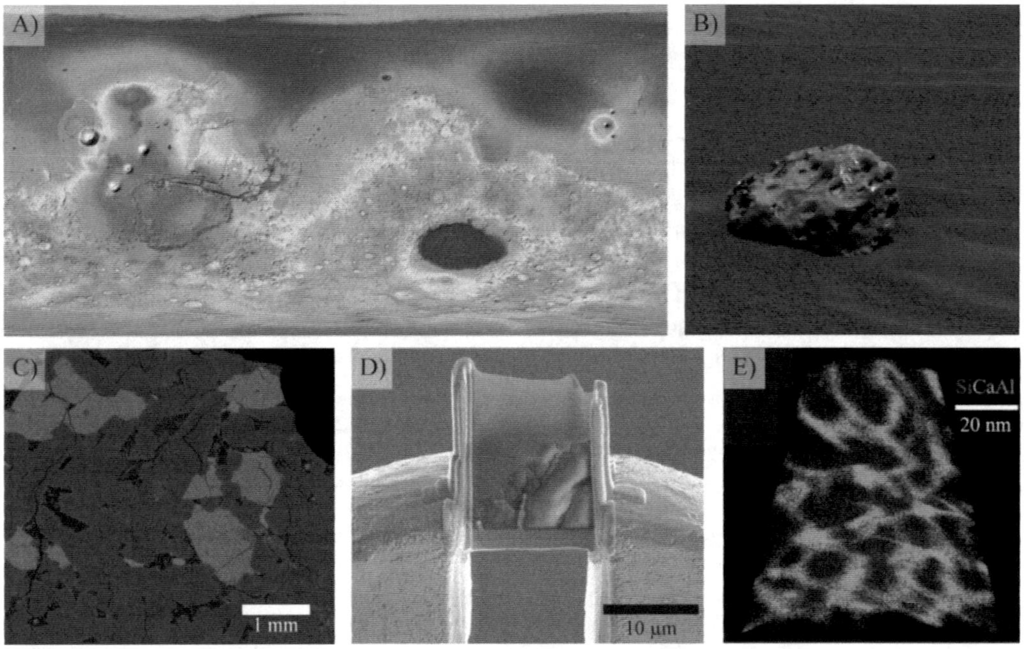

Figure 1. A correlative pipeline across a broad array of length scales to interpret the geological history the planet Mars. a) An elevation map of the Martian surface from the NASA MOLA science team. Image Credit: NASA/USGS Astrogeology Science Centre/Goddard Space Flight Centre. b) An image of an iron meteorite on the surface of Mars taken by NASA's Opportunity rover. Image credit NASA/JPL/Cornell. c) A red (Al), green (Fe), blue (Mg), EDS image of Martian meteorite Yamato 000802. d) A TEM lamella of a section through a magnetite grain in the Tissint meteorite. e) A APT analysis of an atom probe needle extracted from the mesostasis of Martian breccia Northwest Africa 11522 (after [12]).

2. Focussed ion beam TEM / APT sample preparation

The preparation of geological and extraterrestrial samples for TEM and APT usually involves extraction from a polished section using a focussed ion beam (FIB) instrument. Typical TEM sample preparation requires extracting a 1 - 2 µm thick bar of material that is then thinned to electron transparency, usually < 100 nm [13]. TEM samples produced by Ga-FIBs are generally 15 µm × 20 µm as larger samples would be too time intensive to extract. However, with the development of Xe/plasma-FIB instrumentation that can remove 'vast' volumes of material rapidly, the width and depth of a TEM sample are limited only by the cohesive strength of the material. It is now feasible to produce 100 µm wide TEM sample. To avoid bending/warping during thinning of wider samples the structural integrity can be maintained by only thinning 'windows' at regular intervals, with structurally sound 'posts' in between. Wider TEM samples are no longer limited to small regions of interest and can now evaluate heterogeneities in mineralogy, density, porosity, etc., across 100 µm lengths. This is incredibly powerful in meteoritics because most meteorites are heterogeneous over these length scales.

Typical APT sample preparation approaches involve milling out a triangular prism and attaching segments to pre-prepared posts [14]. Sample mounts for APT are Cu or Mo five post grids otherwise used for TEM analysis that have been FIB milled into needles or pre-grown arrays of Si posts. Once attached, the APT samples are thinned using an annular milling pattern to produce a needle with a 100 nm diameter tip [14]. APT samples can also be prepared in situ using the 'satellite dish' method ([15] Fig. 2a). This approach involves cutting a 150 µm diameter wide 50 µm deep circular depression in the sample with APT needle in the centre [15]. A wedge shape 'window' is also cut to view the needle in the atom probe ([15] Fig. 2a). This approach maintains the orientation of the specimen relative to the sample. The satellite dish method combined with electron backscatter diffraction (EBSD) would be a powerful correlative analytical approach. However, this approach is only viable using a Xe-FIB due to the 'vast' amounts of material that need to be excavated and where most of the sample surrounding the region of interest is forfeit and as such is only suitable for specific applications. The 'button method' of FIB APT/TEM sample preparation enables sub-micrometre grains, interfaces, defects and twin boundaries to be targeted ([16] Fig. 2b). This approach becomes particularly powerful when combined with SEM, EBSD, FIB-TOF-SIMS characterisation prior to FIB preparation to pinpoint the region of interest [16]. A 'button' of electron beam deposited platinum is deposited to highlight the feature and locate it after covering with the ion beam protective Pt strip ([16] Fig. 2b). These buttons permit accurate extraction and thinning of TEM/APT samples to ensure the region of interest is contained within the sample. This approach has been successfully applied to extract 100 nm diameter refractory metal alloys in meteorites [17, 18].

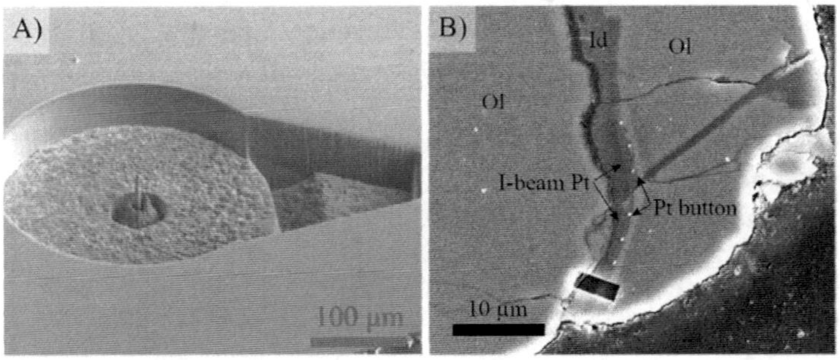

Figure 2. Images of a suite of FIB sample preparation approaches to both TEM and APT. a) APT sample preparation using the in situ 'satellite dish' approach, adapted from [15]. b) The 'button method' of APT sample preparation of dish from Rickard *et al.* [16] used here on to decorate and target the olivine-iddingsite interface in the Lafayette meteorite.

3. Transmission electron microscopy

TEM has been used extensively to characterise water-bearing minerals in Martian meteorites (e.g., [19, 20]), and is complementary to APT. TEM works by irradiating a thin sample – ideally < 100 nm – with a beam of high energy electrons (typically 200 kV) under vacuum. These thin samples are often referred to as 'electron transparent'. Those electrons that pass through the sample can be used to form images, down to atomic resolution. If the beam is relatively broad and static, bright- and dark-field images are formed using the unscattered and scattered electrons, respectively. By scanning a focussed beam over the sample, scanning TEM (STEM) images are formed, again in bright-field (using unscattered electrons) and dark-field (using scattered electrons) modes. Electron diffraction patterns represent the angular range of scattering, and the *d*-spacings obtained from them can be used to identify minerals and determine their crystallographic orientation. The energy that the electrons lose as they pass through the sample can also be measured to obtain information on chemical composition and bonding (i.e., electron energy loss spectroscopy, EELS). An energy-dispersive X-ray spectrometer located above the thin sample enables quantitative and qualitative chemical analysis in the same manner as EDS in an SEM. For a more detailed description of TEM and STEM in planetary science see Lee [21].

4. Atom probe tomography

The APT technique is described in detail in Kelly and Larson [22] and Gault et al. [23]. Therefore, we only provide a brief overview of the technique here. APT is a high-resolution technique capable of analysing small cylindrical volumes (up to approximately 100 × 100 × 1,000 nm) with a spatial resolution of 0.3 nm in three dimensions [22]. The required specimen geometry for APT analysis is needle-like, with a 100 nm diameter tip and a half shank angle < 10° [22]. The needle-like specimen is loaded in the atom probe under high vacuum and situated near an annular electrode [24]. A voltage of ~ 1,000 V is applied to the specimen by the electrode. If the specimen is conductive, the voltage can be pulsed to facilitate field evaporation of individual atoms from the surface of the needle [22, 23]. However, if the sample is non-conductive (i.e., most geological and extraterrestrial samples) field evaporation can be initiated through the addition of heat by applying a high frequency (100 - 250 kHz) ultraviolet laser to the tip of the sample to achieve single atom ionisation [25]. It was the addition of the laser to atom probe instruments that enabled the analysis of non-conductive geological samples [25].

Field evaporated ions are accelerated across a potential difference where they impinge upon a position sensitive detector [22, 23]. The time of flight from laser/voltage pulse to impact provides the mass-to-charge-state-ratio of the ion and thus its elemental and isotopic identity, while the x- and y-coordinates on the detector can be used to reconstruct the initial position of the atom within the sample and thus reconstruct the 3D atomic structure of the sample [22, 23]. The detection efficiency of the detector is dependent on the available surface area on the detector surface, and is generally between 50 - 65 % [22, 23]. However, new generation detectors in atom probes have an increased surface area available for detection and can now collect over 80 % of the ions produced. In principle, the energy imparted to the sample should ionise all elements of the periodic table with equal efficiency resulting in 100 % ionisation of the sample [26, 27]. Therefore, APT does not, in theory, suffer mass dependant fractionation of isotopes or produce biases in the elements detected. The only sources of uncertainty are then derived from counting statistics based on the actual number of ions measured [25]. As such, APT has a high sensitivity and spatial resolution, and so can in principle be used to determine crystallography and atomic lattice spacings, complete geochemical characterisation including trace elements [18, 28-30], depict nanoscale geochemical heterogeneities and clustering of atoms [28, 31-33], and isotope analysis across all elements in the periodic table [34-39].

To extract elemental and isotope abundances from the mass-to-charge-state-ratio-spectrum resulting from the analysis, the total number of counts beneath each peak must be calculated. This is done by 'ranging' the counts beneath each peak [23] using the IVAS software package over a specific unified atomic mass unit (*u*) width followed by a local background subtraction from a nearby region that contains no peaks over the same *u* width. Details about ranging methods and advantages and disadvantages of different approaches can be found in Daly *et al.* [37] and Parman *et al.* [24]. In general, for geochemical analysis the full width of the peak should be used whereas for isotope analysis slightly less than the width of the smallest peak should be used consistently for all isotope peaks of interest [37].

Analysis of geological and extraterrestrial samples is complicated by the formation of multi-ion complexes, i.e., OH^+ or $Si_2O_3^+$. These result in isobaric interferences that need to be deconvolved to extract robust elemental or isotopic abundances. APT can detect trace abundances of elements down to 10 atomic parts per million (appm) for sample volumes over 106 nm^3 [22] while newer atom probe instruments are capable of detection limits as low as 1 appm. This means APT can be utilised to analyse the trace element chemistry of mineral phases. Furthermore, the 3D reconstruction of the atomic positions and the size of the sample mean that it is possible to ensure only atoms emitted from the phase of interest are included in the spectrum. This means an accurate bulk chemistry of the phase of interest can be derived (e.g. [18, 28, 29, 40]). Therefore, APT is ideal for analysing sub-micrometre size phases whose trace element chemistry cannot be obtained through conventional techniques, or where the signal from the phase of interest is contaminated by contributions from surrounding phases. In addition, it is possible to extract nanostructural geochemical information such as planar or point cluster concentrations of elements even in trace amounts within samples. This approach has been demonstrated many times for the geochronological studies of Pb-clusters in zircon, and baddeleyite [31, 32, 34-36] and gold in arsenopyrite [28].

The mass resolving power ($m/\Delta m$), where m is mass and Δm is the variation in mass between atoms, of the atom probe is better than 1,000 [25, 41]. Therefore, APT is capable of discerning peaks separated by 0.15 u [25, 41]. Thus, isotopic peaks from ions of different charge states (i.e., $^{94}Mo^{3+}$ and $^{63}Cr^{2+}$) can be distinguished, but APT cannot currently determine the mass difference between ions that have the same mass (i.e., $^{54}Cr^{2+}$ and $^{54}Fe^{2+}$) [26, 42]. $^{54}Cr^{2+}$ and $^{54}Fe^{2+}$ have a difference in mass of 0:0005 u, which requires a mass resolving power > 1:5 × 10^5 [25, 41] to distinguish these ionic species. These isobaric interferences are beyond the mass resolving power of any technique currently available [25, 41] and require peak deconvolution to estimate the contribution of each isotope to the total peak. Field evaporation during APT analysis is ideal for isotopic analysis of nanoscale materials as the whole sample is ionised, and it is assumed that there is no bias in the detection of elements or between isotopes, i.e., all species are ionised and detected equally [26, 27]. However, unique factors in each analysis, such as voltage, tip shape and temperature are not constant between analyses and their effect on the resulting mass spectrum is not quantified and potentially not quantifiable [24, 32]. Despite these challenges a robust standardised approach for isotope analysis has been proposed for Re/Os isotope analysis and is readily adaptable to other isotope systems and can distinguish isotopic ratio variations between 0.5 – 1 % with substantial room for improvement [37].

5. A brief history of atom probe tomography in planetary science

The first paper on non-conductive geological materials using laser assisted atom probe microscopy was by Valley *et al.* [34] who demonstrated that the $^{207}Pb/^{206}Pb$ isotopic abundances can be measured in zircon using APT. The APT $^{207}Pb/^{206}Pb$ isotope ratio recorded an age consistent with measurements of the same zircon using secondary ion mass spectrometry (SIMS) [34]. This study put atom probe on the map as a powerful geochronological tool for geoscience applications that is capable of reliably reproducing results of other more established high precision isotopic techniques but using a much smaller sample volume.

One of the first uses of APT on extraterrestrial samples was Heck *et al.* [38] who reported APT data for C isotopic ratios in nanodiamonds within primitive chondritie meteorites. These results indicated that the $^{12}C/^{13}C$ ratio is anomalous relative to Solar System values [38]. However, these anomalous isotopic ratios may instead reflect contributions of ^{12}C-hydrides to the ^{13}C-peak [38] and possible underrepresentation of the ^{12}C-peak due to molecular ion formation and detector dead times [40]. This work has been followed by continued APT work on pre-solar grains in Lewis *et al.* [39], unpicking the nanostructures and geochronological information trapped in shocked zircon, baddeleyite, Lunar meteorites [33, 35, 36], and space weathered Lunar soil from the Apollo missions [42], unpicking the magnetic properties of iron meteorites [30], and geochemical analysis of refractory metal alloys [17, 18]. We have only scratched the surface of the insights of APT in planetary science.

The real interpretive power of APT and TEM is through correlative analysis combining several techniques to provide context to these nanoscale analyses (e.g., [16]). Below is an example workflow showing how light microscopy, SEM, EBSD, FIB, TEM, STEM and APT can be successfully combined to locate and characterise water-bearing minerals in Martian meteorites.

IOP Conf. Series: Materials Science and Engineering **891** (2020) 012008 doi:10.1088/1757-899X/891/1/012008

6. Water on Mars

Ashworth and Hutchison [7] first described evidence for water-bearing material in the nakhlite group of Martian meteorites (see [8] for a recent review). These authors described the water bearing material as "iddingsite", which they interpreted to have formed by hydrothermal alteration of olivine. This fluid that produced the iddingsite assemblage was oxidising in nature, low temperature (< 140 °C) [11] and affected these rocks approximately 633 Ma [43]. The heat source that melted these subsurface ices is not currently known, but the primary theories are impact generated hydrothermal activity [44] or a magmatic event [43].

The iddingsite veins can be readily identified using transmitted light microscopy (Fig. 3) and are a few micrometres in thickness. They resemble narrow veins of 'rust' in transmitted light (Fig. 4a) and although more detail can be observed by SEM (Figs. 4b and 4c) their properties remain difficult to determine. EDS shows that these veins have a Mg-Fe silicate composition (~ 40 wt% SiO_2, 40 wt% Fe_2O_3, 5 wt% MgO) and their low analytical totals (~ 85 - 90 wt%) indicate the presence of unanalysed OH/H_2O (i.e., Martian water). EBSD demonstrates that the veins have a preferred crystallographic orientation parallel to [010], and this crystallographic information can be used to help guide the orientation in which the electron-transparent samples can be cut and extracted using FIB (Figs. 2c and 2d). Bright-field TEM images show that the host olivine is microstructurally complex, probably reflecting mild shock metamorphism [20] whereas the iddingsite veins appear to be homogeneous and featureless (Fig. 4d).

Figure 3. Transmitted light image of a thin section of the Nakhla meteorite. An olivine grain in the centre of the field of view has a light brown colour and is cut by several very thin veins of rust-coloured iddingsite. The clear tabular mineral grains are augite. Image from the Open University virtual microscope (www.virtualmicroscope.org/content/nakhla-2).

To characterise OH/H_2O-bearing Fe-Mg silicates, compositional maps were obtained from the area in Fig. 4d using EDS and EELS (Fig. 5). Results highlight the location of siderite and parts of the carbonate have altered to ferryhydrite ($Fe^{3+}{}_{8.2}O_{8.5}(OH)_{7.4}$ + $3H_2O$) (Fig. 5). As well as elemental analysis, EELS is used to determine the valence state of Fe (and other elements) [45]. An Fe valence map of the vein shows that the OH/H_2O-bearing Fe-Mg silicate and ferrihydrite both contain Fe^{3+}, whereas the siderite and olivine have Fe^{2+} as expected (Fig. 5).

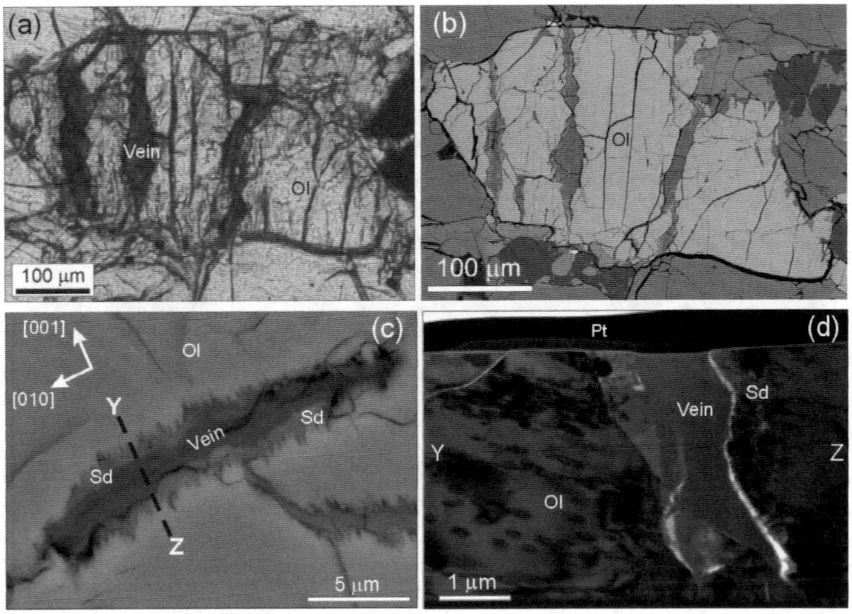

Figure 4. Iddingsite in the Lafayette meteorite, (a and b) [46], and the Nakhla meteorite (c and d) [20], hosted by olivine (Ol). a) Transmitted light image, and b) backscattered electron SEM image of an olivine grain that is cross-cut by three rust-brown coloured iddingsite veins. c) Backscattered electron SEM image of an iddingsite vein. Between the iddingsite and olivine is a selvage of siderite (Sd) with a finely serrated contact with olivine. The crystallographic orientation of the vein relative to the host olivine grain is indicated, as determined by EBSD. The dashed black line Y-Z denotes where a FIB-section was extracted. d) Bright-field TEM image of the sample extracted from (c). The olivine is various shades of grey as its defect-rich microstructure is diffracting electrons to different intensities. At this magnification the vein is homogeneous and apparently free of diffracting crystals. Siderite (Sd) is difficult to distinguish from olivine. The black band along the top of the image is the platinum (Pt) strap that was deposited by the FIB to protect the sample from ion beam damage.

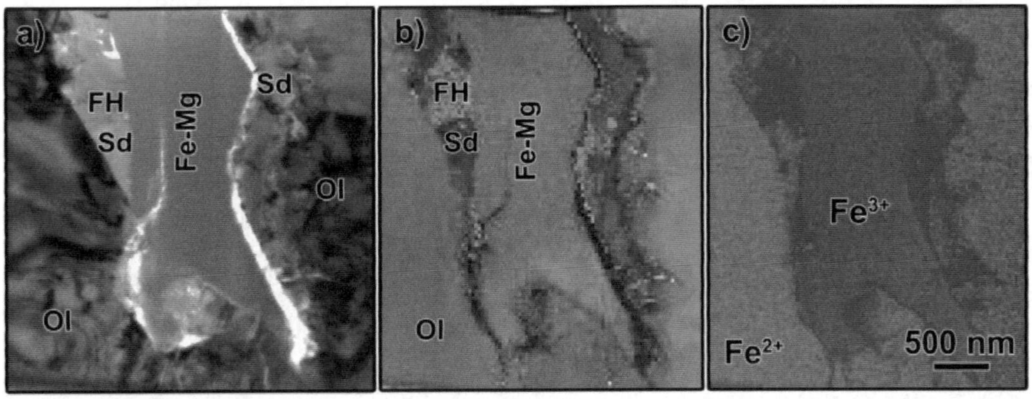

Figure 5. Iddingsite veins from the Nakhla meteorite (from [19], Fig. 3). a) Bright-field TEM image of a vein of OH/H_2O-bearing Fe-Mg silicate (Fe-Mg) that is flanked by siderite (Sd) and hosted by olivine (Ol). The siderite also contains ferrihydrite (FH). b) EDS map of (a) with element combinations coloured so as to render olivine green, siderite blue, ferrihydrite pink, and the Fe-Mg silicate orange. d) EELS valence map of (a) and (b) showing that olivine and siderite both contain Fe^{2+} (coloured green) and the Fe-Mg silicate and ferrihydrite have Fe^{3+} (red).

The identity of the OH/H_2O-bearing Fe^{3+}-Mg-silicate could not be determined by TEM imaging or EELS alone. Thus, higher resolution STEM images and EDS maps were obtained using an aberration-corrected TEM with a probe size of < 2 nm (Fig. 6). STEM images show that the Fe-Mg-silicate is composed of a closely packed mass of particles that are circular/oval and a few nanometres in diameter (Fig. 6a). EDS maps show that the particles are composed of Si and O with a Fe-rich rim (Fig. 6b). Given that this material contains OH/H_2O and its Fe^{3+}, the Fe-Mg-silicate is interpreted to be composed of nanoparticles of opaline silica (amorphous $SiO_2 \cdot nH_2O$) that are rimmed by Mg-bearing ferrihydrite.

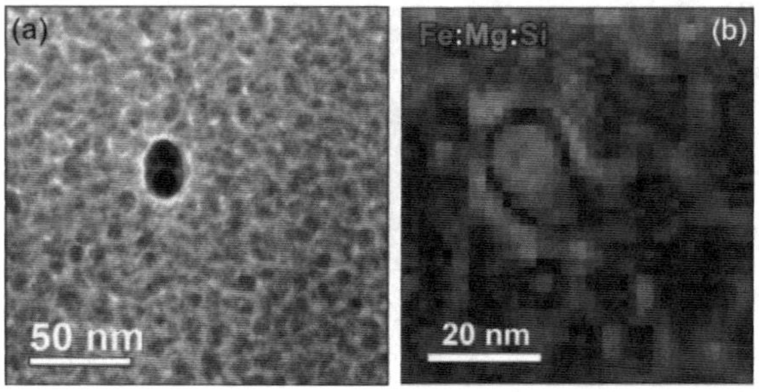

Figure 6. Iddingsite veins from the Nakhla meteorite (from [19] Fig. 4). a) High-angle annular dark-field (HAADF) image of the OH/H_2O-bearing Fe-Mg silicate showing that it is composed of a tightly packed mass of nanoparticles. b) EDS element map of one area of (a) demonstrating that the nanoparticles are composed of silica and have a Fe rich rim.

APT samples have been extracted across the nakhlite olivine-iddingsite interface (Fig. 7) to observe nanoscale geochemical gradients and corroborate TEM observations. APT analysis of iddingsite reveal it had a heterogeneous density consistent with the HAADF images obtained by TEM (Figs. 6a and 7) but homogeneous chemistry. As such on the length scales of an APT needle in this region, we only identified clay minerals. No oxides, carbonates, sulphates or halides [2] or nanoscale chemical domains were found. The density variations in iddingsite are interpreted to represent the preferential ionisation of individual phyllosilicate platelets (high density) followed by 'bursting' removing the clay mineral to leave a void (low density). APT analysis of across the iddingsite-olivine interface reveal that this boundary is gradational at the nanoscale. Towards the interface over 30 - 20 nm steady depletions in the major cations such as Mg are observed. These depletions are complemented by a steady enrichment in H towards the interface (Fig. 7). EBSD analysis of NWA 817 indicate the iddingsite vein is parallel to the (021) plane of the olivine consistent with previous petrographic descriptions of crystal orientation relationships in these meteorites [10]. However, the observation of a 20 nm wide diffuse boundary not a sharp boundary (Fig. 7) is inconsistent with recent models of fluid alteration of glasses, such as interfacial dissolution-reprecipitation [47]. Our APT data agree better with a diffusion coupled hydration model where inward diffusion of hydrogen species is coupled with outward diffusion of cations [48]. In addition, this 20 nm layer may store H more securely than iddingsite whose hydrogen isotopic composition is strongly perturbed by terrestrial weathering [49, 50] making inferences about the fluids origin challenging. However, directly measuring the D/H ratio of hydrogen contained within the olivine across these interfaces may reveal the D/H ratio of the original fluid, and thereby constrain its source (e.g., Martian crust versus atmosphere).

7 Conclusion

We stand upon the edge of a nanorevolution in planetary science. Through measurements at the atomic-scale using TEM and APT we can provide new insights into Martian and Solar System processes that act on length scales over 20 orders of magnitude larger than the sample volume analysed such as action of water on Mars that has implications for past and future habitability of the red planet. APT combined with improvements to the resolution of more traditional analytical approaches like TEM is about to open a whole new suite of nanoscale observations. This is particularly important for samples from the

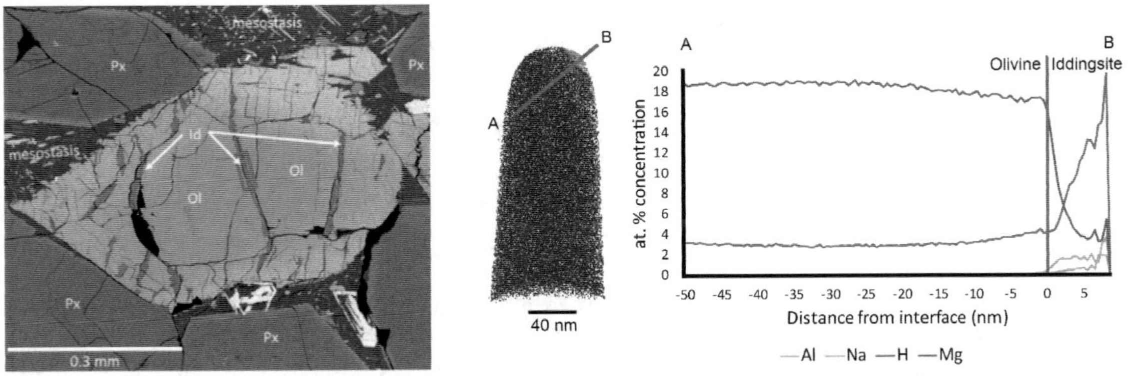

Figure 7. APT analysis of an olivine-iddingsite interface in nakhlite meteorite NWA 817. Image to the left is a backscattered electron (BSE) image of an olivine phenocryst with cross-cutting crystallographically controlled iddingsite veins. A set of APT samples were extracted from the red box so as the iddingsite-olivine interface was present at the apex of the needle. Ol = Olivine, Id = Iddingsite, Px = Pyroxene. Image below is an APT analysis across an olivine-iddingsite interface. The atom map represented by Mg (blue dots) and Na (green dots) atoms defines the interface (red iso-concentration surface). The graph shows a concentration profile of the major elements across the boundary from iddingsite (b) into olivine (A) where a clear depletion of Mg and enrichment of hydrogen is observed up to 20 nm into the olivine. The origin of the majority of the hydrogen signal within APT dataset is likely a combination between inherent hydrogen in the sample and background hydrogen from the sample chamber that is currently challenging to distinguish. However, recent deuteration experiments have shown a clear pathway to conduct these studies in the future [51-53]. In particular, deuteration experiments conducted on olivine by [53] indicate that variations in hydrogen at grain surfaces and reaction interfaces, similar to those detected here are real variations inherent in the sample and as such may provide a secure reservoir for Martian hydrogen. Image after [54].

red planet where meteorite samples are rare and the planned Mars sample return mission will provide at most a kilogram of material. Therefore, it is critical to maximise the research output from these small sample volumes. The combination of APT and TEM form a final set of techniques at the end of a correlated pipeline of analysis that will maximise the information gathered from such small sample volumes.

Acknowledgements
The authors would like to thank proceedings editors as well as one anonymous reviewer for their valuable comments that improved this manuscript. This work was funded by STFC, grant number ST/T002328/1 to MRL.

References

[1] Meteoritical Bulletin Database. 2019 Martian meteorites
[2] Agee C B, *et al.* 2013 Unique meteorite from early Amazonian Mars: Water-rich basaltic breccia Northwest Africa 7034. *Science* **339** 780-785
[3] Nyquist L, *et al.* 2001 Ages and geologic histories of Martian meteorites. in: *Chronology and evolution of Mars*. (New York, NY: Springer) 105-164
[4] McSween H Y 2015 Petrology on Mars. *Amer. Mineralogist* **100** 2380-2395
[5] Williford K H, *et al.* 2018 The NASA Mars 2020 rover mission and the search for extraterrestrial life. in: *From habitability to life on Mars*. (Amsterdam, The Netherlands: Elsevier) 275-308

[6] iMOST, *et al.* 2018 *The potential science and engineering value of samples delivered to Earth by Mars Sample Return.* Mars Exploration Program Analysis Group (MEPAG). White paper, 186.
[7] Ashworth J and Hutchison R 1975 Water in non-carbonaceous stony meteorites. *Nature* **256** 714-715
[8] Cohen B 2017 Taking the pulse of Mars via 40Ar/39Ar dating of a plume-fed volcano. *Nature Comm.* **8** 640
[9] Harvey R P and McSween H Y 1992 Petrogenesis of the nakhlite meteorites: Evidence from cumulate mineral zoning. *Geochim. Cosmochim. Acta*, **56** 1655-1663
[10] Treiman A H 2005 The nakhlite meteorites: Augite-rich igneous rocks from Mars. *Chemie der Erde-Geochemistry* **65** 203-270
[11] Bunch T and Reid A M 1975 The nakhlites Part I: Petrography and mineral chemistry. *Meteoritics Planet. Sci.* **10** 303-315
[12] Daly L, *et al.* 2018 High pressure excursions in the matrix of Martian meteorite North West Africa (NWA) 11522. in: *81st Ann. Meeting of the Meteoritical Society* 2067
[13] Lee M, Bland P and Graham G 2003 Preparation of TEM samples by focused ion beam (FIB) techniques: applications to the study of clays and phyllosilicates in meteorites. *Mineral. Mag.* **67** 581-592
[14] Thompson K, *et al.* 2007 In situ site-specific specimen preparation for atom probe tomography. *Ultramicroscopy* **107** 131-139
[15] Halpin J E, *et al.* 2019 An in-situ approach for preparing atom probe tomography specimens by xenon plasma focused ion beam milling. *Ultramicroscopy* **202** 121-127
[16] Rickard W, *et al.* 2020 Novel Applications of FIB-based ToF-SIMS in Atom Probe Tomography Workflows. *Microsc. Microanal.* doi.org/10.1017/S1431927620000136
[17] Daly L, *et al.* 2017 Crystallography of refractory metal nuggets in carbonaceous chondrites: A transmission Kikuchi diffraction approach. *Geochim. Cosmochim. Acta* **216** 42-60
[18] Daly L, *et al.* 2017 Nebula sulfidation and evidence for migration of "free-floating" refractory metal nuggets revealed by atom probe microscopy. *Geology* **45** 847-850
[19] Lee M, *et al.* 2015 Opal-A in the Nakhla meteorite: A tracer of ephemeral liquid water in the Amazonian crust of Mars. *Meteoritics Planet. Sci.* **50** 1362-1377
[20] Lee M, *et al.* 2013 Evidence for silicate dissolution on Mars from the Nakhla meteorite. *Meteoritics Planet. Sci.* **48** 224-240
[21] Lee M 2010 Transmission electron microscopy (TEM) of Earth and planetary materials: A review. *Mineral. Mag.* **74** 1-27
[22] Kelly T F and Larson D J 2012 Atom probe tomography 2012. *Ann. Rev. Mater. Res.* **42** 1-31
[23] Gault B, *et al.* 2012 *Atom probe microscopy, vol. 160.* (New York, NY: Springer Science and Business Media)
[24] Parman S W, *et al.* 2015 Atom probe tomography of isoferroplatinum. *Amer. Mineralogist* **100** 852-860
[25] Larson D J, *et al.* 2013 *Local electrode atom probe tomography.* (New York, NY: Springer Science)
[26] Kingham D R 1982 The post-ionization of field evaporated ions: A theoretical explanation of multiple charge states. *Surface Sci.* **116** 273-301
[27] Straub H, *et al.* 1999 Absolute detection efficiency of a microchannel plate detector for kilo-electron volt energy ions. *Rev. Sci. Instrum.* **70** 4238-4240
[28] Fougerouse D, *et al.* 2016 Nanoscale gold clusters in arsenopyrite controlled by growth rate not concentration: Evidence from atom probe microscopy. *Amer. Mineralogist* **101** 1916-1919
[29] Pérez-Huerta A, *et al.* 2016 Atom probe tomography (APT) of carbonate minerals. *Micron* **80** 83-89
[30] Einsle J F, et al. 2018 Nanomagnetic properties of the meteorite cloudy zone. *Proc. Nat. Acad. Sci.* **115** E11436-E11445

[31] Piazolo S, *et al.* 2016 Deformation-induced trace element redistribution in zircon revealed using atom probe tomography. *Nature Comm.* **7** 10490
[32] Reddy S M, *et al.* 2016 Mechanisms of deformation-induced trace element migration in zircon resolved by atom probe and correlative microscopy. *Geochim. Cosmochim. Acta* **195** 158-170
[33] Montalvo S D, *et al.* 2019 Nanoscale constraints on the shock-induced transformation of zircon to reidite. *Chem. Geol.* **507** 85-95
[34] Valley J W, *et al.* 2014 Hadean age for a post-magma-ocean zircon confirmed by atom-probe tomography. *Nature Geosci.* **7** 219-223
[35] White L, *et al.* 2017 Atomic-scale age resolution of planetary events. *Nature Comm.* **8** 15597
[36] White L F, *et al.* 2019 Crystallization and impact history of a meteoritic sample of early lunar crust (NWA 3163) refined by atom probe geochronology. *Geosci. Frontiers* **10** 1841-1848
[37] Daly L, *et al.* 2018 Defining the potential of nanoscale Re-Os isotope systematics using atom probe microscopy. *Geostand. Geoanal. Res.* **42** 279-299
[38] Heck P R, *et al.* 2014 Atom-probe analyses of nanodiamonds from Allende. *Meteoritics Planet. Sci.* **49** 453-467
[39] Lewis J, *et al.* 2015 Atom-probe tomography measurements of isotopic ratios of high-field materials with corrections and standardization: A case study of the 12C/13C of meteoritic nanodiamonds. *Microsc. Microanal.* **21** (Suppl. 3) 39-40
[40] Thuvander M, *et al.* 2011 Quantitative atom probe analysis of carbides. *Ultramicroscopy* **111** 604-608
[41] Kelly T F 2011 Kinetic-energy discrimination for atom probe tomography. *Microsc. Microanal.* **17** 1-14
[42] Goupon P, *et al.* 2017 Ultra-reduced pases in Apollo 16 regolith: Combined field emission electron probe microanalysis and atom probe tomography of submicron Fe-Si grains in Apollo 16 sample 61500. *Meteritics Planet. Sci.* **52** 9 1941-1962
[43] Borg L and Drake M J 2005 A review of meteorite evidence for the timing of magmatism and of surface or near-surface liquid water on Mars. *J. Geophys. Res.: Planets* **110** E12S03
[44] Osinski G R, *et al.* 2013 Impact-generated hydrothermal systems on Earth and Mars. *Icarus* **224** 347-363
[45] Garvie L A and Buseck P R 1998 Ratios of ferrous to ferric iron from nanometre-sized areas in minerals. *Nature* **396** 667
[46] Lee, M R, *et al.* 2015 Formation of iddingsite veins in the martian crust by centripetal replacement of olivine: Evidence from the nakhlite meteorite Lafayette. *Geochim. Cosmochim. Acta* **154** 49-65.
[47] Hellmann R, *et al.* 2015 Nanometre-scale evidence for interfacial dissolution–reprecipitation control of silicate glass corrosion. *Nature Mater.* **14** 307
[48] Grambow B 2006 Nuclear waste glasses - How durable? *Elements* **2** 357-364
[49] Lee M R, *et al.* 2018 Aqueous alteration of the Martian meteorite Northwest Africa 817: Probing fluid–rock interaction at the nakhlite launch site. *Meteoritics Planet. Sci.* **53** 2395-2412
[50] Hallis L, *et al.* 2012 Hydrogen isotope analyses of alteration phases in the nakhlite martian meteorites. *Geochim. Cosmochim. Acta* **97** 105-119
[51] Sundell G, *et al.* 2015 Direct observation of hydrogen and deuterium in oxide grain boundaries in corroded zirconium alloys. *Corrosion Sci.* **90** 1-4
[52] Chen Y S, *et al.* 2017 Direct observation of individual hydrogen atoms at trapping sites in a ferritic steel. *Science* **355** 1196-1199
[53] Daly L, *et al.* 2019 Solar wind hydration of Itokawa olivine. in: *82nd Ann. Meeting of the Meteoritical Society* 6273
[54] Daly L, *et al.* 2018 Insights into Martian fluid-rock reactions by atom probe tomography of the interface between nakhlite olivine and iddingsite. in: *50th Lunar and Planetary Science Conference* **50** 2132

IOP Conf. Series: Materials Science and Engineering **891** (2020) 012009 doi:10.1088/1757-899X/891/1/012009

A dictionary indexing approach for EBSD

M De Graef

Carnegie Mellon University, Department of Materials Science and Engineering, 130 Roberts Engineering Hall, 5000 Forbes Avenue, 15213-3890 Pittsburgh, PA, U.S.A.

E-mail: degraef@cmu.edu

Abstract. Dictionary indexing is a new approach to the indexing of electron backscatter diffraction (EBSD) patterns. In this contribution, we will describe the underlying forward model as well as the indexing approach that uses a similarity metric between experimental and simulated patterns instead of feature extraction. Then we provide a number of example applications drawn from the materials and geological communities.

1. Introduction

Historically, electron backscatter diffraction patterns were indexed manually before commercial indexing packages based on the Hough transform became available in the early 1990s. Today, Hough-based indexing can be carried out rapidly (several thousand patterns per second), and produces accurate orientations, provided the detector geometry has been characterised accurately and the patterns have a sufficiently high signal-to-noise ratio. In 2015, we proposed a new indexing approach [1] that does not extract any features from the experimental patterns; instead, the approach matches complete simulated patterns, using a physics-based forward modelling approach, against the experimental patterns, and assigns the crystal orientation of the best matching pattern, using one of many image similarity metrics, to the experimental pattern. In this contribution, we will briefly review the components of the forward model as well as the basic dictionary indexing (DI) algorithm and we will illustrate the approach with a number of concrete examples drawn from both the materials and geological communities. We will illustrate the robustness of the DI approach against pattern noise as well as the ability of the technique to distinguish between materials with nearly identical crystal structure. We will conclude with a short preview of a recently developed, more efficient indexing approach that is based on the forward model combined with spherical harmonic Fourier transforms.

2. Summary of the EBSD forward model

Simulated EBSD patterns can be obtained using a physics-based forward model [2-4] consisting of: (1) a Monte Carlo simulation of the spatial, energy, and depth distributions of the backscattered electrons (BSEs); (2) a dynamical (quantum-mechanical) simulation of the channelling of the BSEs on their way out of the sample; and (3) an accurate geometrical model for the sample-detector configuration. Without going into details, the model produces the BSE yield on a spherical surface (the Kikuchi sphere) surrounding an imaginary spherical single crystal with an electron source with a range of energies at its centre; this intensity distribution represents the EBSD Master Pattern (MP) and is shown in Fig. 1a for Ni at 20 kV. Individual EBSD patterns are then determined by direct interpolation for a given crystal orientation and detector geometry. The spherical nature of the data immediately suggests the use of spherical harmonic functions to generate an indexing scheme; we will briefly describe such an approach at the end of this contribution. To simplify interpolations from the master pattern, we convert the spherical intensity data via a 2D equal-area Lambert projection [5] onto a square (Fig. 1b), which allows for direct bilinear instead of spherical interpolation. Typical pattern generation rates range from a few hundred to several thousand patterns per second, depending on the individual pattern size and the number of compute threads used [6].

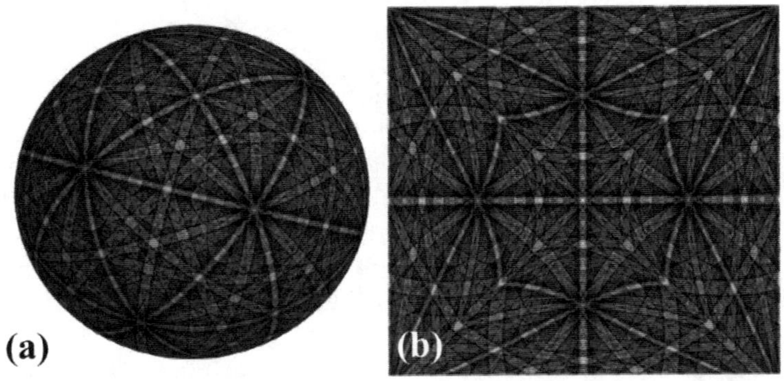

Figure 1. a) Kikuchi sphere for nickel at 20 kV; b) square Lambert projection for the Northern hemisphere of a).

3. The principles behind dictionary indexing

Dictionary Indexing (DI) refers to an algorithm to compare complete experimental EBSD patterns against simulated patterns, using one of many potential image similarity metrics. This approach requires two initial steps, namely the generation of a uniform sample of orientations, and a basic pattern pre-processing step. We will describe both steps in the following subsections, followed by a demonstration of the robustness of this indexing technique against noise. Then we describe the approach as a dimensionality reduction scheme and we introduce a number of useful derived maps. We conclude the section with a brief description of orientation refinement as a post-processing step.

3.1. Orientation space sampling

The EBSD forward model is used in combination with a uniform sampling approach for orientation space *SO(3)* [7] to generate a collection of EBSD patterns for a given angular sampling step size. The sampling algorithm operates in the cubochoric orientation representation, an equal-volume mapping of the quaternion Northern hemisphere onto a cubic grid. Since uniform samples on a cubic grid are trivial to generate, using the cubochoric approach naturally leads to an efficient sampling algorithm. For each point on the grid, the orientation is transformed to a Rodrigues-Frank vector, and this vector is then tested against the geometry of the fundamental zone (FZ) for the particular crystal symmetry under consideration; if the point lies outside the FZ it is discarded from the list, otherwise it remains on the list. Once the list is complete, an individual EBSD pattern is computed for each orientation; the collection of these patterns is referred to as the "pattern dictionary". The number of patterns in a dictionary can become quite large, in particular for low-symmetry crystal structures; for an orientation sampling step size of 1.4°, there are 333,227 unique patterns for cubic crystal symmetry, and 8,000,000 patterns for triclinic symmetry. This is likely the most important drawback of the DI approach and is partially remedied by the new spherical harmonic approach described at the end of this contribution.

3.2. Pattern pre-processing
Before the DI algorithm is applied, both experimental and dictionary patterns undergo two pre-processing steps. The first step is a standard high-pass Fourier filter that eliminates most of the background contrast variations in the patterns. The second step increases the contrast in the patterns by applying an adaptive histogram equalisation (AHE) filter [8] that changes the usually Gaussian-like intensity histogram to an essentially uniform histogram, so that all intensity levels between 0 and 255 are approximately equally present in the pattern. This step increases the differences between pairs of patterns, making it more likely that the indexing process will recover the correct result. Both pre-processing steps are carried out on all experimental and dictionary patterns before the patterns are compared to each other.

3.3. Robustness against noise
Each of the dictionary patterns is then compared to each of the experimental patterns using either a normalised dot product or the mutual information [9] between the two patterns. Since each pattern is represented by a unit vector, simply taking the dot product between two pattern vectors produces a measure for the similarity of the two patterns, in the same way that the dot product between two unit 3D vectors is a measure for the angle between the two vectors (a smaller angle or larger dot product indicates an increased similarity between the vectors). The orientation of the dictionary pattern for which the similarity metric is highest is taken to be the orientation of the illuminated volume. Repeating this process for all experimental patterns then indexes the entire data set [1].

Figure 2 shows an example of the robustness of the dictionary indexing approach for a Ni data set (courtesy of S. Wright) recorded using four different detector settings, with the signal to noise ratio decreasing from top left to bottom right. Even for the noisiest patterns, for which the Hough approach largely fails, the full pattern dot product similarity metric provides a viable means of indexing the patterns, with an 85 % indexing success rate. This rate can be brought close to 100 % by performing a non-local means filter to the data before applying the DI algorithm [10]. Hence, the dictionary approach overcomes limitations that are inherent to the traditional Hough-based indexing algorithm; this robustness indicates that patterns from heavily deformed materials are now no longer impossible to index.

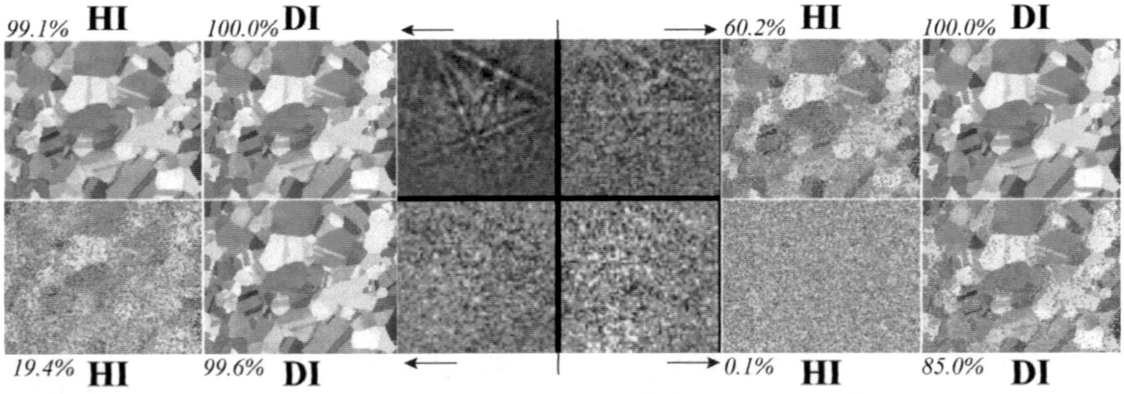

Figure 2. Indexing examples using Hough indexing (HI) and Dictionary Indexing (DI) for four different pattern qualities; the percentages indicate the indexing success-rate of each approach. (Data courtesy S. Wright).

3.4. *DI as a dimensionality reduction algorithm*

An EBSD pattern of *M* by *N* pixels with intensities in the range [0 … 255] can be regarded formally as a point in a discrete *MN*-dimensional space. Indexing the EBSD pattern essentially means that we need to reduce the dimensionality of the pattern from *MN* to 3, since there are three parameters needed to uniquely describe a 3D orientation. This is a significant reduction in dimensionality, even for the smallest patterns typically acquired using 8x binning; if the binned pattern size is 80 by 60 pixels, then the necessary reduction is from 4,800 to 3. This indicates that there should be more than sufficient information in an EBSD pattern, even after 8x or higher binning, to make it possible to index the pattern. Indeed, in our analysis of the performance of the DI approach under a number of different conditions we found that even binning down to patterns of size 15 by 15 pixels does not substantially degrade the indexing success rate or the angular accuracy [11]. Considering EBSD indexing to be a form of dimensionality reduction suggests that the technique may benefit from modern machine learning (ML) approaches, in particular convolutional neural nets (CNNs); the EBSD forward model can be used to generate as many patterns as are needed to properly train the CNN. In our preliminary work, we have shown that reasonable indexing success rates can be achieved by means of machine learning techniques [12]; the main drawback of ML, however, is the lengthy training cycle that is required to set up the neural net (several days of GPU time for a given set of detector parameters and crystal structure). This suggests that machine-learning approaches may have only limited applicability for real-time EBSD indexing.

3.5. *Derived maps*

The DI approach can be used to generate a number of maps that are not readily available using the Hough-based technique. After the pattern pre-processing step, one can compute the average dot product (ADP) map, i.e., a map that shows the average dot product of each pattern vector with its four nearest neighbours (for square grid sampling). The intensity in the ADP map is given by:

$$\alpha_{r,c} = \frac{1}{4}(\hat{\boldsymbol{v}}_{r,c}\cdot\hat{\boldsymbol{v}}_{r+1,c} + \hat{\boldsymbol{v}}_{r,c}\cdot\hat{\boldsymbol{v}}_{r-1,c} + \hat{\boldsymbol{v}}_{r,c}\cdot\hat{\boldsymbol{v}}_{r,c+1} + \hat{\boldsymbol{v}}_{r,c}\cdot\hat{\boldsymbol{v}}_{r,c-1}) \quad (1)$$

where each unit vector represents a pattern and the subscripts indicate the pattern position in the inverse pole figure map. The ADP map (example shown in Fig. 3a for a portion of a 20 kV forsterite data set, courtesy of K. Marquardt) is quite sensitive to the similarity between neighbouring patterns and highlights both microstructural features (e.g., grain boundaries, different phases) and sample artefacts (e.g., surface scratches) [13].

During indexing, one maintains, for each experimental pattern, a list of the dictionary patterns that are similar to it, ranked by decreasing similarity metric. The best matching pattern produces the most likely orientation, and the corresponding similarity metric value can be used to generate a confidence index (CI) map (example for forsterite shown in Fig. 3b). The list of *N* next nearest matches can be used to generate an orientation similarity map (OSM), i.e., a map that shows how similar the top matches list of a given pattern is to those of its nearest neighbours. If $S_{r,c}$ is a set containing the list of *N* top-matching dictionary pattern labels for the pattern at sampling location (r,c), then the OSM value is defined using the following relation (again for a square sampling grid):

$$\eta_{r,c} = \frac{1}{4}\left(\#(S_{r,c} \cap S_{r-1,c}) + \#(S_{r,c} \cap S_{r+1,c}) + \#(S_{r,c} \cap S_{r,c-1}) + \#(S_{r,c} \cap S_{r,c+1})\right) \quad (2)$$

In this equation, # represents the cardinality of the intersection set between the top lists for two neighbouring sampling points. The OSM map provides a remarkably clear microstructure map outlining sharply defined grain boundaries, as is shown in Fig. 3c for the forsterite data set. Figure 3d shows the OSM for a deformed and partially recrystallised Cu-Zn sample (10 kV, data set courtesy M. Taheri); note that the recrystallised grains clearly stand out against the background of deformed grains. The

Figure 3. Forsterite maps: a) inverted ADP map, b) CI map, and c) OSM map. d) OSM map for partially recrystallised CuZn.

contrast between the two sets of grains is readily explained: for a recrystallised grain, neighbouring points have nearly identical top-matching sets $S_{r,c}$, whereas for heavily deformed grains, the list of top matches varies strongly from point to point, thus generating a significantly lower orientation similarity value.

3.6. Orientation refinements
The standard DI algorithm produces the best match on a cubochoric sampling grid of orientations for which the dictionary patterns were computed. The true orientation for a given experimental pattern is likely to lie between sampling grid points, so the final step in the indexing process is a refinement of the orientation. This is carried out using a quadratic bounded optimisation algorithm [14] and results in orientation accuracies on the order of 0.1°. During refinement, one can also carry out the pattern centre correction, since the dictionary patterns are computed using only one single pattern centre for all patterns. This correction is not very important when the sampled region of interest has a size that is similar to the detector pixel size, but for a larger field of view, the correction becomes quite important, as has been shown in [14].

3.7. Background and related information
The DI technique and its foundation and applications have been described in detail in a series of publications which we list here as reference for the interested reader: generation of a reliable orientation space sampling scheme that is uniform in the equal-volume sense [15]; detailed analysis of the accuracy and precision of the DI technique, in particular with respect to pattern size and other detector parameters [14]; introduction of an orientation refinement scheme that allows for accurate indexing of large-area EBSD data sets [14]; introduction of a robust phase differentiation scheme based on the DI approach

[16]; extension of the DI approach to electron channelling patterns (ECP) [15]; and the first application of DI to transmission Kikuchi diffraction (TKD) patterns [17]; successful application of the DI technique to distinguish between dynamically and post-dynamically recrystallised grains in an Inconel 718 sample, showing an orientation angular resolution, after refinement, of about 0.1° [18]; indexing of a three-phase geological sample, consisting of garnet (cubic), clinopyroxene (monoclinic), and an amorphous glassy phase wetting some of the grain boundaries [13].

4. Example applications on metallic and geological materials

4.1. Grain boundaries in shot-peened aluminium

Figure 4 shows a [001] inverse pole figure map for Hough and dictionary indexing (HI and DI) of a portion of a 10 kV shot-peened Al 7075 data set (courtesy T. Burnett), about 30 μm away from the sample surface; the acquisition step size was 23 nm. Red pixels in (a) indicate un-indexed points (using the AZTEC software); they belong to two groups: points near the grain boundaries and points clustered at second phase precipitates. In the dictionary indexing result, all points have been indexed (except for points inside the precipitates) and the grain boundaries are more clearly delineated, as is shown in the magnified views of the region around a triple junction. Occasional points near the boundaries are mis-indexed as belonging to a neighbouring grain; such points can easily be corrected by standard clean-up procedures. This result indicates that the DI approach can handle overlap EBSD patterns without significant issues. The sharpness of the boundaries in the DI result can be improved by performing a statistical analysis on the top N matches, and selecting the orientation that is present the most in this set.

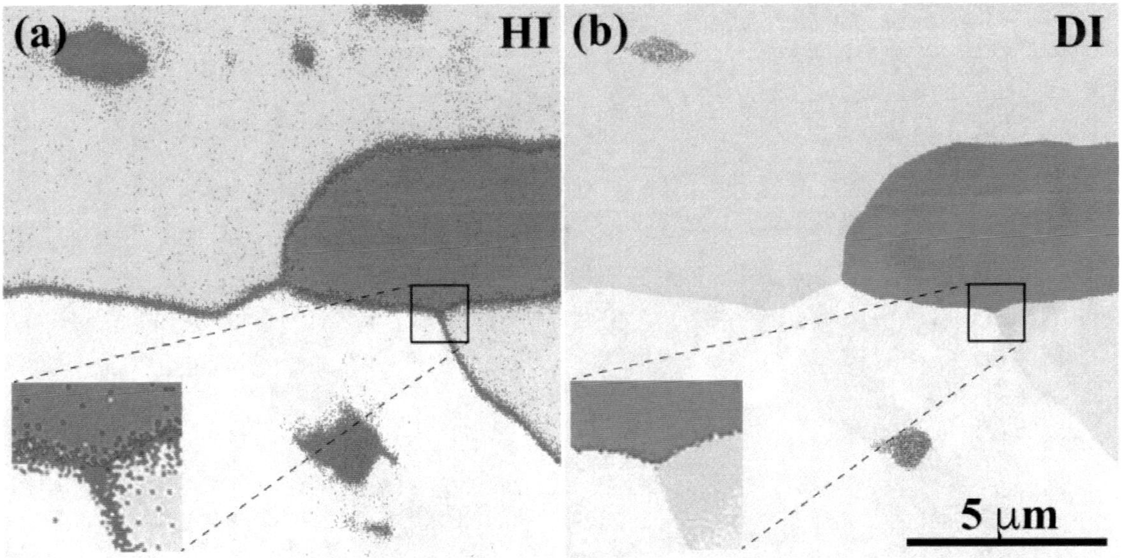

Figure 4. Grain boundaries in a [001] inverse pole figure for a shot-peened Al 7075, indexed using the Hough-based technique (HI) and the dictionary approach (DI). The insets show results near a triple junction.

IOP Conf. Series: Materials Science and Engineering **891** (2020) 012009 doi:10.1088/1757-899X/891/1/012009

4.2. Distinguishing between face-centred cubic phases

Face-centred cubic (fcc) materials produce nearly identical EBSD patterns for a given crystal orientation. A comparison of master patterns for Ag and Ni reveals that there are sufficient differences in intensity along the Kikuchi bands for the DI approach to be able to differentiate between the phases. Figure 5 shows the results of DI analysis of a 7 kV data set for de-wetted Ag on a polycrystalline Ni substrate (sample courtesy D. Chatain; unpublished research); (a) is the ADP map, (b) and (d) shows the Ni and Ag OSM maps that are combined into the phase map in (c) (Ni is red, Ag green). An example experimental 7 kV EBSD pattern is shown in (f) along with the simulated pattern in (e). An analysis of the confidence index values (top dot products) reveals that the differences between the two phases are well above the variation of the top dot product within each individual phase.

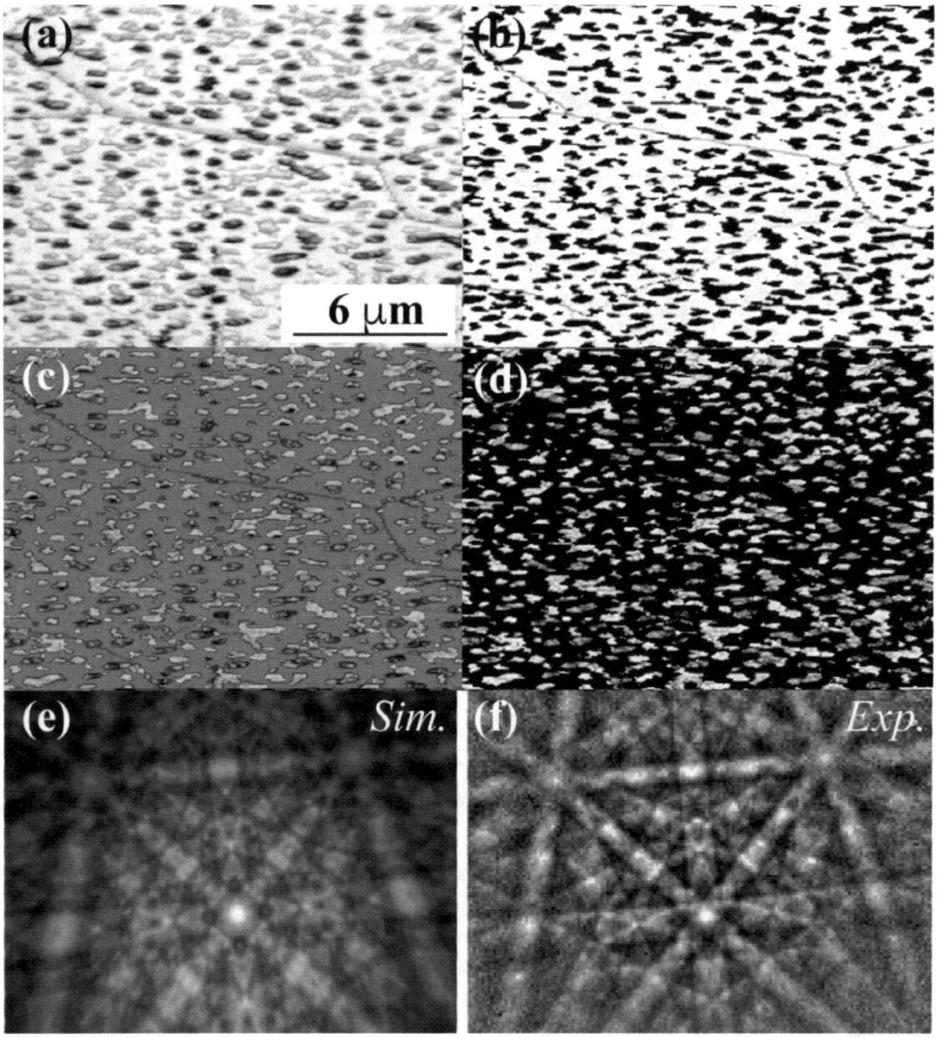

Figure 5. DI results for de-wetted Ag on polycrystalline Ni, 7 kV, 40 nm step size; a) ADP map; b and d) Ni and Ag OSM maps; c) phase map (Ni red, Ag, green); e and f) are simulated and experimental EBSD patters for Ag.

IOP Conf. Series: Materials Science and Engineering **891** (2020) 012009 doi:10.1088/1757-899X/891/1/012009

4.3. A multi-phase geological sample

The final example concerns a multi-phase geological sample consisting of garnet (grt, cubic), clinopyroxene (cpx, monoclinic) and an amorphous melt phase [13]. Since the melt phase does not produce patterns with Kikuchi bands, it is relatively easy to distinguish it from the other two phases simply from the ADP map shown in Fig. 6a.

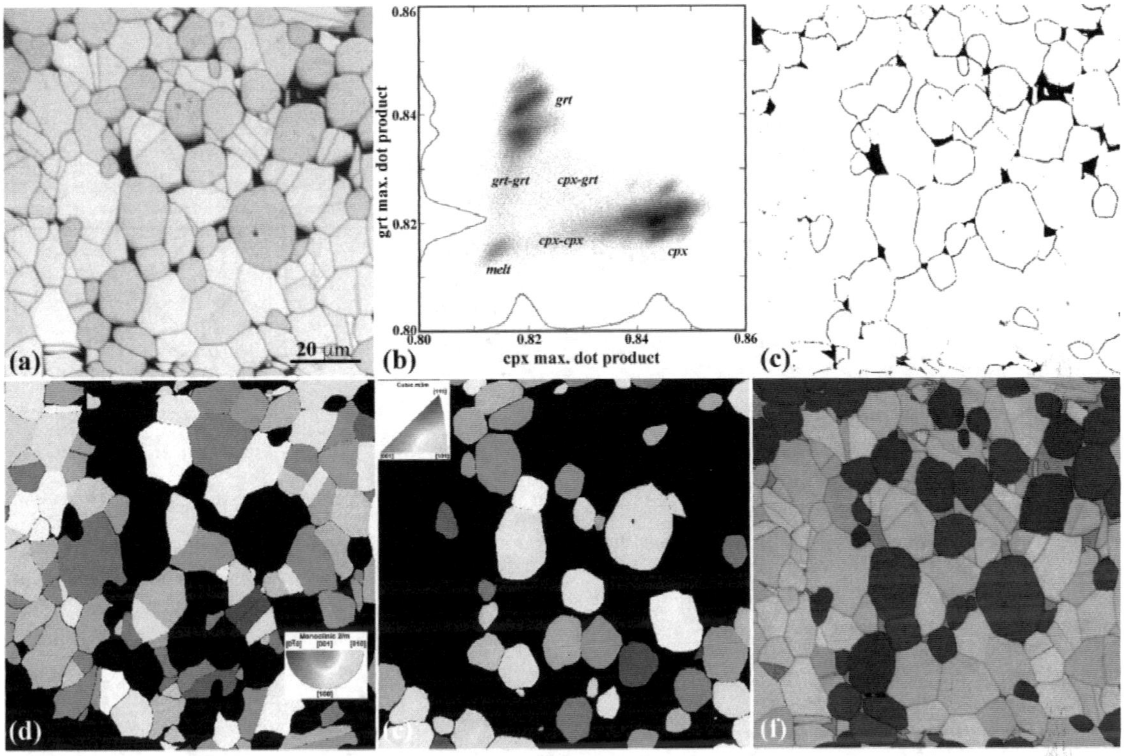

Figure 6. a) ADP map; b) maximum dot product histogram; c) melt phase; d and e) [001] IPF maps for cpx and grt; f) phase map (cpx = green, grt = blue, melt = red).

Application of the DI algorithm using master patterns for each of the phases (cpx and grt) results in two separate best-fit dot product values for each phase; Fig. 6b shows a 2D histogram of all dot product pairs. Several point clusters are clearly identified: patterns that match neither cpx nor grt likely correspond to the melt phase, whereas large clusters are present for each of the two phases. Each large cluster consists of smaller lumps of points, and detailed analysis shows that each lump corresponds to an individual grain. The diffuse area between the large clusters corresponds to grain boundary pixels, and there are three different types: cpx-cpx, grt-grt, and cpx-grt boundaries. Figure 6c shows the locations of the melt, derived from the histogram in (b). The bottom row of Fig. 6 shows the [001] IPF maps for cpx (d) and grt (e) (with colour legends as insets), as well as the combined phase map (f); note that the cpx-grt grain boundaries are wetted by the melt, whereas cpx-cpx or grt-grt boundaries are not wetted.

Figure 7a shows a portion of the OSM map for the cpx-grt-melt sample; the line between A and B crosses a cpx-cpx grain boundary. In Fig. 7b, a total of 40 points is sampled between A and B, and the number of near matches (out of a maximum of 20) that each indexing result has in common with each of the end members A and B is displayed; obviously, the end points have all 20 nearest matches in

IOP Conf. Series: Materials Science and Engineering **891** (2020) 012009 doi:10.1088/1757-899X/891/1/012009

common with themselves. Moving closer to the grain boundary, the number of matches in common with the end point drops slightly until the grain boundary is reached, at which point the number drops to zero. The intersection of the two profiles can be used to determine the location of the grain boundary with an accuracy of better than a single sampling pixel.

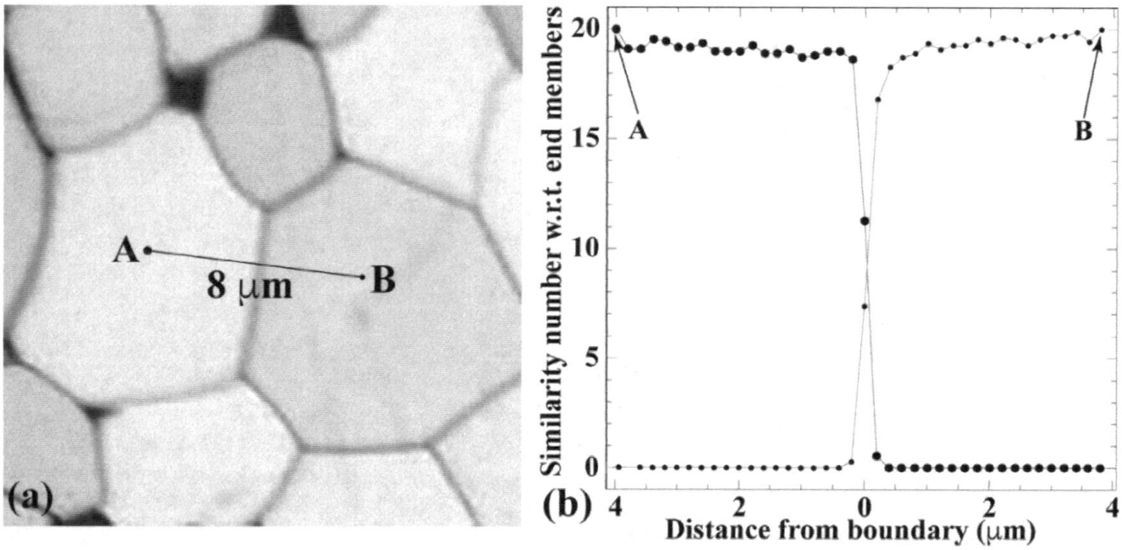

Figure 7. a) Portion of the cpx-grt ADP map with a trace across a cpx-cpx grain boundary between the points A and B; b) number of nearest matches that each sampling point has in common with each of the end members.

5. A brief look at the future

As the dictionary indexing approach becomes more widely used, we anticipate that the algorithm will be optimised for speed by using multiple GPUs combined with smarter pre-processing of the patterns; in the current scheme, each experimental pattern is considered to be fully independent of all other patterns, so that the incorporation of simple clustering schemes could already have a significant impact on the indexing speed. Since much of EBSD research deals with deformed materials, the next step in the development of the DI approach is the incorporation of deformation into the pattern interpolation scheme, so that a refinement of the pattern match with respect to the full deformation tensor becomes a possibility; this work is currently ongoing.

As noted in section 2, the EBSD forward model needs to be further developed to make the result more realistic, in particular with regards to the spectroscopic properties of the patterns. Ideally, this would involve solving the Yoshioka equations for elastic/inelastic electron scattering in combination with appropriate depth and energy integration schemes. This is a challenging problem, but progress in this area would also have a significant impact outside the EBSD field.

We have recently developed an alternative for the DI approach based on the fact that the master pattern resides on the Kikuchi sphere. Simply back projecting each experimental pattern onto the sphere and then carrying out a spherical correlation analysis using a spherical harmonic decomposition of the master pattern results in an efficient indexing algorithm that no longer requires the computation of any dictionary patterns and is thus nearly independent of the symmetry of the crystal structure. We anticipate that this spherical indexing approach, which we call "EMSPHINX", may be capable of near real-time indexing at speeds comparable to the currently used Hough transform algorithm, but with the same increased robustness against noise displayed by the dictionary indexing approach. We are currently exploring the strengths and weaknesses of this new algorithm.

6. Concluding remarks

The dictionary indexing and related algorithms are available in open source form as part of the author's EMSOFT project; all source code (a combination of FORTRAN-95, OPENMP, and OPENCL, with C/C++ wrapper routines) can be downloaded from a GitHub repository (https://github.com/EMsoft-org), and an extensive collection of EMSOFT-related links, including example applications, is maintained at http://vbff.materials.cmu.edu/EMsoft. Test executable packages ("betas") are made available each night for Mac OS X and Windows 10 platforms from the following URL: http://www.bluequartz.net/binaries/EMsoft/experimental. The DI approach is currently being integrated into one of the EBSD vendor packages and will be available to the community for extensive testing and applications. Much progress is being made in quantitative EBSD research as well as in related SEM-based diffraction techniques; it is truly an exciting time to be working in the EBSD field!

Acknowledgements

The author would like to acknowledge numerous colleagues who have over the years contributed ideas or data sets to the EBSD dictionary indexing project, among others S. Singh, F. Ram, W. Lenthe, E. Pascal, P. Callahan, M. Echlin, T. Pollock, M. Jackson, A. Hero, J. Simmons, S. Wright, K. Marquardt, M. Taheri, D. Chatain, P. Wynblatt, and T. Burnett. The author would like to acknowledge financial support from a DoD Vannevar-Bush Faculty Fellowship (#N00014-16-1-2821) as well as the computational facilities of the Materials Characterization Facility at CMU under grant #MCF-677785.

References

[1] Chen Y H, *et al.* 2015 A dictionary approach to EBSD indexing. *Microsc. Microanal.* **21** 739-752

[2] Winkelmann A 2009 Principles of depth-resolved Kikuchi pattern simulation for electron backscatter diffraction. *J. Microsc.* 239 32-45

[3] Maurice C, *et al.* 2011 A method for accurate localisation of EBSD pattern centres. *Ultramicroscopy* **111** 140-148

[4] Callahan P G and De Graef M 2013 Dynamical EBSD patterns. Part I: Pattern simulations. *Microsc. Microanal.* **19** 1255-1265

[5] Roşca D 2010 New uniform grids on the sphere. *Astronomy Astrophys.* **520** A63-A63

[6] Singh S, *et al.* 2017 EMsoft: Open source software for electron diffraction/image simulations. *Microsc. Microanal.* **23** (Suppl. 1) 212-213

[7] Singh S and De Graef M 2016 Orientation sampling for dictionary-based diffraction pattern indexing methods. *Modeling Simulations Mater. Sci. Engng.* **24** 085013

[8] Pizer S M, *et al.* 1987 Adaptive histogram equalization and its variation. *Computer Vision, Graph. Image Process.* **39** 355-368

[9] Gulsoy E B, *et al.* 2009 Application of joint histogram and mutual information to registration and data fusion problems in serial sectioning microstructure studies. *Scripta Materialia* **60** 381-384

[10] Brewick P T W, Wright S I and Rowenhorst D J 2019 NLPAR: Non-local smoothing for enhanced EBSD pattern indexing. *Ultramicroscopy* **200** 50-61

[11] Ram F, *et al.* 2017 Error analysis of crystal orientations obtained by the dictionary approach to EBSD indexing. *Ultramicroscopy* **181** 17-26

[12] Jha D, *et al.* 2018 Extracting grain orientations from EBSD patterns of polycrystalline materials using convolutional neural networks. *Microsc. Microanal.* **24** 497-502

[13] Marquardt K, *et al.* 2017 Quantitative electron backscatter diffraction (EBSD) data analyses using the dictionary indexing (DI) approach: overcoming indexing difficulties in geological materials. *Amer. Mineralogist* **102** 1843-1855

[14] Singh S, *et al.* 2017 Application of forward models to crystal orientation refinement. *J. Appl. Crystall.* **50** 1664-1676

[15] Singh S and De Graef M 2017 Dictionary indexing of electron channeling patterns. *Microsc. Microanal.* **23** 1-12
[16] Ram F and De Graef M 2018 Phase differentiation by electron backscatter diffraction using the dictionary indexing approach. *Acta Materialia* **144** 352-364
[17] Pascal E, *et al.* 2018 Energy-weighted dynamical scattering simulations of back-scattered electron diffraction modalities. *Ultramicroscopy* **187** 98-106
[18] Nicolaÿ A, *et al.* 2019 Discrimination of dynamically and post-dynamically recrystallized grains based on EBSD data. Application to Inconel 718. *J. Microsc.* **273** 135-147

Classification and characterisation of magmatic-hydrothermal tourmaline by combining field observations and microanalytical techniques

K Drivenes[1], W Brownscombe[2], R B Larsen[1], R Seltmann[2], J Spratt[2] and B E Sørensen[1]

[1] Norwegian University of Science and Technology, Department of Geoscience and Petroleum, 7491 Trondheim, Norway
[2] Natural History Museum, Cromwell Road, London SW7 5BD, Great Britain

E-mail: kristian.drivenes@ntnu.no

Abstract. Tourmaline from the St. Byron lobe of the Land's End granite, SW England, was assessed by macroscopic, optical and quantitative microanalytical methods. In total, seven types of tourmaline were distinguished. The seven types reflect different crystallisation environments and stages in the magmatic-hydrothermal transition. Types 1-3 are interpreted to represent a gradual transition from tourmaline crystallising from a silicate melt to precipitation from magmatic aqueous fluids. Types 5-7 crystallised at subsolidus conditions from a different fluid generation than types 1-3. These fluids may be magmatic or mixed with other fluids (e.g., meteoric or formation waters). The Sn-mineralisation in the area is mostly related to the latter fluid generation, and the mineralising potential is reflected by the tourmaline composition.

1. Introduction

Tourmaline is the most common borosilicate mineral in the crust . It is found in magmatic, metamorphic, and sedimentary rocks, but is perhaps best known for the striking, museum-grade crystals found in South America and Central Africa. It has the general formula $XY_3Z_6T_6O_{18}BO_3V_3W$, where the X-site is occupied by Na, K, Ca, Pb, and vacancies; the Y-site by Fe^{2+}, Mg, Mn, Li, Ti, Al, Fe^{3+}, and Zn; the Z-site by Al, Cr, Fe^{3+}, Mg, and Fe^{2+}; the T-site by Si, B, and Al; the B-site by B; the V-site by OH and O; and the W-site by OH, F, and O. Due to a crystal structure that can accommodate a large range of major, minor and trace elements, it has gotten a reputation for being a trash can mineral alongside other complex silicate minerals, such as amphiboles. In addition, tourmaline is strongly resistant to chemical and physical weathering, and is stable over an extended range of P-T-X conditions, hence, it is a good petrogenetic indicator [1] and has potential as a provenance mineral [2-4].

In granitic systems, tourmaline is predominantly found in S-type granites. Boron is generally incompatible in the early-forming granitic mineral assemblage and tourmaline forms late during magmatic differentiation, where it commonly succeeds biotite as the main ferromagnesian mineral. During the magmatic-hydrothermal transition, boron is preferentially partitioned into the aqueous phase ($D^{fluid/melt}$ = 1-3 [5-7]). Within the Cornubian Batholith of SW England, which has served as a field laboratory in this study, evidence of boron-rich fluids migrating through fracture systems in the crystallised granite and country rock can easily be observed as black, tourmaline-rich veins. There is a well-established link between tourmaline veins and tin mineralisation [8]. However, the relationship is not exclusive, and boron is not recognised as a complexing element in the transportation of Sn in ore fluids [9]. A thorough classification and characterisation of tourmaline is important in order to track the magmatic-hydrothermal transition, indicate fluid sources, and distinguish potentially ore-bearing from barren systems. This can only be accomplished by including the full spectrum of analytical techniques; from field observations, through optical microscopy, to quantitative microanalytical studies. This study aims to classify the diversity of tourmaline found in the St Byron lobe of the Land's End granite in Cornwall, SW England, based on field relations, optical properties and mineral chemistry. It will serve as a foundation for future studies trying to interpret the chemical and physical fluctuations within the late stages of granite crystallization and the onset of a potentially ore-forming hydrothermal system; that is, the elusive magmatic-hydrothermal transition.

2. Analytical methods

Quantitative spot analyses of tourmaline were obtained by electron microprobe analysis (EPMA) and LA-ICP-MS at the Natural History Museum, London. Major and minor elements were analysed using a CAMECA SX100 microprobe at 20 kV and 20 nA. Na, Si, Al, and Mg were analysed on a TAP crystal; Mn and Fe on a LLIF; Ca, Ti, and Cl on a PET; K, Sr, and Sn on a LPET; and F on a LPC0 crystal.

Peak counting times varied between 10 (Na) and 30 (e.g., Mn and Sn) seconds, and detections limits typically varied between 170 (Mg) and 800 (F) μg/g. Trace elements, in addition to some major elements (^{7}Li, ^{23}Na, ^{24}Mg, ^{27}Al, ^{29}Si, ^{31}P, ^{43}Ca, ^{45}Sc, ^{49}Ti, ^{51}V, ^{52}Cr, ^{55}Mn, ^{59}Co, ^{60}Ni, ^{63}Cu, ^{66}Zn, ^{69}Ga, ^{72}Ge, ^{85}Rb, ^{88}Sr, ^{89}Y, ^{90}Zr, ^{111}Cd, ^{113}In, ^{118}Sn, ^{121}Sb, ^{133}Cs, ^{137}Ba, ^{139}La, ^{140}Ce, ^{141}Pr, ^{145}Nd, ^{147}Sm, ^{151}Eu, ^{157}Gd, ^{175}Lu, ^{181}Ta, ^{182}W, ^{208}Pb, ^{232}Th, and ^{238}U), were analysed by laser ablation inductively coupled plasma mass spectrometry (LA-ICP-MS) using a 193 nm ESI excimer laser coupled to a 7500cs Agilent ICP-MS. Spots of 45 μm were analysed for 60 seconds using a fluence of 3 Jcm^{-2} and a frequency of 10 Hz, each following 30 seconds of blank measurement. NIST 612 was used as a primary standard with BCR 2g and GSD as secondary standards. Tourmaline results were normalised to known Si values measured by EPMA in the same location prior to LA-ICP-MS. Structural formulae were calculated using WINCLASTOUR [10] normalising for T+Y+Z=15.

3. Classification of tourmaline

3.1. Macroscopic and field observations

In the Land's End granite, tourmaline occurs in a variety of structures and textures: As a trace or minor mineral disseminated in the granite matrix, in pegmatitic pockets, miarolitic cavities, quartz-tourmaline aggregations in the form of orbicules, sheets, and larger bodies +/- K-feldspar phenocrysts (massive quartz-tourmaline (-K-feldspar) rocks – MQT/MQTK), veins, and replacement of former granite. In the field, tourmaline is black, and genetic interpretation must be based on the textural and structural relationship with the granite. The obvious magmatic or hydrothermal textures can easily be distinguished (i.e., disseminations, pegmatites and veins), whereas the field relationship of other textures are ambiguous and their genetic models are controversial [11-14]. An overview of the samples, localities and analyses is listed in Table 1 and Fig. 1.

3.2. Optical microscopy

Optical microscopy reveals a multitude of generations and growth stages of tourmaline. Classification is conveniently based on colour in plane polarised light due to the commonly drastic changes and strong zoning patterns. Seven types are distinguished based on colour and the microtextures within the samples, where the first and final three may be grouped together as two main groups. Examples of the different tourmaline types are shown in Fig. 2.

Figure 1. Samples investigated in this study. See Table 1 for a description and locality references. A: CAR-2. B: KDC-14-18. C: KDC-16-06. D: ES-102. E: KDC-14-14. F: KDC-16-18. G: KDC-16-08. H: KDC-16-12. I: KDC-17-02. J: KDC-16-05. K: KDC-16-19. L: KDC-17-03. M: KDC-17-08: N: KDC-16-13. Letters at the end of sample names in Table 1 refer to specific sections of the sample.

Table 1. Overview samples. See the text for a detailed description of the tourmaline types.

Sample	Lithology/texture	Locality	Tourmaline types	Short description
CAR-2	MQTK	Trevalgan	1,2,3,5	Fine-grained matrix of quartz and tourmaline (type 1,2,3) with K-feldspar phenocrysts. Euhedrally terminated type 3 tourmaline grow into small (few mm) miarolitic cavities. Minor alteration of K-feldspar phenocrysts by type 5 tourmaline.
ES-102	MQT	Nanjizal	1,2,3,5,6,7	Fine-medium grained matrix of quartz and clusters of type 5 tourmaline. Some large grains with multiple tourmaline generations were observed.
KDC-14-14-a	Replacement	Carn Barra	1,5	Fine-medium grained matrix of quartz and clusters of type 5 tourmaline. Minor type 1 tourmaline. Pink K-feldspar phenocrysts are cut by q-tur veins
KDC-14-18-c	Orbicule	Pellitras Point	1,6	Medium-coarse grained type 1 and 2 tourmaline and quartz. Minor type 6.
KDC-14-20-c	Replacement	Pellitras Point	1,2,3,4,5,6,7	Hydrothermal replacement of megacrystic coarse-grained granite. Local variations of tourmaline content. Mostly type 1 and 5
KDC-16-05-c	Q-tur sheet/vein	Carn Barges	1,4	Coarse-grained q-tur sheet. Mostly type 4 with minor type 1
KDC-16-06-aa	Q-tur sheet	Carn Barges	1,2,3,4	Fine-grained q-tur sheet cutting the coarse-grained megacrystic granite. Type 1, 2 and 3 in the sheet and minor type 4 in the contact to the granite.
KDC-16-08-b	Q-tur sheet	Carn Barges	1,2	Quartz-tourmaline sheet with equigranular matrix of type 1 and 2 tourmaline and quartz
KDC-16-12-c	Tur zone in tourmaline granite	Carn Barges	1,2,3,6	Quartz-tourmaline zone in a tourmaline granite. Mainly type 1 and 2, with minor type 3 and 6 tourmaline
KDC-16-13	Tourmline granite	Carn Barges	1	Interstitial type 1 tourmaline in the tourmaline granite matrix
KDC-16-18	Replacement	Pellitras Point	1,2,3,5,6,7	Hydrothermal replacement of megacrystic coarse-grained granite. Local variations of tourmaline content. Mostly type 1, 2 and 5
KDC-16-19-a	Q-tur zone in granite	Carn Barges	1,2,3	Quartz-tourmaline zone in a tourmaline granite with a central tourmaline zone. In the quartz-tourmaline zone, type 1 and 2 tourmaline is commonly overgrown by type 3. Type 3 becomes the dominant tourmaline type towards and in the tourmaline zone.
KDC-17-02-b	Altered FGG/tur zone	Pedn Vounder	1,2,3,5,6,7	Altered (K-metasomatism) fine-grained tourmaline granite with q-tur and tur zones. Great variation of tourmaline types
KDC-17-03	Miarole	Pedn Vounder	1,2,3,5,6	Tourmaline zone in the same rock as KDC-17-02. Highly variable tourmaline, but commonly type 1, 2 and 3, with overgrowths of type 6
KDC-17-08	Miarole	Carn Barges	1,2,3,4	Tourmaline growing in a miarolitic cavity in a fine-grained tourmaline granite. Mainly type 1,2 and 3, with minor type 4
KDC-17-12	MQT	Pedn Vounder	1,2,3	Fine-grained quartz and tourmaline (type 1,2,3) matrix with quartz phenocrysts.

Type 1 tourmaline is brown to orange, often weakly zoned from light to slightly darker brown. In sections normal to the c-axis, the core is commonly lighter in colour than the outer zones. This generation occasionally shows sector zoning. In disseminated tourmaline and orbicules, type 1 is the main, commonly the only, generation, with only minor occurrence of type 6. It is also represented as cores in tourmaline in quartz-tourmaline sheets and bodies (Figs. 2 k and 2 l), and less so in some replacement textures (Fig. 2h)

Type 2 tourmaline is weakly oscillatory zoned. This is observed as concentric, 10 - 200 μm, light brown and light blue bands around the c-axis. The thicker bands may be brown or blue depending on the sample. Type 2 typically occurs as the second growth stage following type 1, with or without resorption of type 1, and is commonly followed by type 3 tourmaline. This generation is common in quartz-tourmaline sheets and bodies (Figs. 2a, 2d, 2e, 2f, 2k and 2l).

Type 3 tourmaline is blue, and typically overgrows type 1 and 2 in tourmaline sheets and bodies. The contact between type 1 and 3 is commonly abrupt, whereas the contact between type 2 and 3 is more gradual. In some samples, type 3 is strongly oscillatory zoned. We found type 3 tourmaline in the same samples as type 2 (Figs. 2a and 2i).

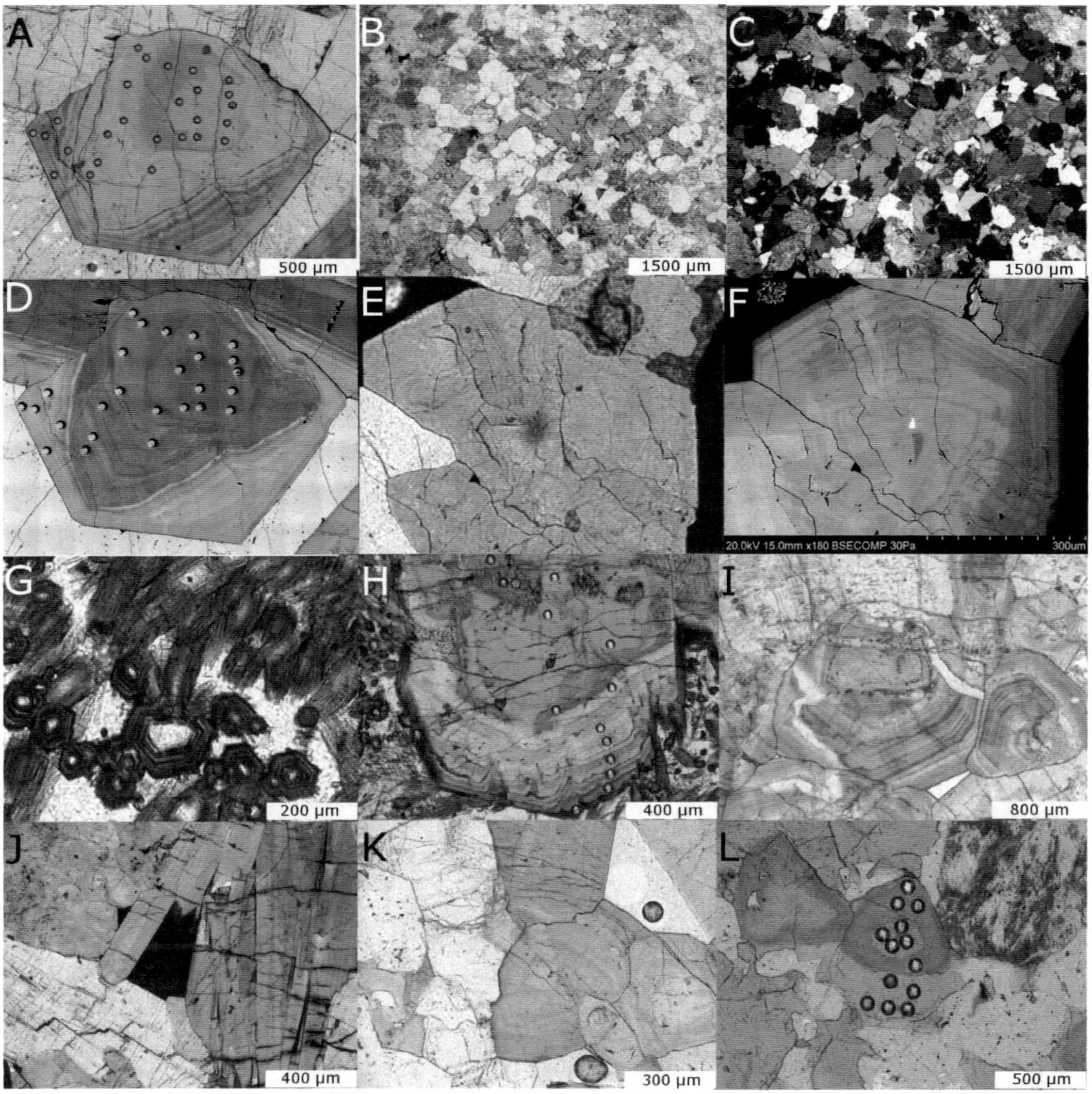

Figure 2. Overview of tourmaline types. All pictures are acquired under plane polarised light except C (cross-polarised), and D and F (backscatter electron). Circular pits are holes from LA-ICP-MS analyses. A: Type 1 (brown), type 2 (weakly oscillatory zoned light blue to light brown, and type 3 (oscillatory zoned dark to light blue) tourmaline from sample KDC-17-03. B: Interstitial type 1 tourmaline from the matrix of a tourmaline granite (sample KDC-16-13). C: Same as B in cross-polarised light showing optical continuity. D: BSE image of A showing minor resorption of type 1 by type 2 and variations in mean atomic number between type 2 and 3. The bright phase surrounding tourmaline is apatite. E: Type 2 tourmaline from the tourmaline zone of sample KDC-16-19-a. F: BSE image of E showing the fine oscillatory zoning pattern not easily observed in the optical microscope. G: Type 5 tourmaline from sample KDC-16-18. Strongly zoned with a colourless core and green and blue rims. This type is typically greenish in sections not normal to the c-axis. H: Type 1 (brown), type 5 (green), type 6 (pale blue-green overgrowth in the upper and right part of the picture), and type 7 (oscillatory zoned colourless-brown rim of the large grain) from sample KDC-16-18. I: Type 1, 2, 3, and 6 from sample KDC-17-02. J: Type 4 tourmaline from sample KDC-16-05. K: Type 2 tourmaline from sample KDC-16-08. L: Type 1 and 2 from sample KDC-17-12.

Type 4 tourmaline is strongly zoned from greenish blue to brown. The zoning is either oscillatory or patchy. Sections normal to the c-axis have a green-blue core and a brown rim. Type 4 tourmaline replace K-feldspar in the granite bordering various quartz-tourmaline structures, and is the main constituent in addition to quartz in a coarse grained quartz-tourmaline sheet (Fig. 2j).
Type 5 tourmaline is strongly oscillatory zoned with a colourless core and a dark blue rim. However, the major growth zones are green, and sections parallel to the c-axis are characteristically pleochroic from colourless to green with an acicular crystal shape. Type 5 tourmaline is ubiquitous in veins and hydrothermal replacement textures (Figs. 2 g and 2h). It is also common in some larger quartz-tourmaline bodies.
Type 6 tourmaline typically occurs as pleochroic colourless to blue overgrowth of other generations of tourmaline. In granite and all quartz-tourmaline aggregations, type 6 tourmaline also occurs as inclusions, likely replacing former feldspar inclusions (Fig. 2h). A stage of brown tourmaline overgrowing all other generations is also included in this type. This is only observed in one sample of miarolitic tourmaline (Fig. 2i).
Type 7 tourmaline covers strongly oscillatory zoned, pleochroic, colourless to brown tourmaline overgrowing type 1. This type was found in replacement textures and a large quartz-tourmaline body (Fig. 2h).

3.3. Mineral chemistry

Tourmaline is divided into primary groups and general series based on their X-site and W-site occupancy, respectively [15]. Most analyses plot (Fig. 3) in the alkali group, with a few type 3 points in the vacancy group, and a few type 5 and 6 in the calcic group. There is a larger spread between the types in the W-site diagram, where types 1, 2 and most of type 3 tourmaline are fluorine-species, and types 5 and 7 are hydroxyl-species. Types 4 and 6 are divided between the fluorine- and hydroxyl-groups. Two smaller clusters of type 3 tourmaline plot in the oxy-field and on the border between oxy- and hydroxyl-groups. A similar grouping is observed in the Fe/Fe+Mg versus X-vacancy plot. Types 1 - 4 are generally schorl, a few points plotting in the foitite field, and types 5 - 7 are generally dravite. From this we can distinguish two main groups of tourmaline in the Land's End granite: F-schorl and dravite. Tourmaline classifying as F-dravite is late brown overgrowths (included in type 6) in sample KDC-17-03 and strongly oscillatory zoned tourmaline (type 7) in sample KDC-16-18.

Substitution of Fe for Mg is the main controlling substitution in tourmaline in granitic systems. The strong chemical zoning in type 5 and 7 tourmaline is largely due to variation in the Fe and Mg content, hence the large spread of these types in the Fe/(Fe+Mg) versus X-vacancy plot. The gradual or stepwise growth zoning shown in tourmaline of type 1 - 3 combined with the gradual increase in Fe may indicate differentiation in a closed system, from a dominantly magmatic (type 1) to a more aqueous fluid-dominated system (type 3). The dravitic types 5 - 7 tourmalines represent a different crystallisation mode. Texturally, these tourmaline types are clearly hydrothermal in origin, and the fluid was more evolved compared to type 3 tourmaline, or possibly mixed with other fluid sources. The latter hypothesis is also evident in the Sn versus Sr plot, where type 1 - 3 show increasing levels of both Sn and Sr, two elements likely to partition into saline aqueous fluids. Types 5 - 7 are the richest in Sr and Sn, giving them a strong hydrothermal signature. Particularly, type 5 is known to be associated with cassiterite mineralisation [11]. The different types showed great separation in the Ni-Co-Zn diagram. Fe and Ti have been proposed to be the main chromophores in tourmaline [16]. However, the clear separation in Fig. 3 may indicate that these elements play also a role.

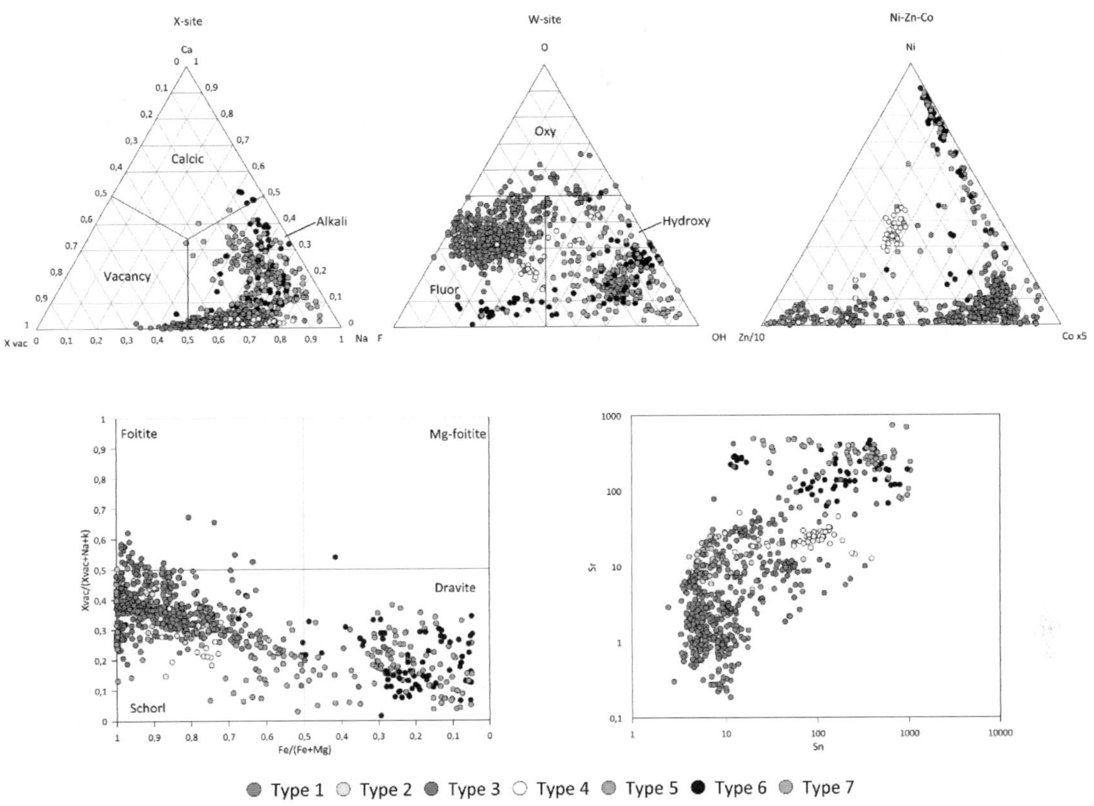

Figure 3. Chemical plots of tourmaline from this study. The colours largely reflect the colour observed in plane polarised light. Most type 2 points are covered by type 1 and 3 points.

4. Conclusions

Seven types of tourmaline were distinguished in samples from the southern part of the Land's End granite. These were primarily discriminated based on colour and growth textures. The different types can also be identified in major and trace element plots. Two main groups, F-schorl and dravite, represent closed magmatic-hydrothermal and open hydrothermal system crystallisation, respectively. A differentiation path can be distinguished within the magmatic-hydrothermal tourmaline from the center of the schorl field towards more Fe-rich composition.

Acknowledgements
We thank Darrell Henry for a constructive review that helped improve the paper.

References

[1] van Hinsberg V J, Henry D J and Marschall H R 2011 Tourmaline: an ideal indicator of its host environment. *Can. Mineralogist* **49** 1-16
[2] Farnsworth-Pinkerton S, McMillan N J, Dutrow B L and Henry D J 2018 Provenance of detrital tourmalines from Proterozoic metasedimentary rocks in the Picuris Mountains, New Mexico, using laser-induced breakdown spectroscopy. *J. Geosci.* **63** 193-198
[3] Henry D J and Guidotti C V 1985 Tourmaline as a petrogenetic indicator mineral- An example from the staurolite-grade metapelites of NW Maine. *Amer. Mineralogist* **70** 1-15

[4] Henry D J and Dutrow B L 1992 Tourmaline in a low grade clastic metasedimentary rock: an example of the petrogenetic potential of tourmaline. *Contr. Mineral. Petrol.* **112** 203-218
[5] Hervig R L, Moore G M, Williams L B, Peacock S M, Holloway J R and Roggensack K 2002 Isotopic and elemental partitioning of boron between hydrous fluid and silicate melt. *Amer. Mineralogist* **87** 769-774
[6] London D, Hervig R and Morgan G V I 1988 Melt-vapor solubilities and elemental partitioning in peraluminous granite-pegmatite systems: experimental results with Macusani glass at 200 MPa. *Contr. Mineral. Petrol.* **99** 360-373
[7] Pichavant M 1981 An experimental study of the effect of boron on a water saturated haplogranite at 1 Kbar vapour pressure. *Contr. Mineral. Petrol.* **76** 430-439
[8] Slack J F 1996 Tourmaline associations with hydrothermal ore deposits. *Rev. Mineral. Geochem.* **33** 559-643
[9] Taylor J R and Wall V J 1993 Cassiterite solubility, tin speciation, and transport in a magmatic aqueous phase. *Econ. Geol.* **88** 437-460
[10] Yavuz F, Yavuz V and Sasmaz A 2006 WinClastour—a Visual Basic program for tourmaline formula calculation and classification. *Comp. Geosci.* **32** 1156-1168
[11] Drivenes K, Larsen R, Müller A, Sørensen B, Wiedenbeck M and Raanes M 2015 Late-magmatic immiscibility during batholith formation: assessment of B isotopes and trace elements in tourmaline from the Land's End granite, SW England. *Contr. Mineral. Petrol.* **169** 1-27
[12] Rozendaal A and Bruwer L 1995 Tourmaline nodules: indicators of hydrothermal alteration and Sn-Zn-(W) mineralization in the Cape Granite Suite, South Africa. *J. African Earth Sci.* **21** 141-155
[13] Perugini D and Poli G 2007 Tourmaline nodules from Capo Bianco aplite (Elba Island, Italy): an example of diffusion limited aggregation growth in a magmatic system. *Contr. Mineral. Petrol.* **153** 493-508
[14] Drivenes K, Larsen R B, Müller A and Sørensen B E 2016 Crystallization and uplift path of late Variscan granites evidenced by quartz chemistry and fluid inclusions: Example from the Land's End granite, SW England. *Lithos* **252–253** 57-75
[15] Henry D J, Novák M, Hawthorne F C, Ertl A, Dutrow B L, Uher P and Pezzotta F 2011 Nomenclature of the tourmaline-supergroup minerals. *Amer. Mineralogist* **96** 895-913
[16] London D, Morgan G B and Wolf M 1996 Boron in granitic rocks and their contact aureoles. in: *Boron : mineralogy, petrology and geochemistry. Rev.Mineral.* **33** 299-330

IOP Conf. Series: Materials Science and Engineering **891** (2020) 012011 doi:10.1088/1757-899X/891/1/012011

Very low energy peak shifts in EDS spectra

F Eggert, P P Camus and F Reinauer

EDAX Inc., AMETEK Materials Analysis Division, 91 McKee Drive, 07430 Mahwah, NJ, U.S.A.

E-mail: frank.eggert@ametek.com

Abstract. The response function of energy-dispersive X-ray spectrometers (EDS) has been a topic for investigation as long as EDS has been used. This work systematically investigates soft X-ray peaks with energies below 150 eV in regard to the detector energy calibration linearity. It was found that Si-L (92 eV) and Al-L (73 eV) lines are not shifted in energy position in comparison to higher energy K-lines of boron, carbon and nitrogen. The reason is simply the silicon detector material has a large absorption jump at 99 eV (Si-L_2/Si-L_3), which causes much higher detection depths of photons with energies below 99 eV. There is also less shift than assumed with older approaches in the energy region 100 to 149 eV (Si-L_1). It is also expected that Li-K radiation will not be affected by an energy shift due to detector artefacts. The authors propose a modified shift-correction which corrects this detector artefact and linearizes the energy calibration of measured EDS spectrum below 600 eV. This energy shift effect of the measured X-ray lines is in principle with any silicon-based X-ray detector. But the shift value depends also on the front contact layer design and quality. The novelty with changed correction is that no shift is applied for X-ray peaks below 99 eV. Older applied correction procedures were found to produce a line energy misplacement in spectrum for this energy region.

1. Introduction

The response function of energy-dispersive X-ray spectrometers (EDS) was a topic of investigation for as long as EDS has been used. Heckel and Scholz [1] and Scholze and Procop [2] developed a full physics-based models, which allows the calculation of the detector response function dependent on detector construction parameters and on the measured X-ray energies. It is known [2] that the low-energy side of the detector response function (peak shapes) depends much on the detected photon energy. It is affected by how deep the X-rays interact with the detector material to create an electron-hole pairs cloud. If the interaction is near the detector surface, electrons can likely escape and then the total measured energy of the photon is with a loss due to different effects of incomplete charge collection (ICC) [2]. With photon energies below 500 eV, the typical low-energy tailing in peak shapes morphs into 'all photons ICC', resulting in a complete peak shift to lower measured energies than the predicted X-ray energy. Usually the photon interaction depth is reduced at lower energies, so the shift is increasing. In [2], the X-rays in the energy range from 100 eV to 10 keV were investigated for three spectrometer systems different in construction, including one being an SDD.

2. Method

An empirically measured numerical shift-correction algorithm is used with EDAX software as a fast processing spectrum correction with energies below 600 eV. Despite being empirical, this shift-correction is derived from principles with the detector physics. The goal of this approach is to correct a measured spectrum for the detector artefact, which results in a non-linear energy calibration due to the detector response. The corrected spectrum is then linear in energy calibration and the low-energy peak positions are correctly measured at their theoretical X-ray energy positions (Fig. 1). All further qualitative and quantitative spectrum analysis can be applied after the linearization of the spectrum energy-axis, including peak deconvolution algorithm with theoretical generated line-series shapes based on well-known X-ray data tables.

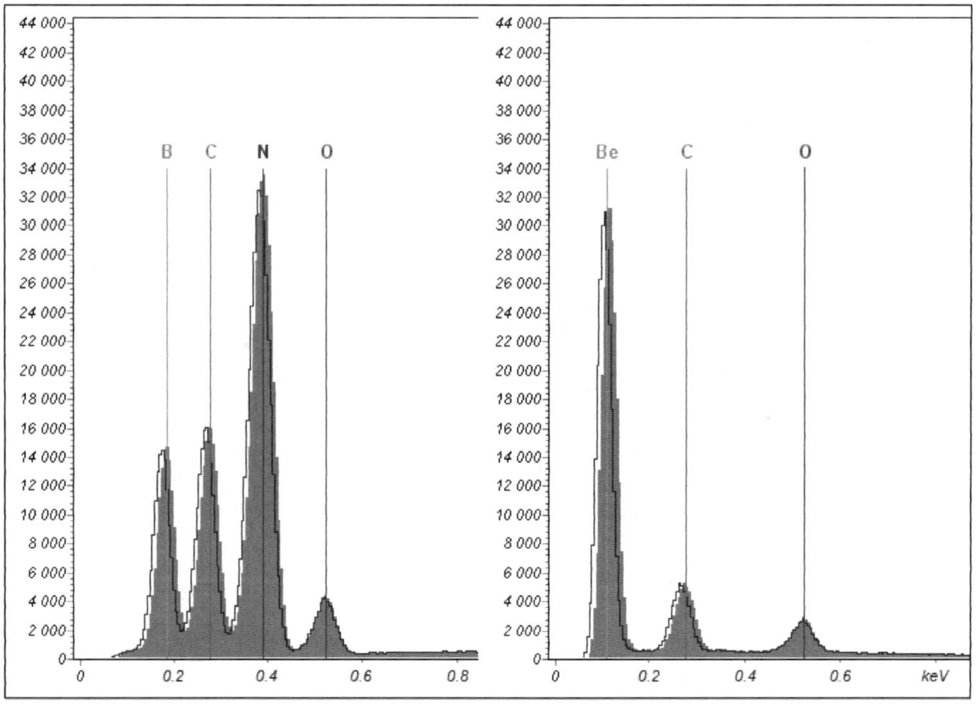

Figure 1. Original measured X-ray spectra (blue lines) versus original shift-corrected spectrum (red bars) in comparison with line-markers which represent the tabulated X-ray energies of several low energy X-ray lines (4.98 eV/channel energy calibration with 10 eV offset).

All peaks on the corrected spectra fit the expected X-ray line energy positions, except Be-K, which looks slightly overcorrected with 5 or 6 eV deviation instead of the expected 108 eV. This discrepancy was never investigated further.

3. Experimental

The more recent solid state detector (SDD) technology [3] and X-ray entrance windows [4] (Fig. 2) allow the investigation of X-ray energies below 100 eV.

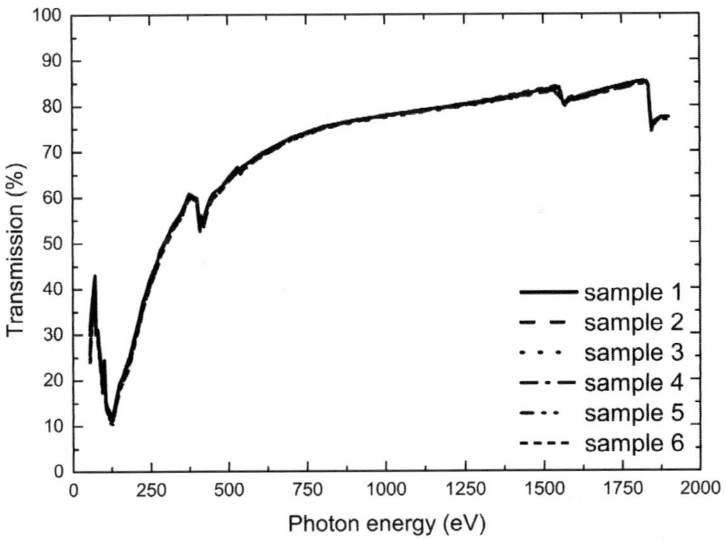

Figure 2. X-ray attenuation of several 40 nm thick Si_3N_4 atmospheric windows measured with synchrotron [4]. The window is additionally coated with a thin Al-layer to protect against visible light.

Whereas the much thicker polymer windows (300 nm) are not, the new atmospheric window type is transparent to soft X-ray below 100 eV. The SDD detector has no metallic contact layer and has only a thin 40 nm thick SiO_2, ideally designed for low-energy detection. Therefore, the EDS spectrum contains X-rays with less energy than 100 eV, and Si-L (92 eV) and Al-L (73 eV) can be clearly detected (Figs. 3 and 4).

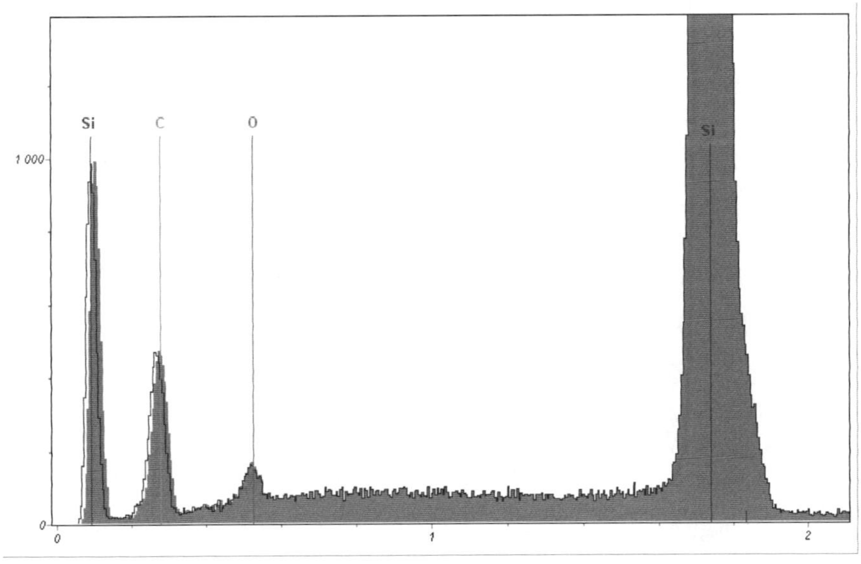

Figure 3. Original measured Si sample spectrum (blue line) versus original shift-corrected spectrum (red bars). The Si-L (92 eV) is with about 10 eV overcorrected with the original shift correction algorithm and misplaced in energy.

4. Theory

There was first a surprise as the X-ray peaks below 100 eV photon energy did not match the theoretical energies after the original shift correction (Figs. 3 and 4). Following the detector physics-based model understanding, the known peak-shift is based on a total incomplete charge collection; all detection events

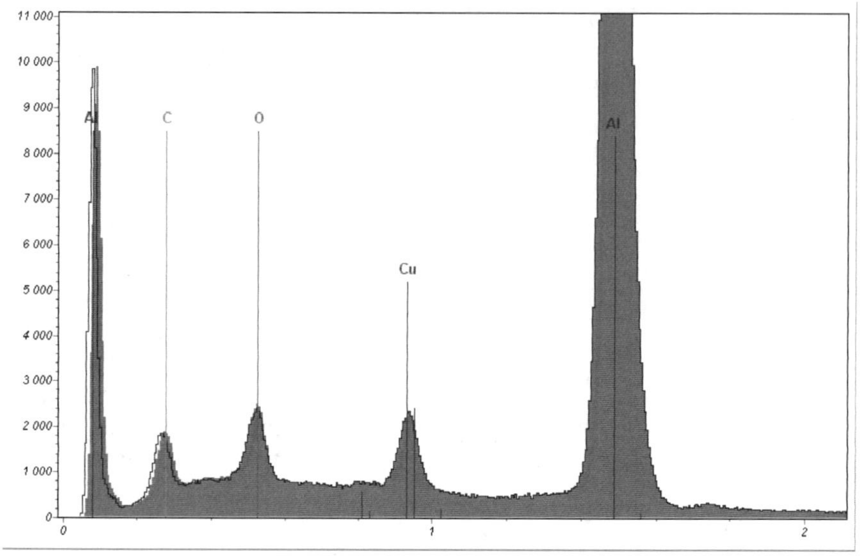

Figure 4. Original measured Al sample spectrum (blue line) versus original shift-corrected spectrum (red bars). The Al-L (73 eV) is with about 11 eV overcorrected with the original shift correction algorithm and misplaced in energy.

with these photon-energies generated electrons are affected by ICC for the voltage jump in the charge-sensitive preamplifier. Therefore, it is required to investigate the excitation depth of photons energy below 100 eV in the detector material (Si).

There is little information about X-ray attenuation length (absorption) in material with energies below 100 eV but can be found at the 'Berkeley Lab' online sources [5] (Fig. 5). The X-ray attenuation length values with energies below 500 eV in silicon supports the original shift correction approach. The depth of photon excitation in SDD detector drops to 100 nm and less, which is the order of magnitude for the contact structures of the silicon detector surface. However, due to the Si-L jumps, the depth of excitation becomes much larger for photon energies below 100 eV than it is, for instance, for carbon Kα (277 eV).

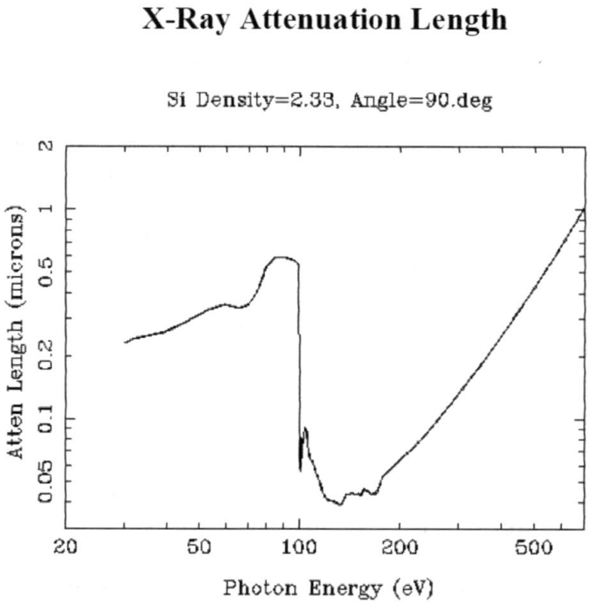

Figure 5. X-ray attenuation length for Si [5], the detector material.

5. Results

Due to the Si-L absorption edge (Si $L_{2,3}$ at 99 eV and Si L_I at 149 eV), the EDAX-used original shift-correction approach needs to be modified for photon energies below 149 eV. The attenuation length in silicon with photons of energies below 99 eV is like the photon energies above 500 eV (Fig. 5), therefore no peak-shifts are normally observed. This is an evidence for the shift effect being due to detector material and design properties, and not due to electronic effects or a non-linearity of the energy calibration.

The original empirical shift-correction applies a moderate linearly decreasing shift correction between 0 and 350 eV and a stronger decreasing correction until 550 eV. We propose a new shift-correction that is reduced below 149 eV to be consistent with the Si L_I edge effects mentioned above, and which drops down to 0 shift below 99 eV (Fig. 6). With this new correction, the measured shifts of Al-L and Si-L are essentially zero, and the Be shift is less than with the original correction. The corrected spectrum with modified shift-correction have finally a perfect linearity in the energy axis (Fig. 7).

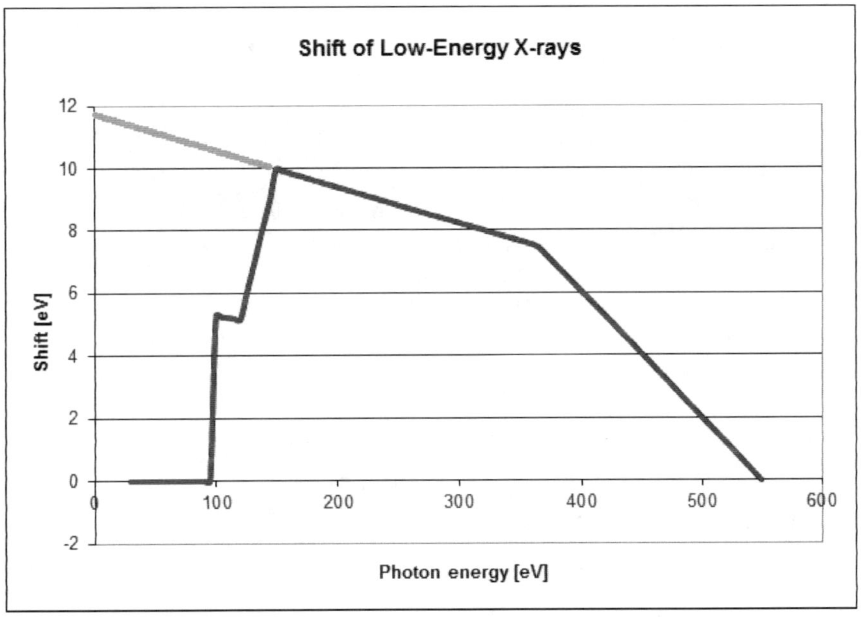

Figure 6. Modified shift correction (blue line) in comparison to the former used curve (grey line). The change is with energies below 149 eV (Si L1) and it drops down to zero shift with energies below 99 eV (Si-L_2/Si-L_3).

6. Conclusions

It is possible to describe the energy shift-effect of the detector response function with an empirical algorithm and to correct this artefact to get linear energy axis spectrum. The new algorithm proposed in this paper accurately corrects the shift effect also below 100 eV. Indeed, no shift-correction is needed with these X-ray energies. Thus, it is not required to modify the known soft X-ray line energies data, if the presented shift-correction modification is used (Fig. 6). The investigation of the low-energy peak shifts requires a clear view of the real zero-point of the energy calibration with the measured X-ray spectrum (the electronic setup of digital pulse processor). The used energy calibration for the linearized X-ray axis was with 4.98 eV/channel with an energy offset of 10 eV. The maximum shift we have

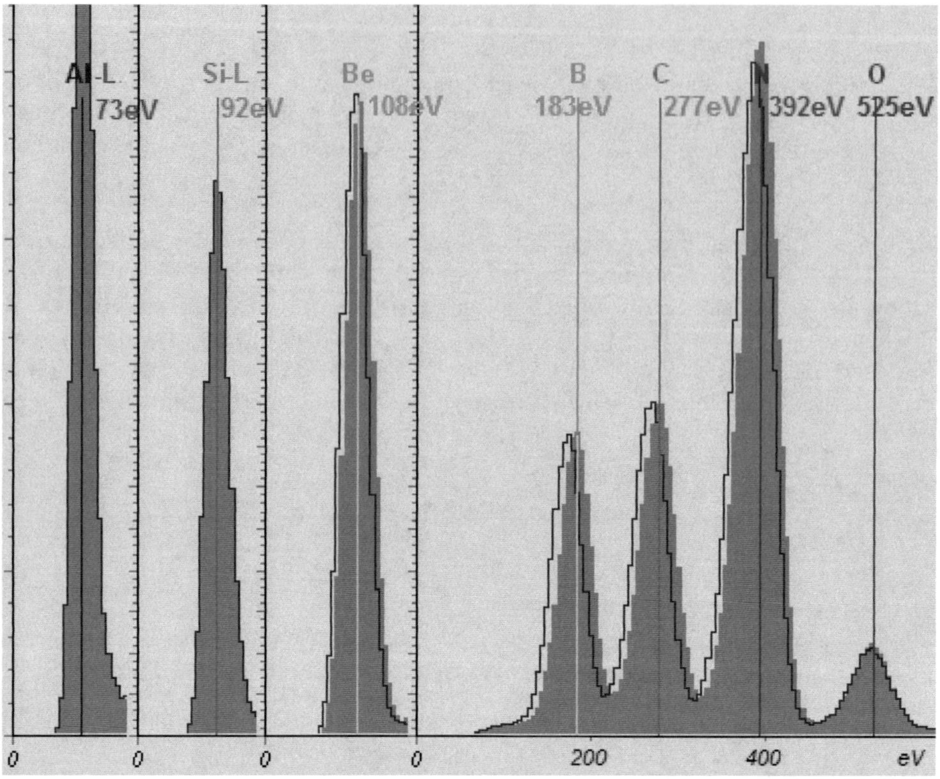

Figure 7. Original measured X-ray spectra (blue lines) versus shift-corrected spectra (red bars) by using the modified approach, in comparison with line-markers which represent the tabulated X-ray energies. The corrected peak positions are now no longer shifted in energy.

observed with boron-K was with about 2 channels. In consequence of this work, it is recommended to use Al-L and Si-L lines for calibration of detectors to determine the real offset (absolute part) of the linear energy calibration.

Acknowledgment
Many thanks to Frank Scholze (PTB) for the enlightening discussions.

References

[1] Heckel J and Scholz W 1987 *X-ray Spectrom.* **18** 181-185
[2] Scholze F and Procop M 2009 *X-ray Spectrom.* **38** 312-321
[3] EDAX June 2014 *Technical Note*
[4] Törmä P T, *et al.* 2014 *IEEE Trans. Nucl. Sci.* **61** 695-699
[5] http://henke.lbl.gov/optical_constants/atten2.html

Identification of Cu-Co-oxide phases of reactive air brazed $Ba_{0.5}Sr_{0.5}Co_{0.8}Fe_{0.2}O_{3-\delta}$-Ag-14CuO joints by EBSD, EPMA and TEM diffraction

L C Ehle[1], S Richter[1], S Herzog[2], C Broeckmann[2] and J Mayer[1]
[1] R.W.T.H. Aachen University, Central Facility for Electron Microscopy, Ahornstrasse 55, 52074 Aachen, Germany
[2] R.W.T.H. Aachen University, Institute for Materials Applications in Mechanical Engineering, Augustinerbach 4, 52062 Aachen, Germany

E-mail: ehle@gfe.rwth-aachen.de

Abstract. The oxygen transport membrane $Ba_{0.5}Sr_{0.5}Co_{0.8}Fe_{0.2}O_{3-\delta}$ (BSCF) was wetted by the reactive air brazing alloy Ag-14CuO. Triple point phases of different cobalt-copper-oxides are formed in the reaction zone and are origin of micro-cracks and stress concentration. Therefore, their identification is necessary for understanding the microstructure evolution and designing the interface structure advantageously by an adapted brazing process. However, as the cobalt-copper-oxide system can form many different oxides with solubility of Co and Cu respectively, different analytical methods are necessary for identification. TEM diffraction is used for the correct identification of the occurring crystal structures, EBSD is used for analysis of composition, arrangement and orientation of the triple point phases and EPMA is used to confirm the crystal structures identified by EBSD and to determine the elemental concentrations. As the triple point phases consist of small grains with unpredictable forms and especially, Cu-Kα can excite Co-Kα by characteristic secondary fluorescence, different evaluation procedures for EPMA are performed on FIB lamellae: 1) Approach based on standard lamellae, where FIB lamellae from standards are used to determine the *k*-factor in dependence of the mass thickness t/λ, and 2) Monte Carlo simulation approach, where an iterative procedure is used to calculate the *k*-ratios, based on bulk standards. The use of FIB lamellae allows excluding the fluorescence effect and incorrect Monte Carlo simulations due to incorrect assumptions of the microstructure below the surface, which is within the excitation volume.

1. Introduction

Ceramic oxygen transport membranes (OTM) allow oxygen separation from air via oxygen ion diffusion during the exposure to an oxygen partial pressure gradient at temperatures above 750 °C [1]. Amongst OTM materials, $Ba_{0.5}Sr_{0.5}Co_{0.8}Fe_{0.2}O_{3-\delta}$ (BSCF) exhibits superior oxygen permeation rates [2] and is therefore object of investigation in many lab-scale and pilot scale membrane modules [3]. For integration into a metallic module environment, the membranes must be joined gastight to metal components. Perovskite-type OTM like BSCF tend to chemically decompose under typical vacuum brazing conditions, i.e., the BSCF crystal structure loose oxygen and therefore its functional properties. The relatively new method reactive air brazing (RAB) operates at ambient conditions and is the exclusive brazing option. Typically, the RAB alloy consists of a precious metal filler (for instance Ag) and a reactive component (for instance Cu). The silver acts as basis of the brazing alloy, because it is stable at high temperatures, does not react and is a ductile metal. However, pure silver as brazing alloy, results in poor joint strengths since the only bonding mechanism is adhesion. The addition of Cu to the silver braze leads to the formation of CuO during heating up. The CuO can subsequently wet both oxide joining partners, the ceramic and the passivation layer of the metal component under formation of reaction phases. In this study, Ag-14CuO (14 at% CuO in silver) was used as brazing alloy on BSCF at 1,000 °C (see Fig. 1). The resulting reaction zone at the BSCF-braze interface exhibits new phases at the grain triple points, which are origin of micro-cracks and stress concentration [4, 5]. Identification of these phases allows firstly to understand the microstructure evolution and secondly, to design the interface structure advantageously by an adapted brazing process. The triple point phases consist of different copper-cobalt-oxides with solubility for Co and Cu respectively.

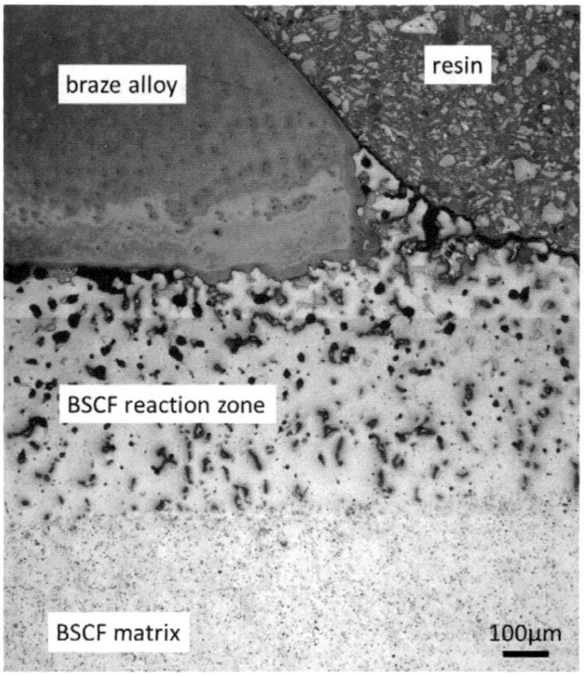

Figure 1. Light microscopic image of the cross-section. Within the BSCF matrix are many small pores, within the BSCF reaction zone are pores and triple point phases.

Different analytical methods are used for identification. Transmission electron microscopy (TEM) diffraction allows determination of crystal structures provided the grains are either large enough (> 1 μm diameter , for easy tilting) to position the grains in different zone axes for selected area diffraction (SAED) of single grains or the grains are small enough (< ~200 nm) to generate ring patterns with clearly distinguishable lattice plane distances. With electron backscatter diffraction (EBSD) the investigation of larger areas is possible with a spatial resolution of about 45 nm (longitudinal resolution for Cu at 10 kV and 1 nA [6]). It provides crystallographic information based on a reference database. To avoid any ambiguity energy-dispersive X-ray spectrometry (EDS) is used as a support for correct indexing of phases with the same crystal structure like BSCF, Co_3O_4 and CoO, which all have a face centred cubic (fcc) crystal structure. Electron probe microanalysis (EPMA) provides accurate quantification of elemental concentrations. Several different phases with solubility for Cu and Co exist in the Co-Cu-oxide system as was investigated by Zabdyr *et al.* [7]. The spatial resolution of EPMA, described by the information volume, depends on the chosen acceleration voltage, chemical composition, density and chosen X-ray line of the analysed phases. It can be well-adjusted to resolve phases with dimensions down to sub-micrometre scale [8]. EPMA corrections are well adapted to handling fluorescence effects within a bulk homogeneous material. But routine EPMA matrix corrections cannot deal with secondary fluorescence from neighbouring materials about which the presence is unknown. There are well established material corrections for thin films, accounting for fluorescence from other layers or for material couples with a simple known geometry [9]. In [10], correction procedures were developed for vertical grain boundaries for instance between pure Co and Cu. However, in the case of an unknown 3D microstructure there exists no fluorescence correction procedure yet. To overcome this problem thin lamellae were cut from each region of interest and EPMA was applied which prevents any influence of neighbouring phases. Disregarding smaller accelerating voltages, using FIB lamellae with thicknesses of less than 150 nm for EPMA is an additional way to reduce the excitation volume, which is then almost limited by the beam diameter [11].

In this work different methods and techniques are used to analyse the phases at the triple points of the reaction zone of BSCF with Ag-14CuO reactive air brazing alloy. Their advantages and disadvantages and their analytical limitations are discussed. In addition, EPMA measurements on Co-Cu-oxide FIB lamellae are performed and evaluated by two different procedures: 1) Approach based on standard lamellae, where FIB lamellae from standards are used to determine calibration factors assuming linear relation between measured X-ray intensities and element concentrations. The dependence on the mass thickness t/λ, which is the lamella thickness divided through the mean free path of the electrons [12] is also considered. 2) Quantification by Monte Carlo simulation, where an iterative procedure is used to get the best fit between calculated and experimental *k*-ratios, i.e., X-ray intensities calibrated by bulk standards (*k*-ratio = background corrected sample intensity / background corrected standard intensity).

2. Experimental

2.1. Sample production

Granulated BSCF-powder (Treibacher) was uniaxially pressed at 180 MPa into a d = 20 mm pellet and subsequently sintered at 1,100 °C for 5 h. Silver and copper powder (Thermo Fisher Scientific) were mixed in the appropriate amounts to achieve the desired molar ratio of 14 mol% CuO in silver and were uniaxially pressed at 200 MPa into braze pellets of d = 6 mm. The wetting of the braze pellet on the BSCF pellet was conducted at 1,000 °C for 6 min under air atmosphere, cooling was performed with 300 K/h. A more detailed description can be obtained from [13]. For manufacturing the reference samples CuO, Co_3O_4 and CoO, the starting powders CuO and Co_3O_4 (ThermoFisher Scientific) were manually mixed with a gel binder (B2-gel, Innobraze) and pressed at 200 MPa into pellets with d = 20 mm. The CuO and the Co_3O_4 pellets were sintered at 760 °C and 850 °C, each for 2 h. To convert the initial powder Co_3O_4 to the high temperature modification CoO [14], a second Co_3O_4 pellet was sintered at 1,080 °C for 48 h, annealed at 950 °C for 24 h and finally quenched in water.

BSCF crystallises in a perovskite structure (ABO_3) with A = $Ba_{0.5}Sr_{0.5}$ and B = $Co_{0.8}Fe_{0.2}$. Wetting the BSCF with alloy Ag-14CuO results in a reaction zone enriched with new phases at grain triple points (see Fig. 2). The triple point phases consist of different copper-cobalt-oxides as well as elemental silver. As reference materials powder-sintered CuO, CoO and Co_3O_4 samples were taken and their stoichiometric composition was confirmed by XRD and TEM diffraction. In order to exclude any kind of influence of neighbouring grains, FIB lamellae were cut from single grains of copper oxide and cobalt oxide as well as a FIB lamella containing several different grains of the triple point phases. The FIB preparation was performed on a Strata 400 by FEI.

2.2. XRD analysis

To confirm the crystal structure of the manufactured reference samples and to exclude the existence of any other phases X-ray interferences were registered on a Seifert XRD 3003 PTS with a step size of

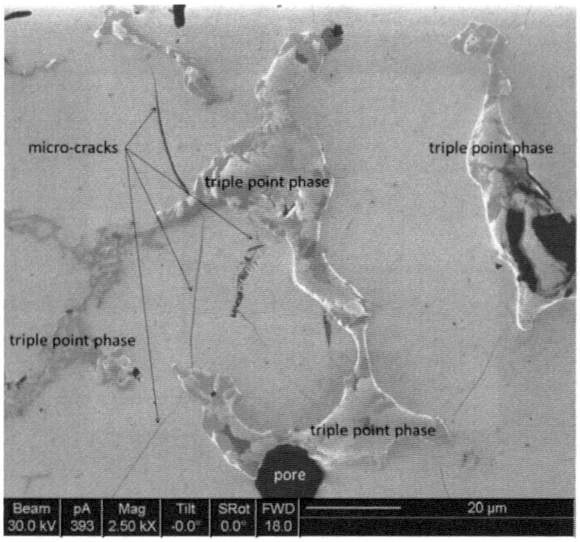

Figure 2. SE image of reaction zone with triple point phases, micro-cracks and pores. Triple point phases consist of different cobalt-copper-oxides.

0.05° and a counting time of 12 s each. Cr-Kα was used while Cr-Kβ was suppressed by means of a vanadium filter. The standard JCPDS-cards #00-043-1003 (Co_3O_4), #01-089-2803 (CoO) and #01-080-1916 (CuO) were taken to confirm the phase purity.

2.3. *TEM analysis*
The selected area diffraction patterns were acquired by a Veleta camera installed at a FEI TEM Tecnai F20. The reference data for identification of the crystal structures are from the online ICSD database with CollCode109462 for BSCF, CollCode16025 for CuO, CollCode36256 for Co_3O_4, and CollCode9865 for CoO [15]. For the acquisition of thickness maps a Gatan GIF camera was used.

2.4. *EBSD measurement*
For the EBSD measurements, a JEOL JSM7000F scanning electron microscope and a Hikari EBSD camera were used. The data collection and the analysis were performed with a combined EDS/ EBSD system from EDAX-TSL consisting of an Octane Plus SSD EDS detector and a "Hikari" EBSD camera. For the measurements, the system was operated at an acceleration voltage of 20 kV and a probe current of approximately 20 nA. The measured areas have been scanned with a step size of 100 nm each. The software used for measurement and data evaluation was OIM DATA COLLECTION and OIM ANALYSIS by EDAX-TSL, both in version 8.0.

2.5. *Electron probe microanalysis (EPMA)*
All measurements were performed by field emission JEOL JXA-8530F microprobe equipped with five wavelength-dispersive X-ray spectrometers. The operating conditions were 15 kV for the acceleration voltage and 40 nA for the beam current. Kα-line X-ray intensities were calibrated by using bulk standards of pure Cu, Co and Fe_2O_3 (for oxygen). For bulk analysis, the wetted sample was embedded in epoxy resin and a polished cross-section was prepared. For conductivity, a copper tape was fixed on the epoxy surface – no carbon coating was used. For EPMA analysis on FIB lamellae, the sample holder in Fig. 3 was used. In order to exclude any kind of influence of neighbouring grains, FIB lamellae were cut from single grains of copper oxide and cobalt oxide. In addition, a lamella containing several different grains of the triple point phases was made for comparison. For validation of quantification algorithms (for the Monte Carlo approach on lamellae) and for determination of k-factors for the approach based on standard lamellae, FIB-lamellae were also cut from reference samples CuO and Co_3O_4.

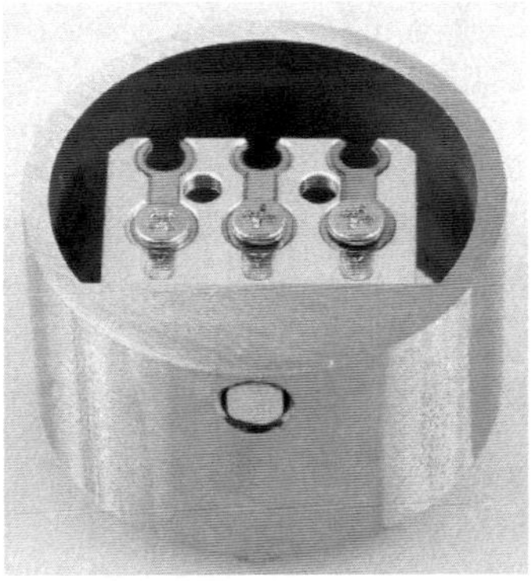

Figure 3. For the EPMA measurements on FIB lamellae, a t-EBSD (transmission-EBSD) holder of the company Micro to Nano was used together with a self-made holder, which fits to the EPMA stage. All components of the holder are made of steel and aluminium to prevent X-rays occurring from the sample holder (the original brass screws and copper plates were replaced by molybdenum plates and steel screws). The grids for the FIB lamellae are made of molybdenum.

2.5.1.1. Approach based on standard lamellae - EPMA on FIB lamellae. Lamellae cut from CuO and Co_3O_4 standards were thinned stepwise to gain areas of different lamella thicknesses. Wavelength-dispersive X-ray spectrometry (WDS) measurements with EPMA were performed on different thicknesses and the background corrected intensities (I) of Cu-Kα and O-Kα for CuO and accordingly Co-Kα and O-Kα for Co_3O_4 were used to determine the k-factors (see Eq. 1) for the known concentrations:

$$\frac{c_{Cu}}{c_O} = k_{CuO}\left(\frac{t}{\lambda}\right) \cdot \frac{I_{Cu}}{I_O} \quad (1)$$

The same equation was applied for the intensities and concentrations of Co_3O_4.

2.5.1.2. Approach based on standard lamellae - Thickness determination of FIB lamellae with TEM. The Gatan Imaging Filter (GIF) camera in the TEM contains a magnetic prism that allows the selection of electrons with different energies after passing the FIB lamella. According to Eq. 2, where t is the thickness of the lamella, λ is the average mean free path for low-energy loss electrons, I_t is the total intensity of electrons with energy between 0 and 50 eV (electrons which suffer an energy loss by passing through the FIB lamella due to plasmon interaction and ionisation) and I_0 is the zero loss peak intensity (electrons which pass through the FIB lamella without energy loss), t/λ can be used as equivalent for the thickness of the lamella provided the mean atomic number of the sample remains similar (since λ also depends on the average atomic number [12]).

$$\frac{t}{\lambda} = ln\frac{I_t}{I_0} \quad (2)$$

If the k-factor is known for different t/λ for the standard lamellae, then the k-factor can be determined for known t/λ for the sample lamellae and with the measured corrected intensities the concentrations can be calculated. The k-factor depends on the lamella thickness, which is usually inhomogeneous within a lamella and different between standard and sample lamella. The standard lamellae were cut with different steps of lamella thicknesses to gain a relationship between k-factor and t/λ. If standard lamellae and sample lamellae have similar mean atomic number, the k-factors determined on the standard lamellae can be used to determine the compositions of the sample lamellae.

2.5.2. Monte Carlo approach. Based on the quantification technique described in [16] a new procedure was developed to quantify thin lamellae under EPMA conditions. Simulations were performed using PYMONTECARLO, a common programming interface written in Python to run different Monte Carlo codes [8]. We used PYMONTECARLO to run WINCASINO 2.5.1.0. We are not using the GUI version because we have direct access to the code, which offers a self-supporting thin film as specimen geometry. Simulations were run for both standards and unknown concentrations to calculate simulated *k*-ratios of Co-Kα, Cu-Kα and O-Kα (bulk standards: Co, Cu and Fe_2O_3) and were compared with the experimental ones. The best fit between experimental and calculated *k*-ratios was chosen to determine the elemental concentrations.

3. Results

In Fig. 3 is a secondary electron (SE) image of the cross-section of the reaction zone between BSCF matrix and Ag-14CuO brazing alloy. Micro-cracks are formed in the reaction zone close to the soldering point starting at the triple point phases. In order to investigate the distribution of the elements present in the triple point phases X-ray intensity maps were acquired by EPMA (Fig. 4). These triple point phases mainly consist of cobalt-copper-oxides and sometimes of silver. According to [7] different combinations of oxides can occur in the Co-Cu-O system like CoO, Co_3O_4, CuO, Cu_2O, and Cu_2CoO_3 with different solubility of Cu and Co. TEM diffraction is used to identify the occurring crystal structures because EBSD can only identify known crystal structures. That means: EBSD does not determine the occurring phases, it only fits the measured Kikuchi patterns to the – from the operator suggested – reference phases. If incorrect phases are chosen or the reference data of occurring phases are missing, the indexing will be wrong. Additionally, the EPMA maps in Fig. 4 reveal that there are also mixed Cu-Co-oxides existing. Especially for small grains, the Kikuchi patterns will be bad (due to overlapping information of neighbour grains) and the EBSD software could fit any phases to these patterns. Furthermore, even if all correct phases were chosen, EBSD is unable to differentiate between different phases with the same crystal structure (for instance: CoO, Co_3O_4 and BSCF are all fcc) unless the EDS signal is acquired during the EBSD measurement as well.

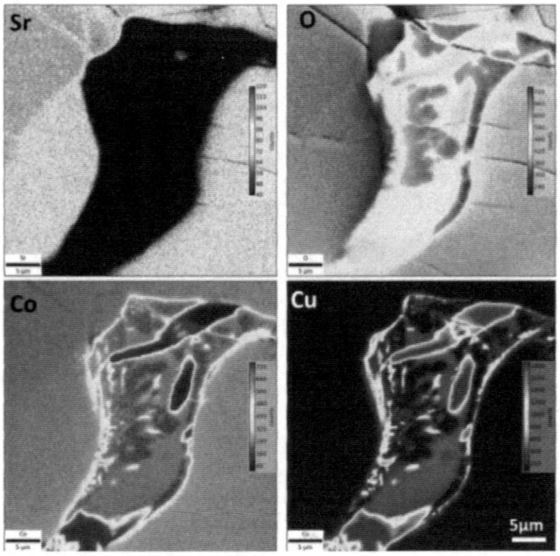

Figure 4. EPMA intensity maps of Sr-Lα, O-Kα, Co-Kα and Cu-Kα.

3.1. XRD measurements

Figure 5 presents the diffraction pattern of the pressed and sintered reference samples. All peaks were identified according to the JCPDS reference data and belong to the desired phase. There are no major peaks from competing phases, for instance Cu_2O in CuO, Co_3O_4 in CoO and CoO in Co_3O_4 which confirms the phase purity of the reference samples.

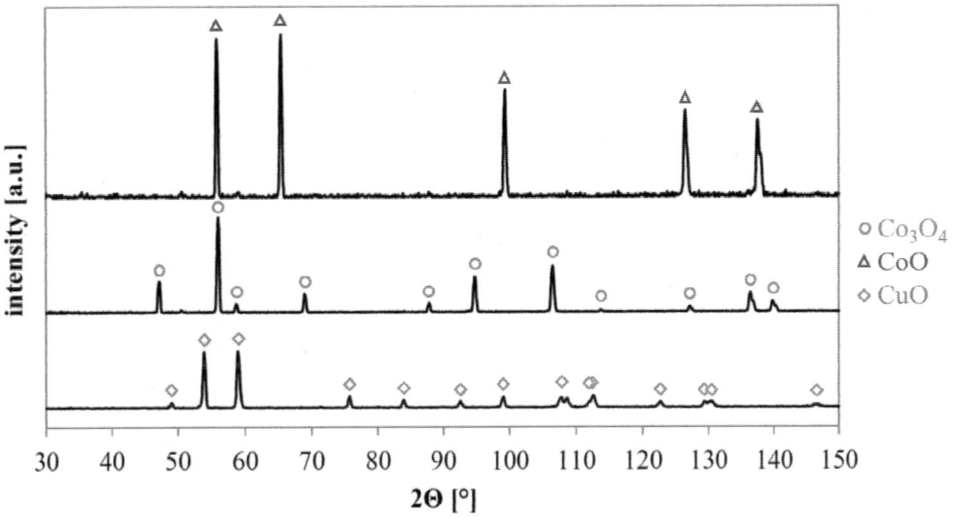

Figure 5. XRD of the reference samples. The markers show the position of the peaks of JCPDS reference data.

3.2. TEM diffraction

A FIB lamella was cut from the triple point phase and selected area diffraction (SAED) was performed on single grains oriented in different zone axes. In Fig. 6, selected diffraction patterns of three different grains of the brightfield image (BF) of the FIB lamella are shown. The determined lattice plane distances as well as the reference lattice plane distances and the dedicated hkl's are listed in the tables below the diffraction patterns. The grains are identified as monoclinic CuO, fcc CoO and fcc Co_3O_4. CoO and Co_3O_4 differ by their lattice parameter, which is 4.263 Å for CoO and 8.072 Å for Co_3O_4.

3.3. EBSD bulk measurements

In Fig. 7, the image quality (IQ) and the EDS maps of the EBSD measurement are depicted. As the expected phases BSCF, CoO and Co_3O_4 are all fcc, the EDS signal is necessary for the automated identification and indexing procedure. However, the EDS signal for Co, Cu and O does not change significantly for CoO and Co_3O_4 respectively. Thus, the more sensitive EPMA maps are used. In Fig. 4 three different phases can be clearly identified in the triple point phase: 1) A copper-oxide phase (identified as CuO), which contains almost no cobalt, 2) a cobalt-oxide phase, which contains almost no copper (identified as Co_3O_4) and 3) a cobalt-oxide phase, which contains some copper (identified as CoO). Within one triple point, the phases CoO and Co_3O_4 have similar orientations. As the lattice parameter of Co_3O_4 is almost twice as big as that of CoO, more bands appear in the Kikuchi patterns as demonstrated in Fig. 8. The different number of counting bands can be used to distinguish between CoO and Co_3O_4 in the EBSD measurement.

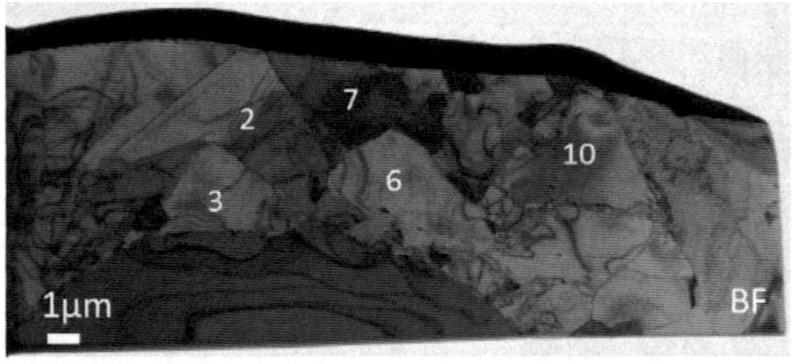

grain	phase	crystal structure
3	CuO	monoclinic
2	Co_3O_4	fcc
6	Co_3O_4	fcc
7	CoO	fcc
10	Co_3O_4	fcc

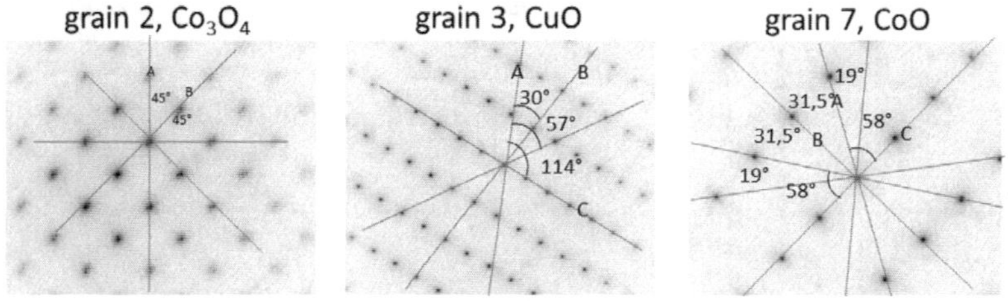

d_{hkl} [Å]	Co_3O_4 d_{hkl} [Å]	hkl
2.8418	2.854	220
2.0135	2.018	400
1.4209	1.427	440
1.2720	1.276	620

d_{hkl} [Å]	CuO d_{hkl} [Å]	hkl
5.045	5.058	001
2.755	2.750	110
2.523	2.523	-111
2.314	2.310	200

d_{hkl} [Å]	CoO d_{hkl} [Å]	hkl
2.4590	2.461	1-11
1.5050	1.507	220
1.2784	1.285	311
1.2295	1.231	222

Figure 6. TEM results; selected area diffraction of single grains in zone axis. The identified phases are monoclinic CuO, fcc CoO and fcc Co_3O_4.

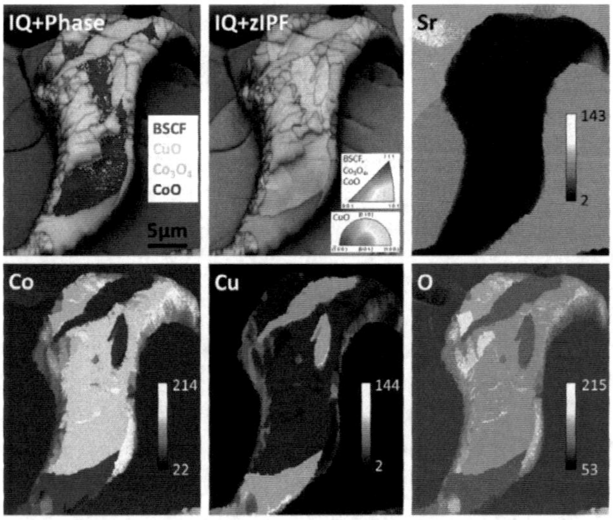

Figure 7. EBSD Image Quality (IQ) with overlaid phase and inverse pole figure (zIPF) map. EDS maps of strontium, cobalt, copper and oxygen. The EBSD map was acquired on the same position as the EPMA maps in Fig. 4.

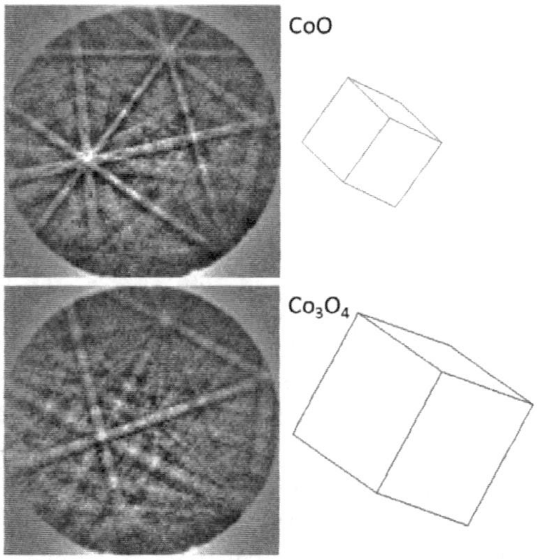

Figure 8. Kikuchi-Pattern for CoO and Co_3O_4 and corresponding oriented unit cell.

3.4. EPMA measurements

In Fig.9, *k*-ratio profiles of Co-Kα, Cu-Kα and O-Kα across the triple point phases of the bulk samples are labelled according to TEM and EBSD results. As can be seen, within the cobalt-oxide phase (Co_3O_4) a small amount of copper X-rays is measured and within the copper-oxide phase (CuO) cobalt X-rays are measured. There appears to be a phase containing a relative high amount of cobalt and copper (labelled CoO?); however, this could be the result of a poor spatial resolution. Different phases are present in the excitation volume.

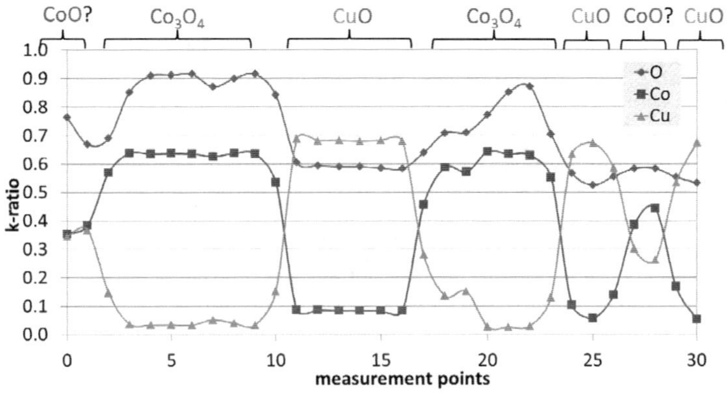

Figure 9. *k*-ratios of Co-Kα, Cu-Kα and O-Kα measured across triple point phases on bulk sample. According to the identified phases by TEM diffraction, the supposed compositions are collated.

As the triple point phases do not have predictable form and size, the influence of excitation volumes including more than one phase (below the surface) and especially, fluorescence effects of Cu-Kα exciting Co atoms by secondary fluorescence X-rays cannot be excluded. That means:

- If the excitation volume includes more than one phase (no matter if it is CoO, Co_3O_4 or CuO) the determined composition will be wrong.
- If the excitation volume includes only CuO, spreading Cu-Kα can excite Co atoms of neighbouring CoO / Co_3O_4 grains resulting in an increased Co content.

Additionally, CoO can solve a wide range of Cu atoms (up to 36 at%). In order to exclude any influence of surrounding grains, FIB lamellae are cut from single phases and – additionally – one lamella that contains several grains (see Fig. 6).

In Table 1 the results of the thin lamella analysis are presented. On the right side, the values were evaluated by using the Monte Carlo simulation programme PyMonteCarlo [8]. For the validation thin lamellae were cut from standards of well-known composition (reference samples), i.e., CuO and Co_3O_4, and measured at the same conditions as the lamellae of the unknown phases of the specimen. It turns out that the measured copper and cobalt concentration is lower than expected. It can be explained with the help of the diagram shown in Fig. 10a. Here the relative deviation of the *k*-ratios with and without contribution of continuums fluorescence is plotted versus the thickness of the lamellae (calculation was performed by using the programme StrataGem [17, 18] assuming a thin film on a hydrogen substrate). For a thickness of 150 nm (density was given) the relative deviation is about 3.3 % for Co-Kα and about 4.5 % for Cu-Kα. Since the *k*-ratios are the measured X-ray intensities of the lamellae calibrated by bulk standards, continuum fluorescence cannot be neglected for the bulk standards.

Table 1. Mass concentrations for Cu, Co and O determined on FIB lamellae with standard lamella based and Monte Carlo based approach. Crystal structures have been confirmed by selected area diffraction. The values show average and standard deviation of more than 6 measurement points each (except for multigrain lamella).

	stoichiometric CuO	*k*-factor CuO		MC CuO		
		single	multi	standard	single	multi
Cu mass. %	79.9	76.12±0.45	78.08±	76.62±0.45	72.16±0.41	73.96±
Co mass. %	0.0	6.61±0.42	3.24±	-	6.85±0.06	3.62±
O mass. %	20.1	17.27±0.28	18.68±	-	-	-
	stoichiometric CoO	*k*-factor CoO		MC CoO		
		single	multi	standard	single	multi
Cu mass. %	0.0	-	15.37±0.08	-	-	14.40±0.03
Co mass. %	78.6	-	62.79±0.22	-	-	61.67±0.24
O mass. %	21.4	-	21.84±0.28	-	-	-
	stoichiometric Co_3O_4	*k*-factor Co_3O_4		MC Co_3O_4		
		single	multi	standard	single	multi
Cu mass. %	0.0	3.13±0.27	2.94±0.13	-	2.99±0.28	2.71±0.04
Co mass. %	73.4	69.84±0.90	69.81±0.41	72.19±0.44	68.93±0.76	67.81±0.72
O mass. %	26.6	27.03±0.98	27.25±0.38	-	-	-

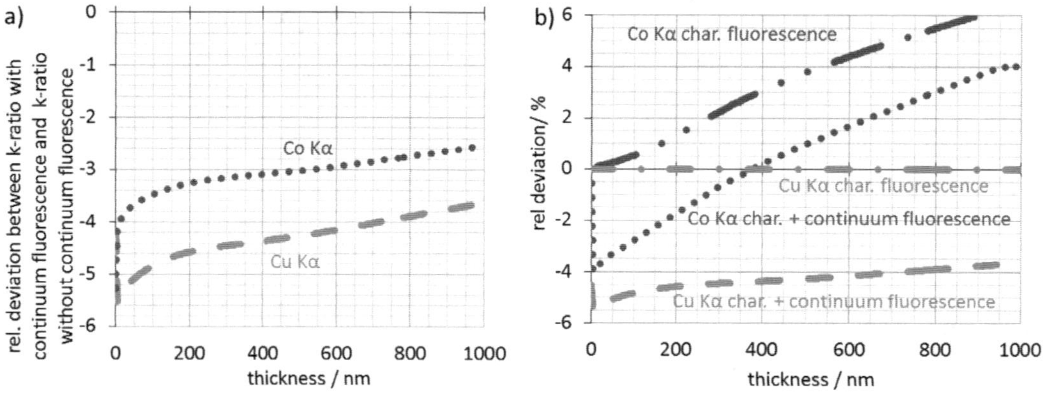

Figure 10. a) relative deviation of the *k*-ratios with and without contribution of continuums fluorescence is plotted versus thickness of the lamella; b) contribution of characteristic and continuum fluorescence for thin lamella consisting of CuO and 7.0 wt% of Co (instead of Cu).

The diagram in Fig. 10b shows the situation for the thin lamella consisting of CuO and 7.0 wt% of Co (instead of Cu). The contribution of characteristic fluorescence is evident for the thin lamellae (red spotted and dashed line) and together with the continuum fluorescence contribution of the bulk standard the relative deviation gets smaller to about 2 % for a thickness of 150 nm (see red spotted line). The small deviation of 2 % is visible in the results shown in Table 1. The measured and by MC calculated concentrations for a thin lamella cut from a single grain (i.e., homogeneous composition) reflect the expected concentrations. Expected concentrations mean that the Cu atoms are substituted by the Co atoms during diffusion of Co into CuO. Figure 10b also shows that the characteristic fluorescence of Co decreases with lamella thickness. If the lamella thickness is below 150 nm – which is usually the case for lamella investigated in the TEM – the contribution of Co-K characteristic fluorescence is below 1 %.

For the approach based on standard lamellae, the *k*-factors are determined on the standard lamellae using Eq. 1. The *k*-factors for CuO and Co_3O_4 are k_{CuO} = 1.93 ± 0.05 and k_{Co3O4} = 1.86 ± 0.03, respectively. There is no clear dependence of the *k*-factors on the lamella thickness identifiable considering that steps of different thicknesses were cut for the standard lamellae. One reason is that the t/λ value is also influenced by bending contours and the exact position of the EPMA measurements on the FIB lamellae cannot be determined. However, the deviation of the *k*-factors is small (see Fig. 11).

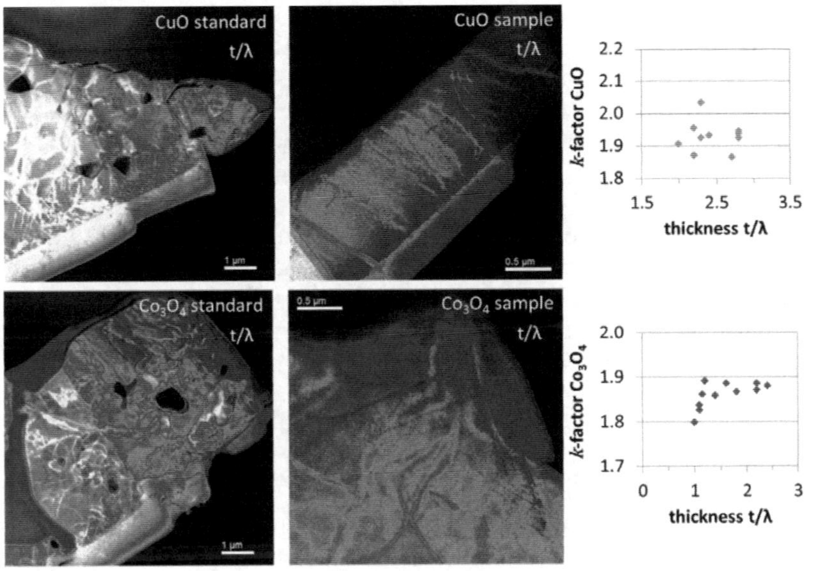

Figure 11. t/λ-maps for CuO and Co_3O_4 standard lamellae as well as CuO and Co_3O_4 single grain sample lamellae. *k*-factors are calculated with the standard lamellae for different t/λ.

Thus, the average could be used for quantification. The concentrations of Cu, Co and O in the sample lamellae can be calculated using the measured intensities on the sample lamellae, the determined average *k*-factors of the standard lamellae, Eq. 1 and the assumption that the total equals 100. Table 1 compares the stoichiometric concentration of the Co-Cu-oxides and the determined concentrations for the *k*-factor and the Monte Carlo method. The average and the standard deviations for more than 6 measurement points are shown (except for the measurements on the multi grain lamella).

As can be seen the MC method and the method based on standard lamellae (*k*-factors) give quite similar results that lie within the standard deviation except for the Cu concentration in the CuO single grain lamella. The results of the multigrain lamella deviate to the determined values of the single grain lamellae, in particular for the CuO lamella, where the Co content of the multigrain lamella is only about half the content of the single grain lamella. As different solubility is possible perhaps a grain of different composition was picked. For the CuO lamellae (standard and single grain lamella), the sum of Cu and Co is significantly higher than the stoichiometric Cu concentration compared to the CoO and Co_3O_4 lamellae, where the sum of Co and Cu almost meet the stoichiometric Co concentration. The reason for this effect can be explained with Fig. 10a, where the influence of continuum fluorescence in the standard bulk sample is higher for Cu than for Co. For CuO in the multi grain lamella the sum of Cu and Co is less than the stoichiometric Cu content, however, as only one measurement point was available no conclusions or explanations can be given.

4. Discussion

As the results showed, TEM diffraction allows the accurate determination of the occurring crystal structures because the grains are big enough for tilting and SAED of single grains in zone axis. Furthermore, the lattice distances of the occurring phases differ enough for unambiguous identification.

For EBSD, the correct identification is much more complicated. As there are three phases with fcc crystal structure (BSCF, CoO and Co_3O_4), unambiguous identification of the phases without EDS signal is impossible. BSCF can be identified by the Sr-signal, but for CoO and Co_3O_4 the differences in Co, Cu and O signal of the EDS data is not strong enough for differentiation (see Fig. 7). WDS is much more sensitive and does clearly show the difference of Co, Cu and O content in CoO and Co_3O_4 (see Fig. 4). However, as the WDS signal is not coupled with the EBSD data it does not help for the automated indexing. As the lattice plane distance of Co_3O_4 is almost twice as large as that of CoO, the Kikuchi patterns of Co_3O_4 contain more bands than the ones of CoO (see Fig. 8), which can be used for the counting bands in the Hough transformation. Within one triple point, CoO and Co_3O_4 have almost the same orientations because the change in crystal structure only depends on small changes in composition.

The WDS measurements on the FIB lamellae of the sample (single grains) clearly showed that a certain amount of Co/ Cu is within the CuO/ CoO and Co_3O_4. Influences of different phases of neighbouring grains as well as characteristic secondary fluorescence effects can be excluded. The lower Cu- and Co-concentrations measured in the standard lamellae could be explained with continuum fluorescence effects. Still, the Co content in CuO is too high (see Table 1) according to the quasi-binary Co-Co-oxide phase diagram, where the Co content is limited to ~2.0 mass% at room temperature [7]. Furthermore, no CoO should exist at room temperature, although it is definitely identified by EBSD, TEM diffraction, and EPMA in the triple point phases. As CoO can be formed above 847 °C, which lies above the brazing temperature of 1,000 °C, its crystal structure might be stabilised during cooling due to oxygen deficiency. However, the oxygen partial pressure that is considered in the ternary phase diagram and especially the oxygen solubility kinetics are not known in the real brazing process and can only be estimated in dependence of the cooling rates [19]. Additionally, small amounts of iron can exist within the triple point phases as well and may change the phase stabilities. Therefore, the quasi-binary Co-Co-oxide phase diagram might not be applicable for this material system and needs to be modified.

The two different quantitative EPMA approaches on FIB lamellae show almost the same values within the standard deviation. It can be concluded that the standard lamella-based approach can be applied for standard lamellae with composition close to that of the sample. The thickness of the lamella does not seem to have a significant effect on the quantification at least for $0.9 < t/\lambda < 3.4$. As the Monte Carlo approach gives comparable results, is much less time-consuming and no standard lamellae are necessary, it is a valuable and easy alternative to get quantitative information about the composition. Together with crystallographic information by TEM diffraction, it also provides unambiguous identification of phases with grain sizes much smaller than the excitation volume in bulk samples.

5. Conclusion

Co-Co-oxide phases have been identified as CuO, CoO and Co_3O_4 by TEM diffraction, EBSD and EPMA on FIB lamellae. For EPMA on FIB lamellae two different approaches were tested: An approach based on standard FIB lamellae with a composition close to the sample lamella, and an approach based on a modified Monte Carlo simulation. For both approaches, quite similar results are gained. The CuO phase contains ~3.24 - 6.85 mass% Co, the CoO phase contains ~14.40 - 15.37 mass% Cu and the Co_3O_4 phase contain 2.71 - 3.13 mass% Cu. Thus, for small phases of unpredictable form and size and element combinations causing secondary fluorescence effects, WDS measurements on FIB lamella allow the quantification of these phases. As the standard FIB lamella-based approach requires standard lamellae with similar composition, it is quite time consuming. However, the presented Monte Carlo approach results in comparable values and is based on bulk standards and, therefore, is a time-saving and reliable alternative to bulk EPMA quantification, especially if TEM diffraction is required anyway.

The comparable high amount of 14 mol% CuO in the brazing alloy was selected for the presented investigations to observe trends and effects of the microstructure evolution that cannot be observed when using only 3 mol% CuO in the brazing alloy. For Ag-3CuO the triple point phases are small and almost no micro-cracks are visible [13]. Nevertheless the BSCF/braze interface is the weakest position in BSCF-Ag-3CuO-steel joints in 4-point-bending tests. The present phase identification of the triple point phases is the first step for future simulations that should answer the following questions:

1. Which kinetic has the solidification of the braze and the precipitation of the triple point phases?
2. When are the micro-cracks formed during cooling?

A thermodynamic dataset for the simulation of solidification will be extended in future by CuO, CoO and Co_3O_4. The subsequent microstructure simulation based on a 2D representative volume element comprising one triple point phase will be added with Young's modulus and coefficient of thermal expansion of CuO, CoO and Co_3O_4 as well as the fracture strength of BSCF to model the crack formation. These simulations allow the choice of the optimal chemical composition of the braze alloy and temperature-time profile for the reactive air brazing of BSCF.

Acknowledgements
The authors thank the German Research Foundation (DFG) for funding the project with the project number 392944287.

References

[1] Sunarso J, *et al.* 2017 *Prog. Energ. Combust.* **61** 57-77
[2] Baumann S, *et al.* 2011 *J. Membrane Sci.* **377** 198-205
[3] Nauels N, *et al.* 2019 *J. Membrane Sci.* **574** 252-261
[4] Kaletsch A, *et al.* 2014 *Adv. Engng. Mater.* **16** 1430-1436
[5] Joshi V, *et al.* 2013 *J. Mater. Sci.* **48** 7153-7161
[6] Chen D, Kuo J C and Wu W T 2011 *Ultramicroscopy* **111** 1488-1494
[7] Zabdyr L A and Fabrichnaya O B 2002 *J. Phase Equilibria* **23** 149-155
[8] Pinard P T and Richter S 2014 *IOP Conf. Ser.: Mater. Sci. Engng.* **55** 012016
[9] Armstrong J T and Buseck P R 1985 *X-ray Spectrom.* **14** 172-182
[10] Llovet X, *et al.* 2012 *J. Phys. D: Appl. Phys.* **45** 225301
[11] Kubo Y 2018 *IOP Conf. Ser.: Mat. Sci. Engng.* **304** 012007
[12] Williams D B and Carter C B (Eds.) 2009 *Transmission electron microscopy.* (New York, NY: Springer) 730 p.
[13] Kaletsch A, *et al.* 2012 *J. Ceram. Sci. Technol.* **3** 95-104
[14] Chen M, *et al.* 2003 *J. Phase Equilibria* **24** 212-227
[15] ICSD database, FIZ Karlsruhe, https://icsd.fiz-karlsruhe.de/search/basic.xhtml
[16] Richter S, *et al.* 2016 *IOP Conf. Ser.: Mater. Sci. Engng.* **109** 012014
[17] Pouchou J-L and Pichoir F 1984 *La recherché aerospatiale* **3** 13
[18] Pouchou J-L and Pichoir F 1991 in: *Electron probe quantitation.* (Heinrich K F J and Newbury D E; Eds.) (New York, NY: Plenum Press) 31-75
[19] Berger D, *et al.* 2014 *Adv. Engng. Mat.* **16** 1475-1480

IOP Conf. Series: Materials Science and Engineering **891** (2020) 012013 doi:10.1088/1757-899X/891/1/012013

Characterisation of super duplex stainless steel by optimisation of EBSD parameters

M Haukali[1], I Westermann[1], M Karlsen[2], Y Yu[1], R de Kloe[3] and J Hjelen[1]

[1] Norwegian University of Science and Technology, Department of Material Science and Engineering, Alfred Getz vei 2, 7491 Trondheim, Norway

[2] Equinor ASA, Stavanger, Norway

[3] Ametek BV, EDAX Business Division, Ringbaan Noord 103, P.O. Box 4144, 5004 JC Tilburg, the Netherlands

E-mail: mohauk.mh@gmail.com

Abstract. Electron backscatter diffraction (EBSD) was used to characterise the microstructure of a super duplex stainless steel (SDSS) alloy Zeron 100 and the intermetallic phases sigma and chi in this material. The material was analysed in three conditions; as-received, heat-treated at 900 °C for 8 min and heat-treated at 750 °C for 4 h. Heat treatments were performed to introduce the intermetallic phases in the steel. It is difficult to get reliable indexing of the sigma and chi phases due to noisy and overlapping patterns. To get the best results, the Hough transformation parameters were optimised and an averaging technique was used. The optimisation was done by removing some of the reflectors in the sigma- and chi-phases in the original TSL material files used for pattern indexing. By changing Hough parameters, considerable improvements in the indexing of the patterns were observed along with the increase of average confidence index (CI) value for both intermetallic phases. The applied SEM, Zeiss Ultra 55 FESEM was equipped with a NORDIF UF-1100 EBSD detector to acquire and stream patterns to HD. Indexing and data processing were performed by TSL/OIM 7.3. In the present study, at the most four different phases are present in the SDSS.

1. Introduction

Super duplex stainless steels are used in petrochemical industry because of their high corrosion resistance and good mechanical properties. These properties come from high content of alloying elements, especially high chromium content which provides high corrosion resistance. However, the wrong heat treatment for these steel grades can lead to embrittlement of the materials through intermetallic phases such as sigma- and chi-phase. This can happen during fabrication, welding processes and prolonged exposure to high temperatures during their service lives. The embrittlement can lead to catastrophic service failure of components.

A way to investigate the intermetallic phases in the scanning electron microscope (SEM) is to apply the electron backscatter diffraction (EBSD) technique. However, the intermetallic phases, i.e., sigma- and chi-phases, can be challenging to index. The aim of this study is to investigate a super duplex stainless steel alloy, UNS S37260 or Zeron 100, focussing on the intermetallic phases. The project is done together with Equinor ASA.

2. Materials and methods

The material used in this study was the alloy Zeron 100, a super duplex stainless steel in annealed condition. The material was provided by Equinor ASA with composition shown in Table 1.

Table 1. Chemical composition (in wt%) of alloy UNS S32760.

Alloy element	Cr	Ni	Mo	Cu	Mn	W	Si	N	C	P	S
wt%	25.55	8.28	3.46	0.72	0.52	0.52	0.42	0.2-0.3	0.018	0.017	0.001

Three samples were chosen for investigation: First one as received as reference sample, a second one was heat-treated at 900 °C for 8 min (~5 wt% of sigma-phase), and a third one heat-treated at 750 °C for 4 h (~30 wt% of sigma-phase). The samples were firstly indexed with default Hough-transform parameters and default material files for all the phases. Thereafter, both Hough parameters and material files for the sigma- and chi-phases were changed in TSL OIM 7.3 software.

2.1. Heat treatments

Heat treatments were performed to introduce different amounts of intermetallic sigma- and chi-phases. Heat treatments were done in a Nabertherm P300 muffle furnace and afterwards the samples were quenched in room temperature water. One of the samples was heat-treated at 900 °C for 8 min, which introduces about 5 wt% of sigma-phase and the other at 750 °C for 4 h, which introduced about 35 wt% of intermetallic phases.

2.2. Sample preparation

The samples were grinded using an ATM Saphir 330 at 150 rpm with SiC-paper up to grit 4,000 using water as lubricant. Thereafter, the samples were polished using a Struers DP-U3 polishing machine at 150 rpm with polishing discs having a diamond suspension with particle sizes 3.0 μm and 1.0 μm, and with an ethanol-based lubricant for 15 minutes. After this step, a vibration polishing with at amplitude of 70 % and a weight of 200 grams was performed with a Buehler Vibromet 2 polisher. After each grinding and polishing step, the samples were cleaned with water and ethanol. Lastly, the samples were plasma-cleaned with Fischione Model 1020 for five minutes.

2.3. Scanning electron microscope (SEM)

The SEM used to study and acquire Kikuchi patterns from the material was a Zeiss Ultra 55 FESEM. The parameters used in the SEM are shown in Table 2.

Table 2. Parameters used in Zeiss Ultra 55 FESEM.

Parameter	**Value**
Accelerating voltage	20 kV
Working distance	25 mm
Magnification	200x
Aperture size	300 μm
Angle	70°
High current mode	ON
Dynamical focus	10 - 15 %

IOP Conf. Series: Materials Science and Engineering **891** (2020) 012013 doi:10.1088/1757-899X/891/1/012013

2.4. Electron backscatter diffraction (EBSD)

EBSD measurements were taken on three investigated samples. A NORDIF UF-1100 EBSD detector to acquire and stream patterns to HD was employed, and NORDIF 3.0.43 software was used. The parameters are listed in Table 3. The step size used was 0.25 μm and the region of interest varied, with the biggest area being 400 × 400 μm. Thereafter, the analysed area was cropped to get a more detailed look at specific phases. For each measurement, five calibration patterns were collected, one in each corner and one in the middle of the area to optimise the pattern centre. A background image was acquired before acquiring the patterns. This was done to get better signal and avoid background noise.

Table 3. Parameters used in NORDIF 3.0.43.

Parameter	Acquisition	Calibration
Averaging	2	5
Speed	400 fps	500 fps
Resolution	120 ×120 px	160 ×160 px
Exposure time	2,450 μs	7,092 μs
Gain	4	1

2.5. TSL OIM analysis

The collected EBSD data were characterised using TSL OIM 7.3 analysis software. At first, default bmt-files for the ferrite, austenite, sigma-and chi-phases were applied. Thereafter, the same acquired patterns were characterised with new bmt-files for sigma- and chi-phases provided by René de Kloe (Ametek). The difference between default and new bmt-material files for the chi-phase lay in the fact that the reflectors [4 2 2] and [6 6 0] were removed in the new, optimised bmt-file. The only change in the sigma bmt-file was the change of symmetry group from tetragonal C4h/[4/m] to tetragonal D4h[4/mmm].

Table 4. OIM data analysis Hough parameters.

Parameter	Before optimisation	After optimisation
Binned pattern size	96	96
Theta step size	1°	0.5°
Rho fraction	85	88
Max peak count	3	3
Hough type	Classic	Classic
Hough resolution	Low	Low
Convolution mask	9 × 9	7 × 7
Mininimum peak magnitude	5	1
Maximum peak magnitude	15	10
Peak symmetry	0.70	0.70
Vertical bias	0 %	0 %

3. Results

3.1. Optimisation of Hough parameters

Optimisation of Hough parameters was done to characterise the sigma- and chi-phases at their best. The optimisation was used on the sample heat-treated at 900 °C for 8 min and the sample head treated at 750 °C for 4 h. Clean up below a confidence index CI = 0.05 was used. The colours of the phases are indicated in Fig. 1.

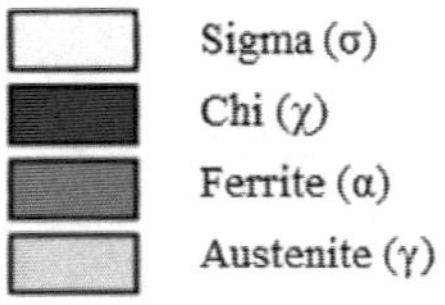

Figure 1. Colours for the different phases.

3.2. Material treated at 900 °C for 8 min

The IQ (image quality) map for this sample is shown in Fig. 2. The area of the measurement was 37.75 µm × 32.50 µm. This is obviously a small area and the phase composition is not representative for the entire sample; however, it was used to show the difference in chosen parameters. The step size used to acquire the signals was 0.25 µm. As can be seen, there are some dark areas on the IQ maps. It can also be observed that the maps are not of the optimal quality. The grain boundaries appear dark but relatively clear.

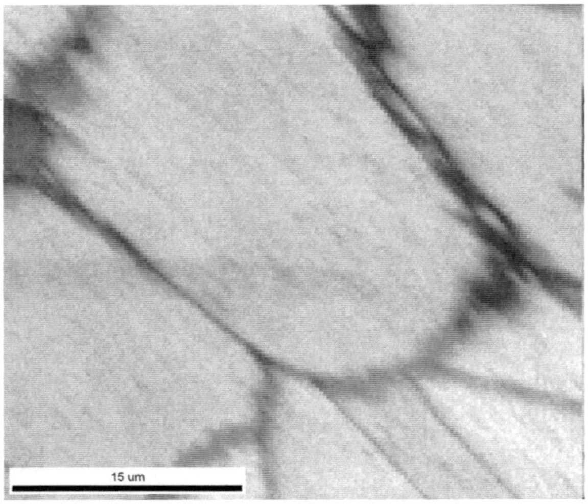

Figure 2. IQ map of sample heat-treated at 900 °C for 8 min

Figure 3 shows the phase maps of the material heat-treated at 900 °C for 8 min before and after Hough parameter optimisation. Austenite is marked in green, ferrite in red and sigma in yellow. It can be observed that the optimisation leads to less noise in the map and less points are characterised as sigma phase. This can also be seen in Table 5, where a higher fraction of austenite after the optimisation is clearly visible.

IOP Conf. Series: Materials Science and Engineering **891** (2020) 012013 doi:10.1088/1757-899X/891/1/012013

Table 5. Phase fraction of the material heat-treated at 900 °C for 8 min before and after Hough parameter optimisation.

Phase	α wt%	γ wt%	σ wt%
Before optimisation	45.4	45.4	7.4
After optimisation	44.2	50.7	5.1

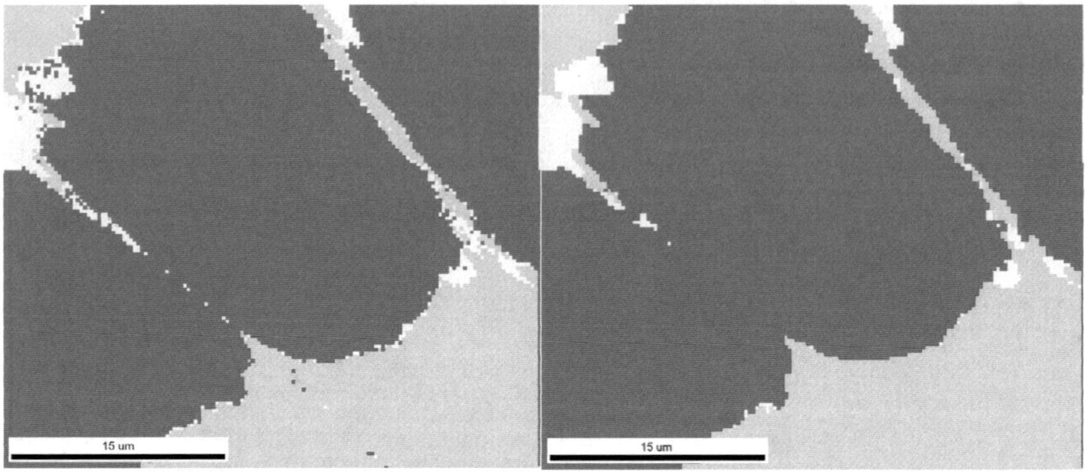

Figure 2. Phase map for material heat-treated at 900 °C for 8 min. Left: before optimisation. Right: after optimisation.

Table 6 shows that the CI value for the sigma-phase increases from 0.08 to 0.14. The CI value for ferrite (alpha) also increases after optimisation, while the value for austenite (gamma) decreases slightly. The IQ values for all the phases increase after optimisation. The fit for all the parameters decreases and gets closer to zero.

Table 6. Average IQ, CI and fit values for the material heat-treated at 900 °C for 8 min. before and after optimisation.

Parameter	γ+α+σ	α	γ	Σ
CI, before	0.64	0.66	0.70	0.08
CI, after	0.73	0.79	0.67	0.14
Fit, before	0.90	0.85	0.93	1.53
Fit, after	0.71	0.66	0.72	1.46

3.3. Material treated at 750 °C for 4 h

In Fig. 4, the IQ map for this sample is shown. The area of the measurement is 39.40 μm × 33.60 μm. Again, the area is not representative for the phase composition of the entire sample. As can be seen, a large part of the IQ map is covered by dark areas. It can also be observed that image is not of optimal quality. Some lighter areas in the upper left and lower right corners of the map can be noticed, as well as some smaller points in between the darker areas. Grain boundaries appear darker.

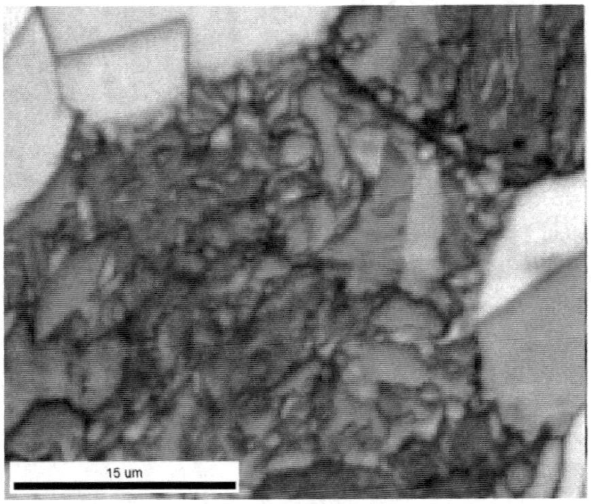

Figure 3. IQ map of sample heat-treated at 750 °C for 4 h.

Figure 5 shows the phase maps of the material heat-treated at 750 °C for 4 h before and after the Hough parameter optimisation. Austenite is marked in green, ferrite in red, sigma-phase in yellow, and chi-phase in blue. It can be observed that the optimisation leads to a considerably lower noise level. It can also be observed that most of the noise comes from the chi-phase. After the optimisation, less phase was indexed as chi-phase and more as ferrite (alpha) phase. This is also shown in Table 7. More points were also indexed as sigma-phase after the optimisation.

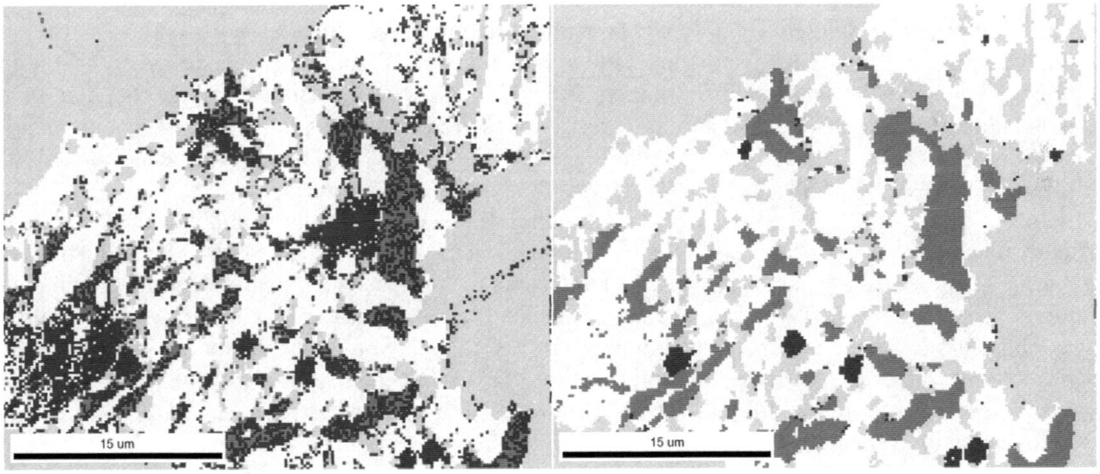

Figure 4. Phase map for material heat-treated at 750°C for 4 h. Left: before optimisation. Right: after optimisation.

Table 7. Phase fraction of the material heat-treated at 750 °C for 4 h before and after Hough parameter optimisation.

Phase	α wt%	γ wt %	σ wt %	χ wt %
Before optimisation	5.6	36.5	41.4	16.5
After optimisation	10.9	38.0	49.1	2.0

Table 8 shows that the CI value for the sigma-phase increases from 0.08 to 0.14. The same does the CI value for chi-phase, mainly from 0.12 to 0.17. Especially worthy of mentioning is the rise of the CI value for ferrite phase from 0.01 before to 0.45 after optimisation. It can be seen, that after optimisation the fit increases for all the phases except the sigma-phase, with the average increasing from 1.04 to 1.09.

Table 8. Average IQ, CI and fit values for the material heat-treated at 750 °C for 4 h before and after optimisation.

Parameter	$\alpha+\gamma+\sigma+\chi$	α	γ	σ	X
CI, before	0.22	0.01	0.47	0.08	0.12
CI, after	0.30	0.45	0.48	0.14	0.17
Fit, before	1.04	0.97	0.73	1.28	1.16
Fit, after	1.09	1.07	0.92	1.21	1.50

4. Discussion

4.1. Hough parameter optimisation
Hough parameter optimisation leads to clear changes in the indexing of different phases, especially the sigma- and chi-phases. The IQ, CI and fit values will be discussed as well as the IQ and phase maps for the two relevant samples. The changes introduced by optimisation will be also discussed.

4.2. Effect on phase characterisation
Optimisation of the parameters leads to a clear improvement of the characterisation of the sigma- and chi-phases. Figures 3 and 5 for the material heat-treated at 900 °C and 750 °C respectively, show that the noise level is much higher before optimisation. This is especially shown in Fig. 5. Most of the wrongly indexed pixels in Fig. 3 assigned to sigma-phase have has been converted after optimisation either to ferrite or austenite. Also, some pixels assigned previously to ferrite have been indexed after optimisation as austenite. This conversion is a result of the clean-up process.

Figure 3 shows the difficulties of characterising the chi-phase. Presumably, there are extra difficulties when there are two phases, which are difficult to be distinguished. This is also illustrated by the optimisation because the areas previously indexed as the chi-phase were after optimisation indexed as sigma-phase (Fig. 5 and Table 7). The chi-phase is reported to have a BCC-structure [1], while the sigma-phase has a tetragonal structure similar to the BCC one. Some of areas previously indexed as chi-phase were converted to ferrite after optimisation. This is because the unit cells of both phases have a lot of common reflectors and the unit cells parameters are almost the same [2]. Optimisation was based on removing reflectors [4 2 2] and [6 6 0] in the chi-phase, which were also used in the ferritic phase.

Optimisation leads to a smaller amount of the sigma-phase in the material heat-treated at 900 °C (Table 5) and a significantly smaller fraction of the chi-phase and a higher amount of the sigma-phase in the material heat-treated at 750 °C (Table 7). Some of the points indexed as the chi-phase in Fig. 5 can be imagined to be ruled out as noise, but some of the points look like clear clusters of the chi-phase. The indexing problem most probably comes from the difficulties of indexing grain boundaries and the overlap of signals acquired from different grains. Grain boundaries were indexed with the help of misorientation angle, so all boundary segments with an angle higher than the defined critical misorientation angle are considered to be grain boundaries [3].

The fraction of austenitic phase increases with the optimisation in both samples (Tables 5 and 7). It is expected that the fraction of austenite remains the same or even increases, compared to the as-received material, while the ferrite phase fraction decreases after the short term heat treatment. This is because sigma- (and chi-) phase is formed where saturation with molybdenum and chromium exists, i.e., in the ferritic phase. Such a saturation also causes new austenite areas to be formed.

The symmetry group of sigma is changed with the optimisation from tetragonal C4h/[4/m] to tetragonal D_{4h}[4/mmm]. In the literature, the space group for the sigma-phase has been identified as $P4_2$/[mnm], which is the same space group as D_{4h} [2]. This corresponds to the optimised case.

4.3. *Effect on CI, IQ and fit values*

Optimisation of the parameters leads to a clear improvement in the CI, IQ and fit values not only for the sigma- and chi-phases but also for the ferritic and austenitic phases. The overall CI value increases after optimisation both for the material heat-treated at 900 °C (Table 6) and for the material heat-treated at 750 °C (Table 8). For the material heat-treated at 900 °C the average CI value for all the phases increases from 0.64 to 0.73 before and after optimisation, respectively. The reason for such an increase in this value is mostly caused by the increase of the CI value for the ferrite (alpha) from 0.66 to 0.79. Sigma-phase also contributes positively with an increase from 0.08 to 0.14. For the material heat-treated at 750 °C that contains a significantly larger amount of sigma- and chi-phases, the average CI value for all the phases increases from 0.22 to 0.30 after optimisation (Table 8). The CI values for sigma-and chi-phase increase from 0.08 to 0.14 and from 0.12 to 0.17, respectively. Another value to be noticed is the increase of ferrite fraction from 0.01 to 0.45. A reason for such an increase might be caused by the difficulty with differentiating between the ferrite and the chi-phase. After removing the strong reflector from the chi-phase HKL list, the number of votes for the chi phase decrease contrary to the CI for ferrite phase. Such an increase in the CI value means that the second solution for phase identification is less likely after optimisation.

The IQ describes the quality of an electron diffraction pattern [4]. The higher IQ value means that the acquired bands are sharper and easier to index. It can be seen on the IQ maps that ferrite grains are darker comparing to austenite ones (Figs. 2 and 4). Comparing Fig. 4 to Fig. 2, it can be stated there are much larger dark areas in the sample heat-treated at 750 °C for 4 h. Therefore, more intermetallic phases are expected. The darkest areas in Fig. 4 can be identified as sigma-phase. The dark areas in the IQ map is due to low IQ value, which results from band edges blurring.

The perfect value for the fit is zero. That means that the angular deviation between calculated and detected bands is also zero [4]. The average fit after optimisation decreases from 0.90 to 0.71 for the material heat)treated at 900 °C and increases from 1.04 to 1.09 for the material treated at 750 °C. One possible reason for such a change was the use of different convolution mask, mainly instead of 9 × 9 the mask 7 ××x 7 was applied (Table 4). The default value is 9 × 9, however for the narrower bands it better to use 7 × 7 mask [4]. For the investigated sigma- and chi-phases, the fit value for sigma is decreasing for both samples, and the fit value for chi-phases is increasing only in the sample heat-treated at 750 °C for 4 h.

The aim of this study was to characterise the sigma- and chi-phases in a better way. From the analysis of the CI, IQ and fit values, it can be deduced that the proposed optimisation resulted in a better sigma-phase evaluation.

5. Conclusion

Changing the Hough parameters as well as the bmt material files in TSL/OIM 7.3 for sigma- and chi-phases led to a better indexing of these intermetallic phases in the SDSS UNS S32760 alloy. This was proven by an increase of the average CI values in both samples from 0.64 to 0.73 and from 0.22 to 0.30 in the materials containing around 5 wt% and 30 wt% of sigma-phase, respectively.

Acknowledgements

The authors would like to thank René de Kloe from Ametek BV, The Netherlands for providing bmt-files for TSL OIM 7.3 analysis software.

IOP Conf. Series: Materials Science and Engineering **891** (2020) 012013 doi:10.1088/1757-899X/891/1/012013

References

[1] Chail G and Kangas P 2016 Super and hyper duplex stainless steels: structures, properties and applications. *Procedia Structural Integrity* **2** 1755-1762 (21st European Conference on Fracture, ECF21, 20-24 June 2016, Catania, Italy)

[2] Warren A D, Harniman R L, Guo Z, Younes C M, Flewitt P E J and Scott T B 2016 Quantification of sigma-phase evolution in thermally aged 2205 duplex stainless steel. *J. Mater. Sci.* **51** 694-707

[3] Oxford Instruments Analytical 2004 *Electron backscattered diffraction*. (High Wycombe, UK: Oxford Instruments)

[4] TSL 1997 *OIM Analysis 7.2 Manual*

The microanalysis of iron and sulphur oxidation states in silicate glass - Understanding the effects of beam damage

E C Hughes[1,2], B Buse[1], S L Kearns[1], R A Brooker[1], D Di Genova[3], G Kilgour[4], H M Mader[1] and J D Blundy[1]

[1] University of Bristol, School of Earth Sciences, Wills Memorial Building, Queens Road, Bristol, BS8 1RJ, Great Britain
[2] California Institute of Technology, Division of Geological and Planetary Sciences, 1200 E. California Blvd., 91125 Pasadena, CA, U.S.A.
[3] Technische Universität Clausthal, Institut für Nichtmetallische Werkstoffe, Zehntnerstrasse 2a, 38678 Clausthal-Zellerfeld, Germany
[4] GNS Science, Wairakei Research Centre, 114 Karetoto Road, RD4, 3384 Taupo, New Zealand

E-mail: ehughes@caltech.edu

Abstract. Quantifying the oxidation state of multivalent elements in silicate melts (e.g., Fe^{2+} versus Fe^{3+} or S^{2-} versus S^{6+}) is fundamental for constraining oxygen fugacity. Oxygen fugacity is a key thermodynamic parameter in understanding melt chemical history from the Earth's mantle through the crust to the surface. To make these measurements, analyses are typically performed on small (<100 μm diameter) regions of quenched volcanic melt (now silicate glass) forming the matrix between crystals or as trapped inclusions. Such small volumes require microanalysis, with multiple techniques often applied to the same area of glass to extract the full range of information that will shed light on volcanic and magmatic processes. This can be problematic as silicate glasses are often unstable under the electron and photon beams used for this range of analyses. It is therefore important to understand any compositional and structural changes induced within the silicate glass during analysis, not only to ensure accurate measurements (and interpretations), but also that subsequent analyses are not compromised. Here, we review techniques commonly used for measuring the Fe and S oxidation state in silicate glass and explain how silicate glass of different compositions responds to electron and photon beam irradiation.

1. Introduction

Magmas are a complex and evolving mix of bubbles, crystals, and melt, where the abundance and composition of these phases changes significantly during ascent. Such changes have implications for how the magma flows and hence eruption style, which has the potential for developing methods to predict eruption style from volcano monitoring data. This requires quantifying the parameters that affect the chemical and physical properties of the magma. Natural silicate melts cover a wide range of compositions, containing ~45 - 75 wt% SiO_2 and varying amounts of TiO_2, Al_2O_3, Fe_2O_3, FeO, CaO, MgO, MnO, Na_2O, K_2O, and P_2O_5 (Fig. 1a). In addition, silicate melts may contain volatiles such as H_2O, CO_2, S, and halogens (e.g., Cl, F, Br, etc.). To cover the range of compositional effects, two

IOP Conf. Series: Materials Science and Engineering **891** (2020) 012014 doi:10.1088/1757-899X/891/1/012014

extreme endmembers are considered (Fig. 1a): one is basaltic, which is low-silica (~50 wt% SiO_2), Fe-rich (~8 wt% Fe), and alkali-poor (<5 wt% Na_2O+K_2O); whilst the other is rhyolitic, which is high-silica (~70 wt% SiO_2), Fe-poor (~2 wt% Fe, except pantellerite which contains ~6 wt% Fe), and alkali-rich (~8 wt% Na_2O+K_2O).

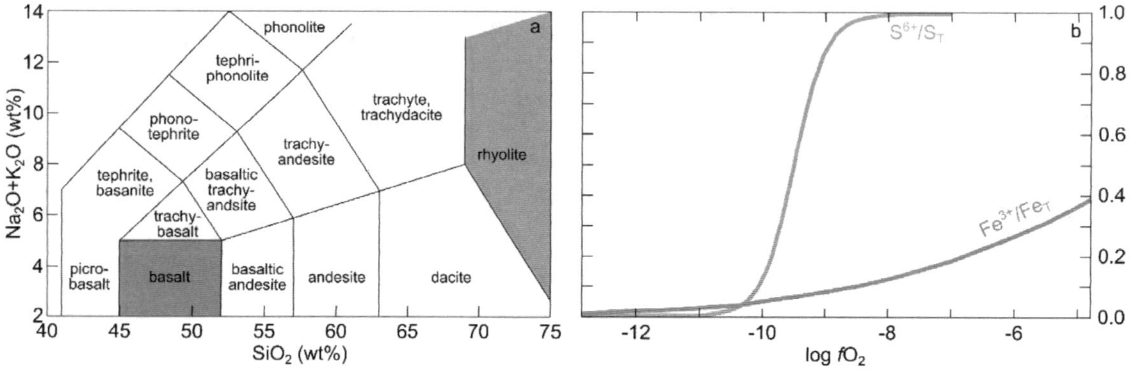

Figure 1. a) Na_2O+K_2O against SiO_2 labelled by composition, where basalt (red) and rhyolite (blue) are shaded. b) Typical effects of oxygen fugacity (fO_2) on melt Fe (Fe^{3+}/Fe_T) [1] and S (S^{6+}/S_T) [2] oxidation state.

Oxygen fugacity (fO_2) is equivalent to the activity of oxygen in a system. It is a key thermodynamic parameter that controls the oxidation state of multivalent elements within the magma (mainly Fe and S, but also Mn, Cr, V, Ce, and Eu; [1, 3, 4]). Figure 1b illustrates how fO_2 typically controls the valency of Fe (Fe^{2+} and Fe^{3+}) and S (S^{2-} and S^{6+}), which can have a major effect on the physical and chemical properties of the magma. The fO_2 affects volatile solubility, mineral-melt equilibria (e.g., phase assemblage), and magma viscosity, as well as controlling the composition of exsolved vapour and even the formation of economic ore bodies (e.g., [5-11]). The fO_2 varies between different magmatic systems, often related to tectonic setting, but can also be perturbed by processes such as fractionation and degassing (e.g., [12, 13]). For these reasons fO_2 is an important parameter to constrain when investigating the evolution of magmatic systems; it can be inferred by measuring the melt Fe or S oxidation state, i.e., the ratio of the reduced to oxidised species (Fig. 1b). Melt inclusions (tiny pockets of glass 1 - 500 µm in diameter in crystals) or matrix glass (final melt composition) are commonly analysed for this purpose (Fig. 2). Melt inclusions form when melt is trapped during crystal growth, providing a unique sample of the melt, which quenches to a glass upon eruption [14].

Multiple microanalytical techniques are often used to analyse melt inclusions and matrix glass due to the wide variety of elements present (at variable concentrations) and range of information required (including concentration, oxidation state, and isotope ratio, e.g., [15]). By necessity, this means that the same area of glass may be analysed multiple times. Interaction of the beam (i.e., electron or photon) with the glass during analysis can cause changes to the composition and structure of the glass, which is broadly referred to as 'beam damage'. As a result, it is important to understand if, and to what extent, these measurements affect the glass and whether subsequent analyses are compromised. The cause of beam damage and the subsequent effects depend on the type and conditions of analysis, as well as the sample (e.g., composition and thickness). Hence, each section covers a separate technique commonly used to analyse Fe and S oxidation state in silicate glass (Fig. 3; Sections 2-6). Sections begin with a brief description of how the technique is used to quantify the Fe (and S if applicable) oxidation state, followed by the observations and possible mechanisms of beam damage. Mitigation strategies and recommendations for multi-technique analytical protocols are suggested in Section 7.

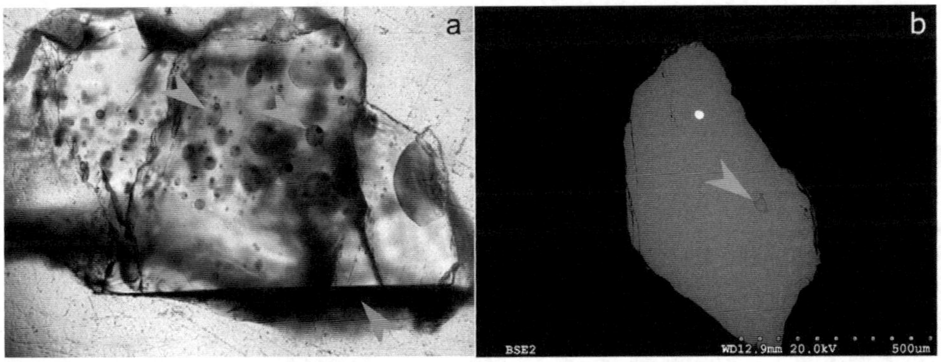

Figure 2. a) Photomicroscope (transmitted light) image of an olivine crystal containing melt inclusions (blue, bottom-right arrows – not all melt inclusions are indicated for clarity) and matrix glass (red, top-left arrow), both of which are brown. b) Backscattered electron image of a pyroxene crystal containing a melt inclusion (blue, bottom-right arrow), which is dark grey.

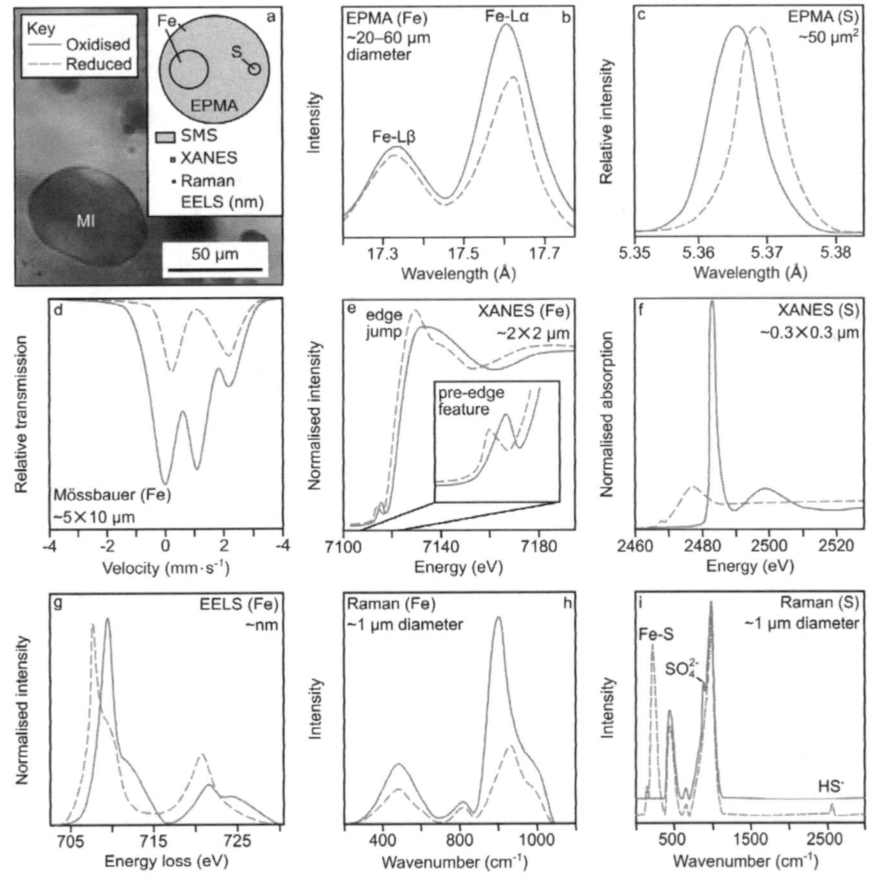

Figure 3. a) Olivine-hosted melt inclusion (MI) with the spatial resolution of the different techniques shown for comparison (spatial resolutions are also quoted on each panel). Schematic spectra for oxidised (solid, blue line) and reduced (dashed, red line) samples for b) Fe-L using EPMA [16], c) S-Kα using EPMA [17], d) Fe using Mössbauer [18], e) Fe K-edge using XANES [18], f) S K-edge using XANES [2], g) Fe L-edge using EELS [19], h) Fe using Raman [20], and i) S using Raman [21].

2. Electron probe microanalysis (EPMA)

2.1. Technique

Measuring Fe oxidation state with EPMA uses the Fe-L X-ray lines, which are produced by the 3d → 2p electron transition (Fig. 3b). The change in position and intensity of the Fe-Lα (~17.6 Å) and Fe-Lβ (~17.3 Å) peaks are a function of Fe oxidation state, as well as the concentration of Fe and its coordination [22, 23]. This is caused by changes in the energy of the 3d electron shell due to bonding, as well as changes in self-absorption of the emitted X-rays [24, 25]. For instance, the intensities of the Fe-L peaks are higher if the glass contains more Fe or if the Fe is more oxidised (i.e., Fe^{2+}/Fe_T decreases) [16]. Additionally, the Fe-L peak positions occur at higher wavelengths for higher Fe concentrations or if the Fe is more reduced (i.e., Fe^{2+}/Fe_T increases) [16]. Nonetheless, careful calibration of these effects can enable the quantification of Fe^{2+}/Fe_T, with an error of ± 0.1 absolute on Fe^{2+}/Fe_T [16, 25, 26].

There are two methods to exploit these changes: the 'peak shift' and 'flank' methods [27]. The peak shift method correlates the Fe-Lα peak position (measured by scanning the spectrometer over the X-ray peak of interest, commonly referred to as a 'wavescan') at a given Fe concentration with Fe^{2+}/Fe_T [28]. The flank method correlates the change in the intensity ratio of positions on the low wavelength flank of Fe-Lα (Fe-$L\alpha_f$) and the high wavelength flank of Fe-Lβ (Fe-$L\beta_f$) (Fe-$L\beta_f$/Fe-$L\alpha_f$) with the Fe^{2+} content of the glass (which is converted to Fe^{2+}/Fe_T using known Fe_T) [23, 27, 29]. The flank method has greater sensitivity than the peak shift method because it utilises both changes in peak position and intensity [27]. Silicate glass standards of similar composition (e.g., basalt or pantellerite) or garnets (andradite and almandine) can be used for calibration [16, 26]. The presence of Fe-bearing nanosized crystals (i.e., nanolites) in the glass prevents quantification of Fe oxidation state using EPMA, likely because the signal is a mix of the Fe in the glass and the nanolites (Fig. 4).

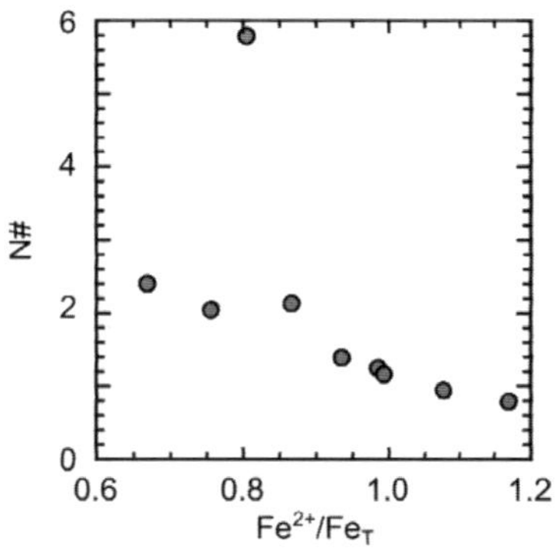

Figure 4. Quantity of nanolites measured by Raman (degree of nanolitisation, N#, defined by [30]) against Fe^{2+}/Fe_T measured using EPMA on natural melt inclusions. The negative correlation between N# and Fe^{2+}/Fe_T (excluding one outlier in red) implies the presence of nanolites changes the intensity and/or peak positions of the Fe-L lines compared to silicate glass. This prevents quantification of the Fe oxidation state using EPMA and can lead to spurious values (i.e., $Fe^{2+}/Fe_T > 1$).

The S-Kα peak (2p → 1s) shifts to higher wavelengths with increased proportion of reduced sulphur due to the lower bond energies of reduced sulphur (Fig. 3c) [31]. The peak shift method is used to quantify S oxidation state (S^{6+}/S_T), although it cannot distinguish S^{4+} from a combination of S^{2-} and S^{6+} [32-35]. Mineral end-members (e.g., barite and troilite) are used to define the peak positions of S^{2-} and S^{6+} and a linear variation in S^{6+}/S_T is assumed between these two positions. An error of ±3.5 % on sulphate can be achieved using this technique [34].

2.2. Beam damage

Electron probe beam damage has been well documented in a variety of studies, as summarised below (beam conditions of 15 - 30 kV accelerating voltage, 4 - 30 μm beam diameter, 2 - 500 nA beam current). Silicate glass is an insulator, hence during EPMA incident electrons are trapped in the band gap within the sample [36]. This generates a region of negative charge below the interaction volume, which produces an electric field within the glass [37]. This electric field causes a variety of changes to the glass, such as cation migration and oxidation/reduction of Fe and S (Fig. 5). In all cases, increasing the electron beam density (i.e., decreased beam diameter, increased beam current, or decreased accelerating voltage) increases the rate or severity of the beam damage as this increases the magnitude of the electric field produced [16, 38-40].

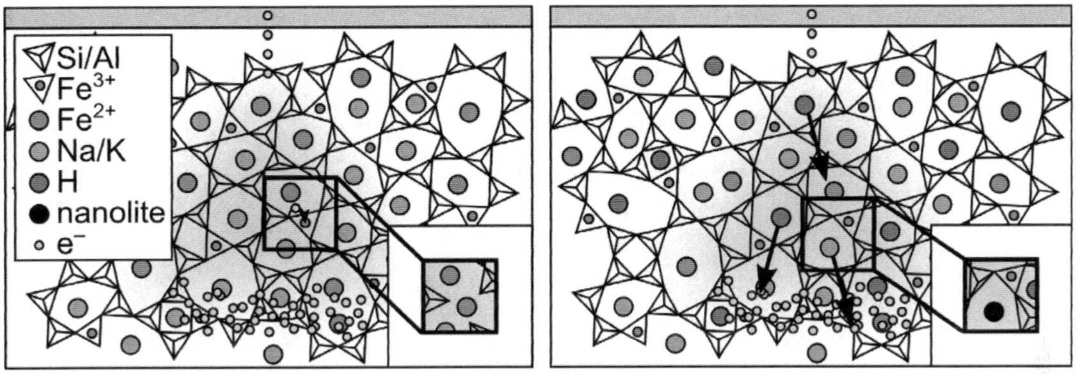

Figure 5. Schematic diagram of the processes occurring during electron beam irradiation in an electron probe. The silicate glass is carbon-coated (shaded grey) in a vacuum and contains mobile elements (e.g., H – blue circle, Na and K – green circles), variably oxidised Fe (Fe^{3+} – red circle in triangle, Fe^{2+} – red circle), and other glass components (Si and Al – triangles) (glass structure based on [41]). During electron beam irradiation, electrons (small yellow circle) are trapped building up a region of negative charge above which an electric field (yellow region) is present. a) In anhydrous silicate glass, electrons are transferred from O to Fe^{3+} causing reduction if there is sufficient Fe. b) In hydrous silicate glass, positively charged mobile ions migrate towards the negative charge at depth, which can oxidise Fe and precipitate Fe-bearing nanolites (black circle).

The electric field results in the migration of mobile cations (e.g., Na^+, H^+, K^+) towards the region of negative charge (Fig. 5b). This causes the concentration of mobile cations (e.g., H, Na, and K) to decrease with time, which can be observed using time-dependent intensity (TDI) measurements of X-rays using EPMA (e.g., decreasing K-Kα intensity; Fig. 6a) and Raman maps of the analysed area (decrease in height of H_2O peak at ~3,500 cm^{-1}, Fig. 6d) [39, 42]. These ions are known to migrate into the sample as subsequent secondary ion mass spectrometry (SIMS) depth profiles show a decrease in the concentration of mobile elements (e.g., H^+, Li^+, Na^+, and K^+) towards the surface (Fig. 6e) [38]. The exception to this is H in rhyolite, which increases in concentration towards the surface, implying OH^- (which is negative and hence would migrate away from the region of negative charge) rather than H^+ migration [38]. Immobile ions remain static and therefore their concentration increases, as observed using EPMA TDI measurements (e.g., Si, Al, Fig. 6b) [39, 42]. At cryogenic temperatures (-190 °C) the mobile cations are unable to migrate in response to the electrostatic field [40]. The rate or severity of migration increases with decreasing atomic radius of the mobile element (i.e., H > Li > Na > K), because smaller ions migrate faster, and increasing SiO_2 content of the glass (rhyolite > dacite > andesite) due to increasing polymerisation [38-40].

The migration of mobile cations leaves behind O which can oxidise Fe, hence hydrous basalts and alkali-rich glasses are able to oxidise (Fig. 6c) [16, 43]. With continued oxidation, nanolites of magnetite and even haematite can precipitate, as observed using Raman spectroscopy (increase in height of magnetite peak at ~670 cm^{-1}, Fig. 6d) [16]. Additionally, secondary electron scanning electron microscopy imaging has shown changes in the surface morphology of irradiated glass, which is interpreted to result from oxygen bubbles outgassing from the sample [43]. The rate of oxidation increases with increasing H_2O concentration because the rate of diffusion of mobile cations is faster at higher H_2O concentrations [16, 38].

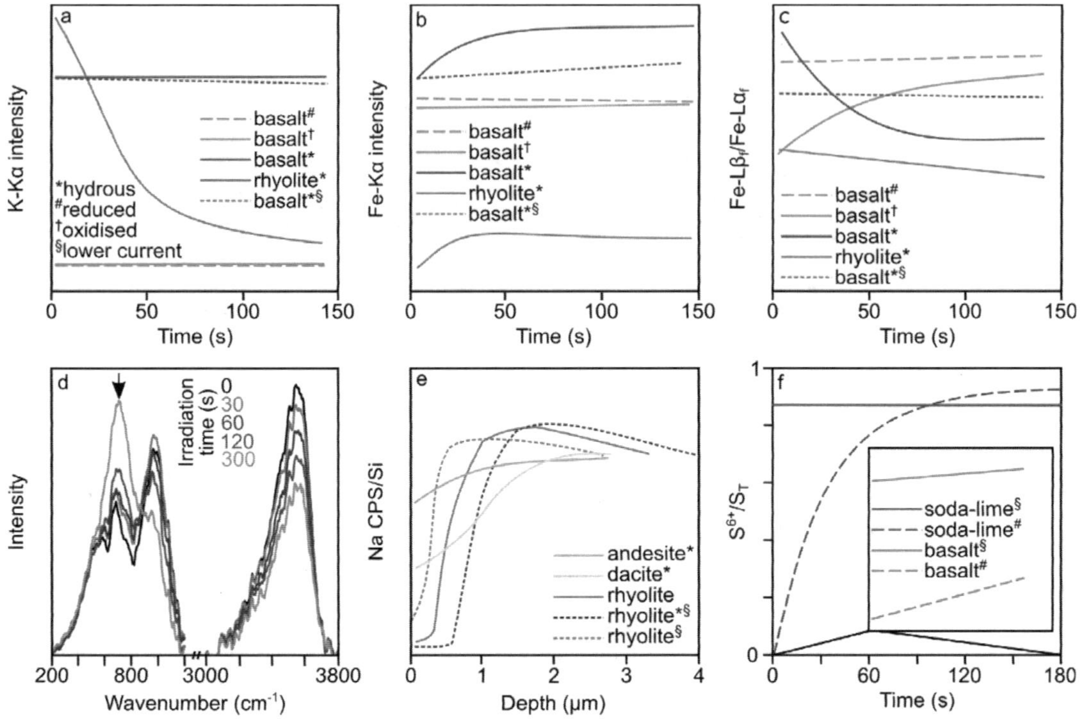

Figure 6. Changes in glass composition and structure during electron beam irradiation in an electron probe. EPMA TDI data from [16] for a) migration of a mobile element (K), b) grow-in of an immobile element (Fe), and c) changes in Fe oxidation state. d) Raman data over time for a nanolite-bearing hydrous basalt showing the silicate peak region (200 - 1,400 cm^{-1}, magnetite is the peak at ~670 cm^{-1} shown by the arrow) and H_2O peak (3,000 - 3,800 cm^{-1}). e) SIMS data from [38] for Na. f) EPMA TDI data from [21, 34] for oxidation of S. Line colour indicates glass composition (basalt = red, andesite = green, dacite = yellow, rhyolite = blue, and Fe-free soda-lime = purple), where darker hues indicate the glass is hydrous*; dashed line indicates reduced# glass, solid line indicates oxidised† glass, and dotted line indicates lower beam current§ used for analysis.

Fe reduction likely occurs in the glass via electron transfer from O to Fe^{3+} but is only observed when the rate of oxidation is slow (Fig. 5a) [16, 44]. Anhydrous basalts can reduce because there are insufficient mobile cations (e.g., H^+) to migrate and leave behind oxygen [16]. The rate of reduction is controlled by the concentration of Fe^{3+} remaining, resulting in an exponential trend in oxidation state with time, and increases with lower initial Fe^{2+}/Fe_T (Fig. 6c) [16].

Anhydrous and hydrous rhyolites (with low and high Fe), and hydrous basalts containing nanolites, do not appear to change oxidation state using TDI EPMA under beam densities that cause damage in hydrous basaltic glass [16]. On the other hand, there was an increase in the quantity of magnetite and haematite nanolites observed using Raman after EPMA for pantelleritic and nanolite-bearing hydrous basaltic glass, which implies oxidation did occur but was too fast to be observed using EPMA [16].

No nanolites were observed for hydrous rhyolitic glass with low Fe, hence these did not oxidise [16]. Therefore, either OH^- rather H^+ migration occurred (as suggested by SIMS), which would not leave oxgyen behind for oxidation, or their Fe contents are too low for oxidation to proceed [16, 38].

The mechanism for changes in S oxidation state is probably similar to Fe, although oxidation and reduction proceed via the formation of S^{4+} as observed using XANES analysis of the S K-edge (Fig. 6f) [33]. Less data are available on the compositional controls on the rate and direction of redox change. TDI measurements of S oxidation state using EPMA show that the rate of oxidation appears to increase with decreasing Fe concentration, increasing S concentration, and decreasing initial S^{6+}/S_T [17, 21, 34, 35]. Extremely rapid oxidation has been observed for very S-rich but Fe-free glass, which implies these redox changes are coupled and that Fe more readily changes redox state than S [17, 21]. Conversely, reduction occurs for initially oxidised glasses [45].

3. X-ray absorption near-edge structure (XANES) spectroscopy

3.1. Technique

XANES is an X-ray absorption spectroscopy technique that uses changes in the pre-edge structure of the Fe K-edge (1s → 3d) that are caused by changes in the Fe concentration, coordination, and oxidation state (Fig. 3e). Broadly, with decreasing Fe^{2+}/Fe_T, the edge jump (~7,120 eV) and crest move to higher energies and the higher energy pre-edge feature increases (and the lower energy pre-edge feature decreases) in intensity [18]. There are a variety of ways of quantifying the Fe oxidation state from XANES spectra, such as the centroid position or height ratio of the pre-edge features (after fitting the spectra with a linear function and damped harmonic oscillator) and principal component analysis [18, 46, 47]. Very high precision on Fe^{2+}/Fe_T measurements (± 0.0045 - 0.013 absolute [18, 47]) can be obtained using well-characterised standards, but data processing must be consistent between different datasets for robust comparison. Additional complexity, such as contamination from the host-mineral, crystallinity within the glass, or differences in the matrix glass composition, can severely compromise oxidation state quantification.

Similarly, the S oxidation state can be quantified using the S K-edge. In addition to S^{2-} and S^{6+}, S^{4+} that is produced by beam damage can also be detected (Fig. 3f) [2, 17]. The proportion of each species can be quantified by comparing to reference spectra, resulting in ± 5 % total error on S^{6+}/S_T [21].

3.2. Beam damage

Fewer studies have investigated beam damage during XANES compared to EPMA, and hence less is known about the compositional controls. Data described below are for typical beam conditions, which vary depending on the synchrotron used for data collection (10^3 - 10^{11} photons·s^{-1}·μm^{-2} flux). Time-dependent measurements at a single energy (e.g., the second pre-edge multiplet) or repeated measurements on the same spot during XANES can be used to monitor changes in Fe or S oxidation state. The mechanisms for beam damage during XANES are different to EPMA, resulting from the energy transferred from the photon beam to the glass rather than in response to an electric field (Fig. 7).

Reduction of Fe is thought to occur by charge transfer from localised defects in the glass to neighbouring Fe^{3+}, causing reduction to Fe^{2+} (Fig. 7a) [48]. The region of localised defects created by the high energy photon beam causes cation displacement [48]. The resulting anionic vacancies produce visible grey marks in the glass after irradiation [48]. This was inferred by Gonçalves Ferreira *et al.* [48] as silica-rich (soda-lime) glass with low Fe concentrations (< 5,000 ppm Fe) reduced over time with no change in Fe coordination (Fig. 8a). The rate of reduction increases with decreasing Fe concentration, but the absolute amount of Fe reduced is the same regardless of the total Fe content (Fig. 8a). At elevated temperatures these defects relax, therefore the reduction of Fe can be reversed by heating the sample and no Fe reduction occurs when the measurements were made at higher temperatures (~500 °C) [48].

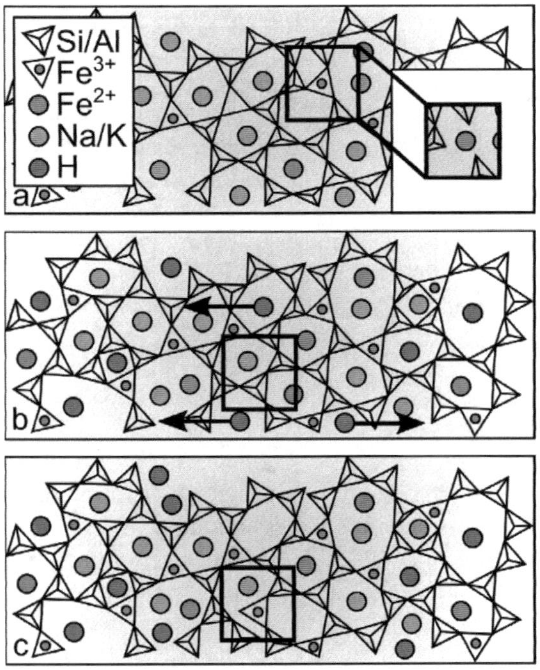

Figure 7. Schematic diagram of the processes occurring during photon beam irradiation in a synchrotron. The silicate glass sample contains H (blue circle), variably oxidised Fe (Fe^{3+} – red circle in triangle, Fe^{2+} – red circle), and other glass components (Si and Al – triangle, Na and K – green circle) (glass structure based on [41]). a) The photon beam creates defects in the glass, which provides the charges to reduce Fe^{3+} to Fe^{2+}. b) The photon beam irradiation breaks O–H bonds and H migrates faster out of the analysis area than O, and c) this O oxidises Fe^{2+} to Fe^{3+}.

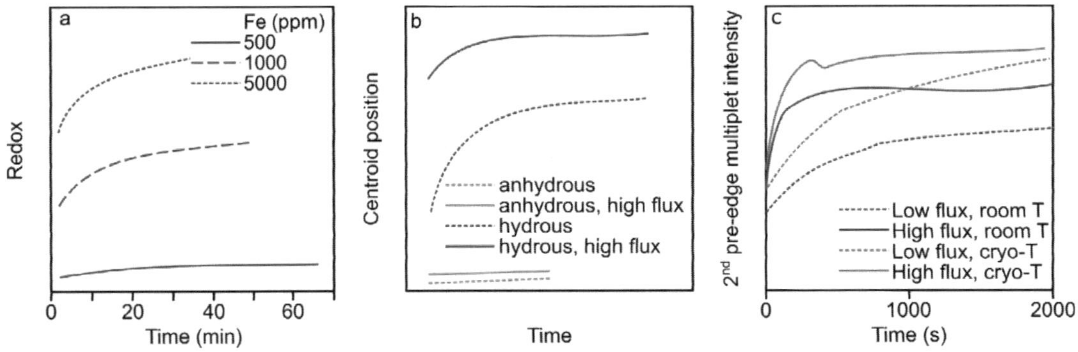

Figure 8. Changes in glass composition during photon beam irradiation in a synchrotron during XANES spectroscopy: a) Fe reduction with time from [48] for Fe-poor, soda-lime glass, b) Fe oxidation with time from [49] for anhydrous and hydrous basaltic glass at low and high photon fluxes, and c) Fe oxidation with time from [49] for hydrous basaltic glass for low and high photon fluxes at room and cryogenic temperatures.

On the other hand, anhydrous basaltic glass (i.e., high Fe concentration) shows no change in Fe oxidation state with time (Fig. 8b) [18, 46, 50]. Anhydrous and hydrous dacitic to rhyolitic glass also show no change in Fe oxidation state with time [51]. The latter was deduced from the lack of change between repeated analyses, rather than time-dependent measurements, and therefore the change may already have occurred within the first analysis [49, 51].

Cottrell *et al.* [49] investigated beam damage to hydrous basaltic glass (Fig. 8b). They observed that during analysis the Fe^{3+} multiplet increases in intensity, the Fe^{2+} multiplet decreases, and the white line shifts to higher energy, which implies oxidation of Fe^{2+} to Fe^{3+} rather than increased 3d-4p hybridisation. The mechanism proposed for Fe oxidation involves H_2O ($2FeO_{(glass)} + H_2O_{(glass)} \rightarrow Fe_2O_{3(glass)} + H_{2(gas)}$)

[21, 49]. Photon radiation breaks the O–H bonds (Fig. 7b), H then outgases from the sample leaving behind O to oxidise Fe (Fig. 7c). Oxidation was faster at higher photon fluxes, lower temperatures (cryogenic, -190 °C), higher H_2O concentrations or more reduced samples, and was permanent (Fig. 8c). Fourier transform infra-red (FTIR) spectra of the analysed area showed a reduction in H_2O concentration [49]. The increased rate of oxidation observed at cryogenic temperatures for Fe suggests that O cannot diffuse at such low temperatures and hence the oxidation rate is increased [49].

The data for S are more complicated, as different beamlines appear to produce different results, likely due to their different beam intensities [17]. Initially oxidised glass reduces over time and S^{4-} is observed, where the rate is controlled by the beam conditions and also increases with H_2O present, whereas partially or completely reduced samples do not change oxidation state [17, 33]. On the other hand, S oxidation is seen at higher intensity beamlines, especially when H_2O is present in the glass, and a similar reaction to Fe oxidation is proposed ($S^{2-}_{(glass)} + 4H_2O_{(glass)} \rightarrow SO_4{}^{2-}{}_{(glass)} + 4H_{2(gas)}$) [33]. The oxidation state of S becomes more stable with decreasing S/Fe ratio of the glass, again suggesting coupled redox changes as for EPMA [21].

4. Electron energy loss spectroscopy (EELS)

4.1. Technique
EELS in a transmission electron microscope (TEM) affords the opportunity to measure Fe oxidation state at sub-nanometre scale. The Fe L-edge (2p → 3d) has two distinct sub-edges, whose shape and position are dependent on Fe oxidation state, symmetry, and coordination state (Fig. 3g) [19]. Gaussian fits to the spectra could provide a way to quantify the Fe oxidation state, but it can prove difficult to find standards that are homogeneous at the analysis scale, which currently precludes quantification [19].

4.2. Beam damage
Burgess *et al.* [19] studied beam damage of silicate glass during EELS (beam conditions of 60 - 100 kV accelerating voltage and $0 - 7 \times 10^6$ $e \cdot nm^{-2} \cdot s^{-1}$ dose). They observed glass density changes and the concentration of most elements decreases in the analysed region, except for Si which is conserved and therefore increases in its relative proportion. Fe oxidation and reduction were observed, but there is no clear relationship between rate of redox change and sample thickness, electron dose rate, or total dose. Generally, reduced samples become more oxidised with increasing dose, whilst at the highest doses reduction occurs. Magnetite nanolites were observed in the glass, but it is not certain whether these were present before analysis. Additionally, during sample preparation and loading into the TEM, surface oxidation occurred.

These observations illustrate that the damage is caused by an electric field in the glass, but via a different mechanism than during EPMA. For EELS, as the sample is very thin, the emission of secondary and auger electrons generates a region of positive charge in the analysis area, generating an electric field (Fig. 9a) [19, 37]. Electrons can hop between Fe^{2+} and Fe^{3+}, which may cause both reduction and oxidation (Fig. 9b) [19]. The region of positive charge causes the cations to migrate out of the analysis area, reducing their concentration (Fig. 9c) [19]. Fe^{2+} can migrate faster than Fe^{3+}, hence this causes oxidation (Fig. 9d) [19]. During sample preparation, initially oxidised samples have a thinner oxidised layer as there is less driving force for oxidation [19].

5. Raman spectroscopy

5.1. Technique
Raman spectroscopy measures the shift in wavelength of monochromatic light due to inelastic scattering of the incident photons when they interact with molecular vibrations, where the electron cloud is deformable. Uniquely, Raman (in confocal mode) can focus the beam at depth on unexposed areas of glass, which means melt inclusions do not have to be exposed at the surface for analysis [52].

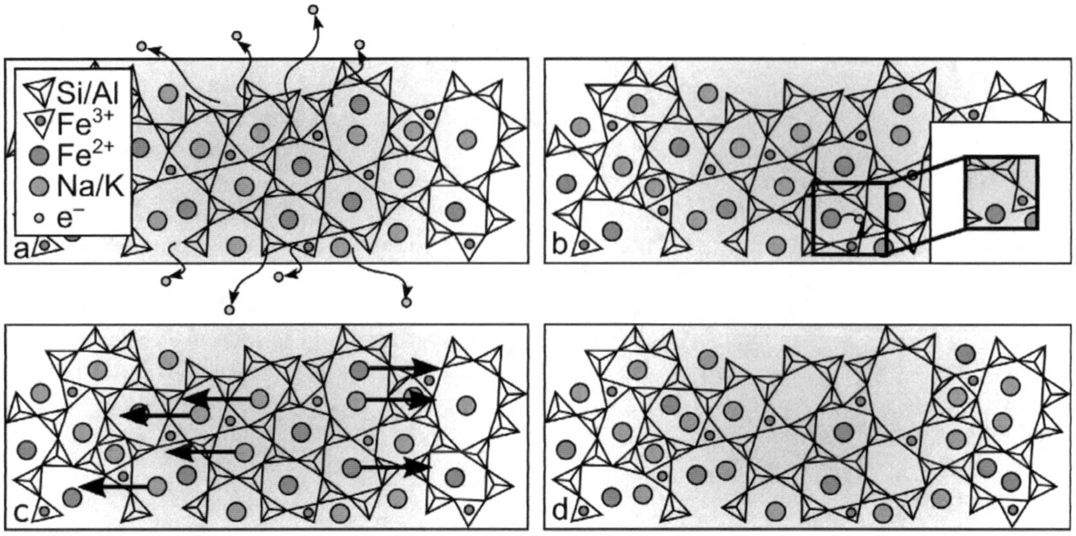

Figure 9. Schematic diagram of the processes occurring during electron beam irradiation in a TEM. The silicate glass sample is sufficiently thin to be transparent to electrons and contains variably oxidised Fe (Fe^{3+} – red circle in triangle, Fe^{2+} – red circle) and other glass components (Si and Al – triangles, Na and K – green circle) (glass structure based on [41]). a) Electron beam irradiation creates secondary electrons (small, yellow circle), which generates a region of positive charge (yellow region). b) Electrons hop between Fe^{2+} and Fe^{3+} causing changes in oxidation state. c) Most cations migrate away from the positive charge, and (d) as Fe^{2+} migrates faster than Fe^{3+} oxidation occurs.

Unfortunately, fluorescence can cause difficulties in processing spectra [21]. There are various bands sensitive to the Fe oxidation state of silicate glass, which increase in intensity with increasing Fe^{3+} content at constant Fe concentration [20, 53] (Fig. 3h). Both spectral deconvolution [53] and empirical calibrations based on ideal mixing equations [20] have been used to quantify Fe^{2+}/Fe_T. The sensitivity of the technique and calibration are composition dependent, with pantellerite being more sensitive than basalt, and the H_2O content also being important.

For sulphur, specific peaks in the glass spectra are produced due to different bonding environments associated with S such as Fe-S (372 cm^{-1}), SO_4^{2-} (990 cm^{-1}), and SH^- (2,574 cm^{-1}) (Fig. 3i) [21]. This provides greater information about the speciation of S than EPMA or XANES as information on the cation the sulphur is bonding to is also given. Unfortunately, Raman is relatively insensitive to sulphur, therefore, significant quantities of S may be required to see these peaks above the noise or other silicate-related vibrations. For Fe-free glasses (i.e., no peak at 372 cm^{-1}), the S oxidation state can be quantified by using the height of the 990 and 2,574 cm^{-1} peaks after the spectra have been deconvolved. The peak heights are assumed to be proportional to the S concentration in the glass but not using the same proportionality factor, but this can be accounted for by creating a calibration curve [21].

5.2. Beam damage

Di Genova *et al.* [54] investigated beam damage during Raman spectroscopy (maximum beam conditions of 11 mW laser power and 26 min per analysis). Beam damage causes oxidation (increase in the intensity of the band associated with Fe^{3+} structural bonds at ~970 cm^{-1}), decrease in intensity of the H_2O (OH) peak, and nanolite formation (with a new peak at ~670 cm^{-1}). The change increased with increasing acquisition time or at higher laser powers and was more prominent in alkali- or Fe-rich samples. The intensity of the photon beam is far less than for XANES, hence laser heating is thought to cause the same reaction to occur as for XANES [54].

6. Synchrotron Mössbauer source (SMS) spectroscopy

6.1. Technique

Mössbauer spectroscopy uses the recoil-free resonant absorption of gamma rays to probe changes in the energy levels of an atomic nucleus in response to the surrounding electric and magnetic environment. Gamma rays over a narrow energy range are transmitted through the sample and the resulting absorption spectra give information on the coordination and oxidation state (Fig. 3d). Careful fitting of the spectra allows quantification of the Fe oxidation state and knowledge about the number of different environments of Fe within the glass structure [55]. This is a relatively simple mathematical procedure which is not directly dependent on comparison with standards [55]. In conventional Mössbauer spectroscopy for iron (^{57}Fe), a ^{57}Co radioactive source is moved over a range of velocities to produce gamma rays over the required energy range. These gamma rays are difficult to focus and relatively weak, therefore long measurement times (days) are needed and the resulting spatial resolution is poor. SMS spectroscopy uses synchrotron radiation, which gives a focussed (down to 10 μm), high brilliance gamma source allowing measurements in minutes [56].

6.2. Beam damage

There are currently no published studies investigating beam damage on glasses during SMS. SMS and XANES both use synchrotron-generated, highly-focussed photon beams for analysis. Hence, the mechanisms of beam damage and resultant changes to the glass would likely be similar (e.g., oxidation and reduction of Fe and H_2O loss). In contrast to XANES, SMS at a beamline such as ID18 at ESRF uses ^{57}Fe resonant gamma radiation generated by passing the synchrotron beam through an iron borate crystal. This mechanism also reduces the photon flux arriving at the sample by ~8 orders of magnitude, which negates the chance of beam damage that would result from XANES (pers. comm. Valerio Cerantola). For instance, the ID18 beamline at ESRF is reduced to a photon flux of 2×10^4 photons s^{-1} by the time it arrives at the sample. This is four orders of magnitude less than that the maximum flux observed to avoid beam damage on hydrous glass (~10^8 photons s^{-1}) at the I18 beamline at the Diamond Light Source [47, 56]. If higher photon fluxes are used for SMS, then similar problems to XANES might be expected.

7. Mitigation strategies

There are two approaches to mitigating beam damage: (1) change the analytical conditions such that beam damage is minimised or (2) modify data collection and processing techniques such that the analyses correct for any beam damage effects. In all cases, beam damage is more severe using higher beam densities (e.g., smaller beam size, higher beam current for electrons, higher intensity for photons). Hence, using lower beam densities reduces beam damage, but this sacrifices spatial resolution, increases analytical time, and/or decreases precision. In some cases this is acceptable, for instance using <5 mW laser power and <3 minutes analysis time for Raman spectroscopy can still produce good results with modern detectors [54]. For XANES, the photon flux must be heavily attenuated to eliminate beam damage in hydrous glasses, but reasonable errors on Fe oxidation state have been achieved by doing this [47]. For EPMA, mobile major elements are often analysed for short-times at the start of the analysis, when they have had little time to migrate, as their concentration is sufficiently high that precision is not compromised (e.g., [15]). Using a cold stage during EPMA can eliminate element migration [40], but has not been proven to prevent redox changes and has been shown to exacerbate these problems during XANES [48, 49].

If the glass sample size permits, the sample can be moved around constantly to analyse undamaged areas, an approach that has been used for Fe and S oxidation states [26, 34, 35]. Alternatively, TDI measurements can be extrapolated to time zero prior to beam damage, but this requires knowing what will change and how it will change. This approach has been applied to mobile element concentrations and Fe oxidation state determinations using EPMA and could perhaps be applied to XANES in the future [16, 42].

As beam damage results in changes to glass composition and structure, the order of analyses is critical to ensure reliable results. Ideally, the least damaging technique would be used but instrument access may not allow this. Raman can be used in such a way that beam damage does not occur, and hence can be used first even before polishing to reveal melt inclusions at the sample surface. H_2O easily migrates during EPMA and XANES, hence techniques such as FTIR and SIMS should be applied next as subsequent concentration and isotope measurements could be compromised. Although SMS and XANES are less damaging measurement techniques, the need for a thin, doubly polished wafer for analysis (not for all beamlines in the case of XANES) can make subsequent analyses challenging. Oxidation state measurements using EPMA should be carried out last (after standard EPMA), due to the high beam density required for analysis.

8. Conclusions

EPMA, XANES, Raman, and EELS can all cause reduction or oxidation of Fe and S in silicate glass. The rate and direction of redox change depends on the primary beam conditions (e.g., beam type, accelerating voltage, beam current, photon flux, laser power, etc.) and measurement time, but also on glass composition (e.g., Si, Fe, and H_2O). For EPMA alkali-poor anhydrous glass reduces, whilst alkali-rich or hydrous glass oxidises. Conversely, for XANES Fe-poor glass reduces, whilst Fe-rich glass oxidises. Broadly, increased H_2O, alkalis, or Fe increases the rate of redox change with both electron and photon beams. Oxidation is caused by cation migration in response to: electric field formation due to electron addition (EPMA) or removal (EELS), photon-induced bond breaking (XANES), or sample heating (Raman). The beam conditions for SMS appear to be too low to cause beam damage. More data are required to understand Fe redox changes in EELS, Raman, and SMS; and S redox changes in EPMA and XANES. The simplest way to mitigate beam damage is to reduce beam density during analysis, although moving the sample and time-dependent intensity methods can also be used. Careful consideration of analytical sequence and conditions can allow the glass to be analysed multiple times using a variety of techniques to fully characterise silicate glass samples.

Acknowledgements
We would like to thank Valerio Cerantola and Alexandr Chumakov for their comments on the SMS section, Jon Wade for his helpful review that greatly improved the manuscript, and Bjørn Sørensen for his editorial handling.

References

[1] Kress V C and Carmichael I S E 1991 The compressibility of silicate liquids containing Fe_2O_3 and the effect of composition, temperature, oxygen fugacity and pressure on their redox states. *Contrib. Mineral. Petrol.* **108** 82-92
[2] Jugo P J, Wilke M and Botcharnikov R E 2010 Sulfur K-edge XANES analysis of natural and synthetic basaltic glasses: Implications for S speciation and S content as function of oxygen fugacity . *Geochim. Cosmochim. Acta* **74** 5926-5938
[3] Wilke M 2005 Fe in magma - An overview. *Ann. Geophys.* **48** 609-617
[4] Vicenzi E P, Green T and Sie S 1994 Effect of oxygen fugacity on trace-element partitioning between immiscible silicate melts at atmospheric pressure: A proton and electron microprobe study. *Chem. Geol.* **117** 355-360
[5] Tattitch B and Blundy J D 2017 Partitioning of Cu and Mo between felsic melts and saline magmatic fluids: influence of salinity, fO_2 and fS_2. *Appl. Earth Sci.* **126**. 99
[6] Cukierman M and Uhlmann D R 1974 Effects of iron oxidation state on viscosity, lunar composition 15555. *J. Geophys. Res.* **79** 1594-1598
[7] Dingwell D B and Virgo D 1987 The effect of oxidation state on the viscosity of melts in the system Na_2O-FeO-Fe_2O_3-SiO_2. *Geochim. Cosmochim. Acta* **51** 195-205
[8] Hamilton D L, Burnham C W and Osborn E F 1964 The solubility of water and effects of oxygen fugacity and water content on crystallization in mafic magmas. *J. Petrol.* **5** 21-39

[9] Carmichael I S E and Ghiorso M S 1990 The effect of oxygen fugacity on the redox state of natural liquids and their crystallizing phases. *Rev. Mineral. Geochem.* **24** 191-212
[10] Bouhifd M A, Richet P, Besson P, Roskosz M and Ingrin J 2004 Redox state, microstructure and viscosity of a partially crystallized basalt melt. *Earth Planet. Sci. Lett.* **218** 31-44
[11] Kolzenburg S, Di Genova D, Giordano D, Hess K U and Dingwell D B 2018 The effect of oxygen fugacity on the rheological evolution of crystallizing basaltic melts. *Earth Planet. Sci. Lett.* **487** 21-32
[12] Humphreys M C S, Brooker R A, Fraser D G, Burgisser A, Mangan M T and McCammon C 2015 Coupled interactions between volatile activity and Fe oxidation state during arc crustal processes. *J. Petrol.* **56** 795-814
[13] Kelley K A and Cottrell E 2009 Water and the oxidation state of subduction zone magmas. *Science* **325** 605-607
[14] Roedder E 1979 Origin and significance of magmatic inclusions. *Bull. Mineral.* **102** 487-510
[15] Blundy J D and Cashman K V 2008 Petrologic reconstruction of magmatic system variables and processes. *Rev. Mineral. Geochem.* **69** 179-239
[16] Hughes E C, *et al.* 2018 High spatial resolution analysis of the Iron oxidation state in silicate glasses using the electron probe. *Amer. Mineralogist* **103** 1473-1486
[17] Wilke M, Klimm K and Kohn S C 2011 Spectroscopic studies on sulfur speciation in synthetic and natural glasses. *Rev. Mineral. Geochem.* **73** 41-78
[18] Cottrell E, Kelley K A, Lanzirotti A and Fischer R A 2009 High-precision determination of iron oxidation state in silicate glasses using XANES. *Chem. Geol.* **268** 167-179
[19] Burgess K D, Stroud R M, Dyar M D and McCanta M C 2016 Submicrometer-scale spatial heterogeneity in silicate glasses using aberration-corrected scanning transmission electron microscopy. *Amer. Mineraogist* **101** 2677-2688
[20] Di Genova D, Hess K U, Chevrel M O and Dingwell D B 2016 Models for the estimation of Fe^{3+}/Fe_{tot} ratio in terrestrial and extraterrestrial alkali- and iron-rich silicate glasses using Raman spectroscopy. *Amer. Mineralogist* **101** 943-952
[21] Klimm K, Kohn S C, O'Dell L A, Botcharnikov R E and Smith M E 2012 The dissolution mechanism of sulphur in hydrous silicate melts. I: Assessment of analytical techniques in determining the sulphur speciation in iron-free to iron-poor glasses. *Chem. Geol.* **322-323** 237-249
[22] de Groot F 2001 High-resolution X-ray emission and X-ray absorption spectroscopy. *Chem. Rev.* **101** 1779-1808
[23] Hofer H E and Brey G P 2007 The iron oxidation state of garnet by electron microprobe: Its determination with the flank method combined with major-element analysis. *Amer. Mineralogist* **92** 873-885
[24] Smith D G W and O'Nions R K 1971 Investigations of the $L_{II,III}$ X-ray emission spectra of Fe by the electron microprobe Part I: Some aspects of the Fe $L_{II,III}$ spectra from metallic iron and haematite. *J. Phys. D: Appl. Phys.* **4** 320
[25] Gopon P, Fournelle J, Sobol P E and Llovet X 2013 Low-voltage electron-probe microanalysis of Fe-Si compounds using soft X-rays. *Microsc. Microanal.* **19** 1698-1708
[26] Zhang C, *et al.* 2018 Electron microprobe technique for the determination of iron oxidation state in silicate glasses. *Amer. Mineralogist* **103** 1445-1454
[27] Hofer H E, Brey G P, Schulz-Dobrick B and Oberhansli R 1994 The determination of the oxidation state of iron by the electron microprobe. *Eur. J. Mineral.* **6** 407-418
[28] Fialin M, Bézos A, Wagner C and Humler E 2004 Quantitative electron microprobe analysis of $Fe^{3+}/\Sigma Fe$: Basic concepts and experimental protocol for glasses. *Amer. Mineralogist* **89** 654-662
[29] Hofer H E 2002 Quantification of Fe^{2+}/Fe^{3+} by electron microprobe analysis - New developments. *Mössbauer Spectrosc.* **144-145** 239-248

[30] Di Genova D, Caracciolo A and Kolzenburg S 2018 Measuring the degree of 'nanotilization' of volcanic glasses: Understanding syn-eruptive processes recorded in melt inclusions. *Lithos* **318-319** 209-218
[31] Carroll M R and Rutherford M J 1998 Sulfur speciation in hydrous experimental glasses of varying oxidation state: Results from measured wavelength shifts of sulfur X-rays. *Amer. Mineralogist* **73** 845-849
[32] Wallace P J and Carmichael I S E 1992 Sulfur in basaltic magmas. *Geochim. Cosmochim. Acta* **56** 1863-1874
[33] Wilke M, *et al.* 2008 The origin of S^{4+} detected in silicate glasses by XANES. *Amer. Mineralogist* **93** 235-240
[34] Rowe M C, Kent A J R and Nielsen R L 2007 Determination of sulfur speciation and oxidation state of olivine hosted melt inclusions. *Chem. Geol.* **236** 303-322
[35] Métrich N and Clocchiatti R 1996 Sulfur abundance and its speciation in oxidized alkaline melts. *Geochim. Cosmochim. Acta* **60** 4151-4160
[36] Bonnelle C 2004 Charge trapping in dielectrics. *Microsc. Microanal.* **10** 691-696
[37] Cazaux J 1996 Electron probe microanalysis of insulating materials: Quantification problems and some possible solutions. *X-ray Spectrom.* **25** 265-280
[38] Humphreys M C S, Kearns S L and Blundy J D 2006 SIMS investigation of electron-beam damage to hydrous, rhyolitic glasses: Implications for melt inclusion analysis. *Amer. Mineralogist* **91** 667-679
[39] Morgan G B and London D 2005 Effect of current density on the electron microprobe analysis of alkali aluminosilicate glasses. *Amer. Mineralogist* **90** 1131-1138
[40] Kearns S L, Steen N and Erlund E 2002 Electron probe microanalysis of volcanic glass at cryogenic temperatures. *Microsc. Microanal.* **8** (Suppl. 2) 1562-1563
[41] Henderson G S, Calas G and Stebbins J F 2006 The structure of silicate glasses and melts. *Elements* **2** 269-273
[42] Nielsen C H and Sigurdsson H 1981 Quantitative methods for electron microprobe analysis of sodium in natural and synthetic glasses. *Amer. Mineralogist* **66** 547-552
[43] Fialin M and Wagner C 2012 Redox kinetics of iron in alkali silicate glasses exposed to ionizing beams: Examples with the electron microprobe. *J. Non. Cryst. Solids* **358** 1617-1623
[44] Nishida T 1995 Mössbauer effect in inorganic glasses. *Hyperfine Interact.* **95** 23-39
[45] Jugo P J , Luth R W and Richards J P 2005 Experimental data on the speciation of sulfur as a function of oxygen fugacity in basaltic melts. *Geochim. Cosmochim. Acta* **69** 497-503
[46] Shorttle O, Moussallam Y, Hartley M E, Maclennan J, Edmonds M and Murton B J 2015 Fe-XANES analyses of Reykjanes Ridge basalts: Implications for oceanic crust's role in the solid Earth oxygen cycle. *Earth Planet. Sci. Lett.* **427** 272-285
[47] Moussallam Y, *et al.* 2019 Mantle plumes are oxidised. *Earth Planet. Sci. Lett.* **527** 115798
[48] Gonçalves Ferreira P, de Ligny D, Lazzari O, Jean A, Cintora Gonzalez O and Neuville D R 2013 Photoreduction of iron by a synchrotron X-ray beam in low iron content soda-lime silicate glasses. *Chem. Geol.* **346** 106-112
[49] Cottrell E, *et al.* 2018 A Mössbauer-based XANES calibration for hydrous basalt glasses reveals radiation-induced oxidation of Fe. *Amer. Mineralogist* **103** 489-501
[50] Moussallam Y, *et al.* 2014 Tracking the changing oxidation state of Erebus magmas, from mantle to surface, driven by magma ascent and degassing. *Earth Planet. Sci. Lett.* **393** 200-209
[51] Fiege A, *et al.* 2017 Calibration of Fe XANES for high-precision determination of Fe oxidation state in glasses: Comparison of new and existing results obtained at different synchrotron radiation sources. *Amer. Mineralogist* **102** 369-380
[52] Thomas R 2000 Determination of water contents of granite melt inclusions by confocal laser Raman microprobe spectroscopy. *Amer. Mineralogist* **85** 868-872

[53] Di Muro A, Métrich N, Mercier M, Giordano D, Massare D and Montagnac G 2009 Micro-Raman determination of iron redox state in dry natural glasses: Application to peralkaline rhyolites and basalts. *Chem. Geol.* **259** 78-88

[54] Di Genova D, Sicola S, Romano C, Vona A, Fanara S and Spina L 2017 Effect of iron and nanolites on Raman spectra of volcanic glasses: A reassessment of existing strategies to estimate the water content. *Chem. Geol.* **475** 76-86

[55] Mysen B O, Carmichael I S E and Virgo D 1985 A comparison of iron redox ratios in silicate glasses determined by wet-chemical and ^{57}Fe Mössbauer resonant absorption methods. *Contrib. Mineral. Petrol.* **90** 101-106

[56] Potapkin V, *et al.* 2012 The ^{57}Fe synchrotron Mössbauer source at the ESRF. *J. Synchrotron Radiat.* **19** 559-569

Using complementary microanalytical techniques to analyse diamond anvil cell experiments

E S Jennings

Department of Earth and Planetary Sciences, Birkbeck, University of London, Malet Street, London WC1E 7HX, Great Britain

E-mail: e.jennings@bbk.ac.uk

Abstract. Diamond anvil cell (DAC) experiments are being used with increasing frequency to examine deep planetary processes. A variety of analytical techniques are available to characterise the products of high-pressure, high-temperature DAC experiments. Recent developments in both sample preparation and analytical techniques allow in-situ measurements made during experiments (typically synchrotron-source X-ray techniques) to be supplemented and supported by ex-situ measurements made of the extracted experimental sample after the experiment. The combination of complementary microanalytical techniques facilitates the verification of analytical results and allows additional information to be obtained from these technically-challenging experiments. Techniques with differing spatial resolutions can also be effectively used together. This contribution describes some techniques used to analyse the products of DAC experiments, and discusses example case studies from the Earth and planetary sciences where the combination of several complementary techniques has resulted in important additional insight into the interpretation of DAC experimental results.

1. Introduction

Diamond anvil cell (DAC) experiments are used to access the extremely high pressure and temperature conditions that are present in deep planetary interiors. These pressures are achieved by compressing samples between the truncated tips of a pair of diamonds, and very high temperatures can be reached by irradiating the sample with powerful infrared lasers [1]. DAC experiments are incredibly versatile: not only can a huge range of physical conditions be accessed, but the experimental products can also be analysed by a wide variety of techniques. In-situ measurements can be made through the diamonds, for example using X-ray methods (X-ray fluorescence, XRF; X-ray diffraction, XRD; imaging), FTIR (Fourier-transform infrared spectroscopy) and Raman spectroscopy. Following a high-PT experiment the sample can be extracted for additional measurements e.g., electron microbeam techniques. Multiple analytical techniques can be used in a complementary way to fully interpret the experiments. For example, in-situ XRD structural measurements of high-PT phases can be interpreted with the aid of follow-up electron beam imaging and compositional measurements.

In this contribution, I begin by presenting an overview of DAC experiments. This is followed by a brief review of a selection of techniques that are used to analyse DAC experiments. I then present a series of case studies that highlight the benefits of using multiple complementary analytical techniques to interpret DAC experiments. Examples here have a focus on Earth and planetary sciences, where DAC experiments have become a popular technique to answer a wide variety of questions, ranging from phase relations and melting in the deep Earth through to mineral physical measurements that inform the quantitative analysis of geophysical measurements of planetary interiors.

2. Experiments

To generate the pressure conditions of planetary interiors in DAC experiments, samples of a material of interest are placed between the truncated tips (culets) of two gem-quality diamonds (Fig. 1). The diamonds are then mechanically forced together, e.g., by manually tightening screws that connect diamond backing plates, or by pressurising a membrane [1]. The force is focussed over a tiny area (around 0.01 to 0.1 mm^2, depending on culet diameter), generating static pressures from a few GPa up to several hundred GPa (dependent on geometry), with maximum pressures far in excess of those accessible through larger-volume techniques such as the multi-anvil press. The diamonds are mounted in such a way that light can pass into, and out of, the sample, i.e., with a cut-out in the backing plate. Pressures of up to ~ 80 GPa are routinely used; higher pressures still can be reached with very tiny truncations or through specialist tip geometries (e.g., [2]). The culets are often 100 to 300 μm in diameter, with a significantly smaller sample (e.g., 50 to 100 μm) contained within a bored-out chamber in a metal gasket (e.g., Fig. 1).

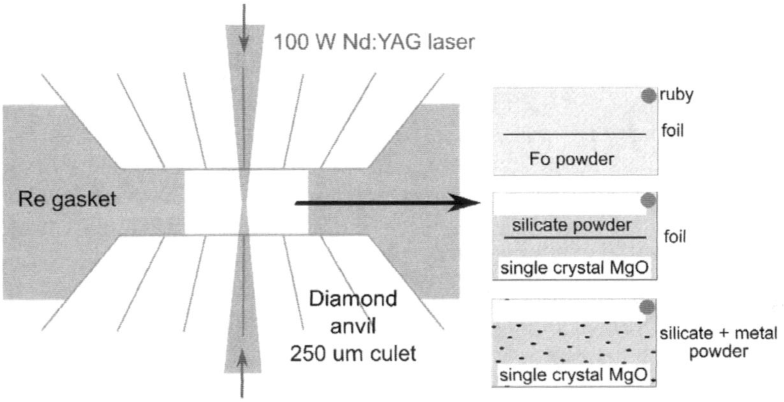

Figure 1. Example of a DAC experimental setup, designed to examine elemental distribution between molten metal and silicate (the result of such an experiment is shown in Fig. 2). Three possible sample configurations are shown on the right-hand side (not to scale). Note the ruby chip for pressure measurements and the MgO disk to insulate the diamond from the hot spot.

The pressure reached can be measured through the diamond using a range of in-situ microanalytical techniques. Often, a sphere or chip of ruby is included within the sample chamber. The ruby is fluoresced by laser, and the wavelength of the emission lines shifts with pressure [3]. Alternatively, pressure can be measured by making confocal Raman spectroscopy measurements of the diamond culets themselves (e.g., [4]), where the diamond singlet peak shifts in response to stress in the (001) direction. Pressure determinations can also be made during experiments by making XRD measurements of materials with well-characterised equations of state such as Au, MgO or NaCl.

DAC experiments can be heated restively using electrical parts to temperatures of up to around 1,000 K, although they are more commonly heated by infrared laser in order to reach the much higher temperatures of deep planetary interiors (temperatures of up to 5,000 K are sometimes reported). The laser-heated spot will usually be focussed on just a small area within the experimental sample of some 10's of μm diameter. Samples may need to be doped with a metal that can interact efficiently with the

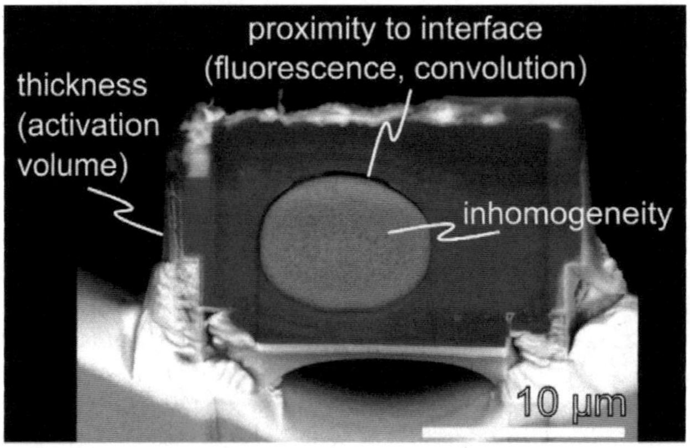

Figure 2. 15 kV BSE image showing the product of a laser-heated DAC experiment, with analytical difficulties highlighted. The central ~8 µm wide paler sphere is the metal, and the surrounding darker material is silicate glass. Metal and silicate were heated together, forming two immiscible liquids. Their compositions can be measured after the experiment is quenched, extracted and prepared for analysis by focussed ion beam (FIB), with an aim to define the distribution of various elements between the two phases at various PT conditions

laser. Temperature is determined in-situ through spectro-radiometric analysis of the thermal emission spectrum over a 1D or 2D profile across the sample. Large temperature gradients result from the small laser spot and spatial variations in laser coupling efficiency, which, combined with assumptions made about emissivity, result in significant temperature uncertainties of hundreds of degrees. When resistive heating is used for lower temperatures, temperature can be measured more precisely with a thermocouple (e.g., [5]).

Different structural and chemical measurements can be made using a range of techniques, and complementary information can be gained by combining multiple techniques. In-situ measurements are made whilst the experiment is at high pressure, and sometimes temperature, contained between the diamonds. Such measurements include XRD, XRF and X-ray imaging because X-rays are energetic enough to penetrate the diamonds, although high-temperature measurements are significantly more technically challenging than low-temperature ones (reviewed by [6]). A synchrotron light source is usually required, and because of the difficult geometries associated with laser heating and X-ray pathways, only specialised beamlines can make measurements during laser heating [7].

Following an experiment, the sample can be carefully removed and prepared for a wider range of microbeam measurements such as electron beam methods, nanoSIMS (secondary ion mass spectrometry), and atom probe measurements, in addition to X-ray techniques (e.g., Fig. 2). Such measurements can corroborate or complement in-situ measurements: for example, combining in-situ XRD and ex-situ compositional measurements can confirm phase identities, P and T conditions, and then identify the compositions of the phases and their mixing behaviours.

DAC experiments have a wide range of applications in Earth Sciences and beyond. They are used to examine phase relations within the deep silicate Earth, including its melting behaviour, composition and physical properties (e.g., [4, 8-10]). They are also used to examine metallic systems (e.g., [11-13]) and metal-silicate mixtures [12, 14-16]; Figs. 1 and 2) in order to determine the composition and properties of planetary cores and to investigate their formation conditions. Rapid quenching by switching off the laser means that equilibrium compositions of phases at high temperature are effectively "frozen in" to the experimental products.

Because of the small dimensions of the heated experimental products, there are some unique difficulties, which must be considered when analysing DAC experiments. Some techniques will be used close to their analytical limits, and care must be taken to mitigate against analytical artefacts. Effects from fluorescence, thickness and small-scale inhomogeneity must be considered (Fig. 2).

3. Analysing diamond anvil cell experiments: methods and limitations

In this section, a range of micro-analytical techniques that have been used, or can be used, to analyse DAC experimental products are reviewed. These focus specifically on applications suitable for the Earth and planetary sciences.

3.1. In-situ analytical techniques

3.1.1. Synchrotron X-ray diffraction (XRD). X-ray diffraction measurements are made to characterise the lattice structure and unit cell parameters in crystalline materials, including the ways in which they change with pressure and temperature. It is widely used in Earth and planetary sciences, and is a suitable in-situ technique for identifying and characterising materials in DAC experiments. X-ray diffraction measurements can be made using a lab X-ray source, but the far brighter X-rays generated by a synchrotron light source are preferable given that a proportion of the direct beam is absorbed by various DAC components other than the sample. The shorter counting times afforded by a synchrotron source allow multiple and near-continuous in-situ measurements to be made of DAC experiments, e.g., through heating and/or pressure cycles. A narrow beam will be required such that X-rays only diffract through the portion of the sample corresponding to the laser-heated spot. Typically, a diffracted monochromatic beam is detected by 2D imaging plates.

DAC experiments from which XRD measurements are made usually focus on solid materials, e.g., equation of state characterisation, synthesis experiments, and experiments to examine phase relations. Samples are usually contained within a chemically inert pressure medium (e.g., a noble gas or salt) within the sample chamber, which transmits quasi-hydrostatic pressure: measurements of this pressure medium can be used to determine experimental pressure from material volumes. In some cases, the powdered sample itself is also the pressure medium.

Single-crystal XRD measurements can be made in the DAC: technological developments are reviewed in detail by Boffa Ballaran *et al.* [5]. DACs designed with wide opening angles on the backing plates can be used to maximise the angular range of detected X-rays. Laser heating of single crystals in DAC experiments is not usually possible due to geometric constraints (but see [5] for developments), although resistive heating allows single crystal XRD conditions at elevated temperatures. Examples of applications include the characterisation of newly synthesised phases and quantification of compression behaviour (e.g., [17-19]).

Powder XRD is frequently used to examine phase relations, equation of state and other aspects of high-PT behaviour of various materials (e.g., [20]). Some specialised beamlines allow the simultaneous laser heating and X-ray analysis of DACs, meaning that phase assemblages can be characterised during heating cycles [7]. XRD can also be performed ex-situ after the experiment at many beamlines, where experimental samples are extracted from the DAC and measured directly. Assuming that the high-pressure phases were preserved on quench and depressurisation, they can still be identified.

There are many examples of application, particularly in the Earth Sciences, where phase relationships, including melting behaviour, are examined in materials representing the deep Earth's interior. The phase identities, density profiles and melting curves of pyrolytic mantle, subducted oceanic crust, and simple analogues shed light on the formation and evolution of the deep Earth and inform the interpretation of seismic data [21-25]. Phase relations in metals at planetary core conditions inform models of the composition and physical conditions of inner core crystallisation [11, 13].

Finally, developments in experimental design and data processing have allowed the quantification of the densities, local structure, and compression mechanisms of amorphous materials, especially liquids, in the DAC from direct measurements of X-ray diffuse scattering. Materials range from silicate melts [26] to molten metals [27], with application to Earth's deep interior. XRD measurements are frequently combined with other techniques, such that phase identifications and physical properties are complemented by fully quantified compositions and spatial context (section 4).

3.1.2. Synchrotron micro-X-ray fluorescence (μ-XRF). In XRF, the fluorescence of characteristic X-rays from a material is induced by irradiation with high-energy primary X-rays. Similarly to electron beam methods, the energy spectrum of these X-rays is used to quantify the elemental composition of a material. The X-ray beam can be focussed to a small (micrometre or sub-micrometre scale) spot. Synchrotron radiation provides the high flux typically required for the high spatial resolution and sensitivity needed. Detection limits can be lower than for EPMA analysis. As with XRD, high-energy X-rays can penetrate through diamond, allowing XRF to be used for in-situ measurements in the DAC at high-PT conditions: This is particularly powerful for the measurement of fluids, as quenching is not required [28, 29].

Matrix corrections must be made for X-ray absorption by the sample, and secondary effects occur such as the absorption of fluoresced X-rays and subsequent emission of additional secondary X-rays (known as enhancement). The volume from which X-rays can be generated through enhancement is potentially much larger than the focussed beam spot size. This should be considered in the context of measuring small or heterogeneous materials in the DAC. Additionally, the interaction volume for each X-ray is dependent on its characteristic energy: this makes the resolution of an XRF map dependent on the energy of the X-ray and on complicated matrix corrections.

μ-XRF has been used for the in-situ measurement of geologically-relevant samples in DAC experiments where there is a requirement to physically retain a fluid phase at high-PT conditions, e.g., mineral solubility studies [28, 30]. Andrault *et al.* [31] used μ-XRF to make elemental maps with ~ 1 μm spatial resolution of a quenched DAC experiment to examine partitioning of FeO between solid phases and liquid at lower mantle conditions. μ-XRF has also been pioneered in the analysis of quenched liquid metal – liquid silicate DAC partitioning experiments, where detection limits of 2 to 5 ppm allow the analysis of strongly siderophile (affinity for iron) or lithophile (affinity for silicate/rock) elements which have trace concentrations in one phase [32].

3.2. Ex-situ analytical techniques

3.2.1. Focussed ion beam (FIB) sample preparation. DAC experiments can be prepared for microbeam analysis by FIB, Ar ion milling, or by conventional polishing. FIB gives fine control over the sample site and shows 3D context because the lamella can be imaged front and back [33]. For EPMA analysis, a lamella of at least several μm thickness is required – this can be subsequently thinned for TEM analysis. A dual-beam instrument allows simultaneous imaging with an electron beam and milling with a Ga^+-beam. The process of cutting the site, extracting the sample, welding it to a Cu grid and thinning to the final geometry [33] is shown in Fig. 3, using a DAC experiment with a molten metal sampling region as an example. The lamella must be cleaned of implanted Ga+ and redeposited material using a low-power Ga^+-beam. For TEM analysis, these final stages must be carefully performed to avoid beam damage [33].

A related procedure can be used to produce an APT (atom probe tomography) needle tip: 1) a wedge-shaped chunk is cut from a sample and lifted out; 2) the end of this chunk is welded to a truncated Si needle (a post) of a sample holder; 3) an annular beam raster is used, that becomes successively smaller and lower current/voltage, until an extremely low power beam is used for the final sharpening to a ~30 nm point. A very low power beam removes traces of implanted Ga^+.

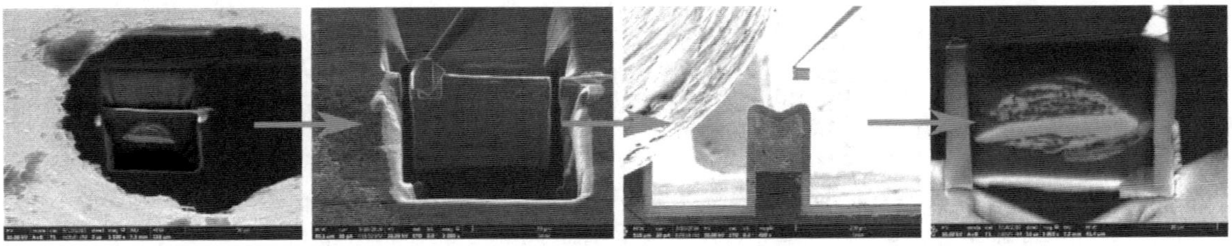

Figure 3. The process of preparing a sample as a lamella for EPMA, SEM or TEM by FIB.

3.2.2. Electron probe microanalysis (EPMA). EPMA is by far the most commonly used technique to perform non-destructive, small spot-size compositional analysis of solid samples, and is particularly relied upon in the Earth Sciences. It offers rapid, fully quantified analyses at µm-scale spatial resolutions, combined with low relative errors. A flat polished surface is required, which is traditionally achieved by mechanical polishing. Because of the small dimensions of DAC samples, a polished surface is more usually achieved through extraction of the analytical area and FIB milling.

Although EPMA is considered a routine technique, it was not originally envisaged for very small samples such as DAC samples, where multiple phases co-exist within a 10's of µm diameter area. Users may be unaware of the limitations of EPMA for tiny samples. Difficulties stem from the fact that although the beam can be focussed to a sub-micrometre diameter on a sample surface, electrons and X-rays are able to travel and spread within the sample. The volume within which electrons can travel, interact with atoms and excite X-rays (and so generate a primary signal), the interaction volume, is ~ 1 µm diameter at typical operating conditions of 15 to 20 kV. X-rays themselves can excite secondary X-rays and are less impeded than electrons, so can travel much further in a material. The generation of secondary signal from 10's to 100's µm from the analytical point thus represents a minor, but potentially important, source of analytical error. In tiny samples, thin samples, or measurements made very close to grain boundaries, such as when analysing high pressure experimental samples, these represent a limitation of the technique [34, 35].

There are many examples of EPMA used for the measurement of DAC samples. These include partitioning studies, where two or more phases are equilibrated at temperature, quenched, and analysed [15, 36-40], with many studies focussed on planetary core formation. These experiments usually involve equilibration of metals with silicates at > 30 GPa in a laser-heated DAC. The highly heterogeneous nature of these experiments, coupled with their small size, are challenging for the analyst. Subsolidus DAC experiments usually produce sub-micrometre crystals that are too small for easy EPMA characterisation.

An EPMA equipped with a field emission gun (FE-EPMA) rather than a W-filament may be more appropriate for the measurement of tiny samples, as the beam can be better focussed at low accelerating voltages (kVacc), reducing the interaction volume. For example, Tateno et al. [41] measured the composition of quenched melt in DAC melting experiments performed at lower mantle conditions using a 10 kVacc beam, and Sinmyo and Hirose [42] mapped Soret diffusion (discussed in section 4.2.5) in a DAC using this method, also at 10 kV_{acc}. However, whilst this offers benefits in spatial resolution, it presents difficulty with measuring higher energy characteristic X-rays such as Fe-Kα_1. Furthermore, instrumental effects such as carbon deposition must be minimised for analysis to be meaningful.

Finally, energy-dispersive X-ray spectrometry (EDS) measurements have also been used for measuring DAC experiments, on an SEM or EPMA instrument. For example, Chidester *et al.* [43] measured DAC experiments examining the high-PT metal-silicate partitioning of U by EDS. However, spectral resolution and detection limits are significantly better in wavelength-dispersive X-ray spectroscopy (WDS) than EDS, so WDS is preferred for fully quantitative analyses. EDS is convenient and a useful tool for elemental mapping and, when fully calibrated, can be very useful for analysing major elements.

3.2.3. Transmission electron microscope (TEM). TEM involves transmitting a focussed beam of electrons through a thin sample, and the possible range of modes of use and types of analyses are dictated by the instrumentation. Spatial resolution is at the nm- scale, so is much higher than SEM (also EPMA), allowing crystallographic defects to be imaged (for example), and crystallographic data from nm-scale domains can be obtained by electron diffraction. A very high accelerating voltage (e.g., 200 kV) is frequently used relative to SEM/EPMA (e.g., 15 kV_{acc}) for higher resolution, since it is assumed that electron scattering is negligible during the beam's journey through the sample.

A particular adaptation of TEM, STEM (Scanning TEM), involves rastering the finely focussed beam over the sample. When combined with an analytical TEM technique (e.g., EDS, electron energy-loss spectroscopy EELS), this can be used to obtain elemental and/or chemical maps of a sample. As with SEM/EPMA, a spectrum of emitted characteristic X-rays can be measured by an EDS detector: This can be used to obtain either point analyses or maps of relative elemental concentrations of a sample at a nm-scale spatial resolution. Measurement is usually standardless and requires knowledge of *k*-factors and sample thickness to calibrate the counts.

EELS relies on the fact that primary electrons that pass through the sample will lose some energy through inelastic scattering interactions with electrons from within the sample. This technique is complementary to EDS, as it is more sensitive to lower atomic number elements. Both of these techniques give better spatial resolution, but worse precision and accuracy, than WDS by EPMA, and may also be affected by fluorescence artefacts as discussed for EPMA in section 3.2.2.

In order to be analysed by TEM, samples must be electron-transparent (e.g., ~ 100's nm thick, material- and accelerating voltage-dependent). Geological samples can be prepared by FIB or by ion milling [33, 44]. Sample preparation for TEM is generally more difficult and time consuming than for EPMA/SEM.

STEM with EDS and EELS has long been used to measure elemental concentrations in many tiny natural geological samples and DAC experiments. Frost *et al.* [45] analysed DAC experiments to examine O distribution between quenched liquid iron veins and solid magnesiowüstite. Fischer *et al.* [16] analysed quenched liquid metal – liquid silicate DAC partitioning experiments using a broader spot size of 0.5 - 2 μm (to integrate fine quench textures). In both cases, STEM was used, with relative elemental concentrations determined by standardless EDS, and relative O ± C concentrations determined by EELS.

3.2.4. Atom probe tomography (APT). APT is a powerful technique, allowing the identification and 3D mapping of individual atoms in a sample (reviewed elsewhere e.g., [46]). However, it is also amongst the most difficult, expensive and time-consuming techniques in terms of sample preparation and analytical success. It is best suited to one-off measurements, and is more successful for conductive materials, which better withstand the analytical conditions. Samples are on the order of 100's nm and features at the sub-nm (atomic) scale are resolvable, so this is best suited for when extremely high spatial resolution is required. A sample must be prepared as a sharp needle tip. The evaporated atoms are ionised and identifiable by their mass/charge ratio (i.e., isotopic identification is possible). Detection efficiency is very high, and using the position and relative timing of counts, a 3D image is reconstructed.

APT is not routinely used in the Earth Sciences because insulating materials are not best suited to the high field gradients. It has, however, occasionally been applied to natural samples where nm-scale 3D resolution and compositional identification are required. Some notable examples include: i) characterising Pb nanoclusters within Earth's oldest zircon [47]; ii) the U-Pb dating of impact and crystallisation events from discrete nanostructural domains in baddeleyite [48]; iii) developments in APT are being made towards the characterisation of the C isotopic composition of pre-solar nanodiamonds (e.g., [49]). In section 4.2, an APT measurement of exsolution within quenched metal from a DAC experiment is presented.

3.2.5. Nanoscale secondary ion mass spectrometry (nanoSIMS). SIMS relies on a primary ion beam (e.g., O^- or Cs^+) that bombards the sample and sputters ionised material; these secondary ions are analysed by a mass spectrometer. It can be effective for both chemical mapping and spot analyses. NanoSIMS is a variant of SIMS optimised for nanoscopic scale spatial resolution, with theoretical resolutions of around ~100 nm [50]. Like SIMS, it has the advantage of making isotopic as well as elemental measurements. However, using a very small spot size compromises on analytical sensitivity. SIMS techniques depend on appropriate standardisation, as the ionisation efficiency of various elements is matrix-dependent, so closely matrix-matched standards are needed to convert counts to composition. In nanoSIMS, even variations in surface topography and sample morphology can cause offset relative to standards: analyses have a reputation for being difficult. SIMS/nanoSIMS require a highly polished surface for analysis.

NanoSIMS is a well-known technique for tiny samples, and is particularly used for its isotopic analysis capability: Instrumentation and examples of used are reviewed in detail elsewhere [50]. Examples include cosmochemical studies of extra-terrestrial materials such as presolar grains [51]. NanoSIMS has also been used in the chemical analysis of DAC experiments when a high spatial resolution or light element measurement is required [52]; recently, Suer *et al.* [53] quantified the sulphur content of run products in DAC metal-silicate partitioning experiments. However, the trade-off between sensitivity and analytical resolution is ever-present, whilst matrix matching of exotic materials, and in particular metals derived from very high pressure experiments, is not trivial.

4. Combined approaches

4.1. Introduction
Each technique described in section 3 can be used in isolation to analyse DAC experiments. However, when used in combination, complementary techniques are hugely powerful in extracting more information from experiments and verifying observations of these tiny samples. For example, whereas XRD provides material structural data (phase identification, structure, unit cell parameters), it cannot provide compositional data. Other techniques such as electron microbeam methods and XRF provide compositional data, but not structural. Finally, different resolutions can be accessed through different techniques, such that "bulk" measurements that are assumed to be homogeneous on the μm-scale can be further interrogated for spatial variations at the nm-scale using TEM, APT and nanoSIMS. In the following section, some case studies are highlighted where complementary techniques were required to fully interpret DAC experiments.

4.2. Case studies

4.2.1. Chemistry and mineralogy of Earth's lowermost mantle. A seismic discontinuity, known as D″, exists near the base of the Earth's mantle, a few hundred kilometres above the core-mantle boundary. It is seismically anisotropic and is characterised by a step increase in seismic velocity, although the lowermost mantle more generally is known to be seismologically complex [54]. Following the discovery of a new deep-mantle phase transition by Murakami *et al.* [23], this discontinuity is now explained by the transformation of bridgmanite to a post-perovskite structure. Post-perovskite was observed and characterised by Murakami *et al.* [23] using in-situ synchrotron XRD measurements of laser-heated DAC experiments up to 134 GPa and 2,600 K. Soon after, Murakami *et al.* [55] used similar high-PT DAC experiments on a pyrolite composition to both better characterise phase relations, and to make compositional measurements of the phases. Using in-situ XRD followed by Ar ion thinning and TEM EDS analysis, they determined that Fe is significantly less compatible in post-perovskite than in perovskite, shifting the distribution of Fe and Mg between these phases at the transition. Additionally, the authors observe that the Mg/Fe ratio in bridgmanite relative to periclase is higher when measured by electron beam methods than when inferred directly from unit cell changes measured by XRD [56], highlighting the benefit of combining multiple analytical methods to verify results.

4.2.2. Melting and crystallisation in a deep magma ocean. Melting and crystallisation are a fundamental mechanism by which planets differentiate into their component layers, where deep magma ocean crystallisation and/or melting could drive chemical fractionation. Additionally, there may be seismological evidence for the presence of low-degree melts in present day Earth's lower mantle [54]. In order to understand the compositional consequences of deep melts and their physical plausibility, high pressure experiments are needed to examine phase relations, melt compositions and melt properties.

Tateno *et al.* [41] used high-PT DAC experiments to generate melts in equilibrium with pyrolytic mantle up to 179 GPa. They analysed phases in quenched samples by FE-EPMA and additionally verified phase identifications in some samples by XRD. It was found that partial melts became increasingly Fe-rich with pressure, which was mirrored by a drop in Fe/Mg in bridgmanite and an even lower Fe-content in post-perovskite. The resultant high Fe-content predicted for low-degree melts means that they would be dense and negatively buoyant deeper than 2,000 km, i.e., they would be trapped at the base of the mantle. This conclusion contrasts that of Andrault *et al.* [31], who ran similar experiments but analysed recovered samples by creating simultaneous XRF and XRD maps, allowing them to identify the phases present in each pixel and assign XRF compositions accordingly. Andrault *et al.* [31] found liquid compositions that were lower in FeO than those of Tateno *et al.* [41]. For a more accurate perspective on mantle melting temperatures, Nomura *et al.* [57] used ex-situ X-ray micro-tomographic imaging to identify the onset of melting, where the Fe-rich nature of the melt (as confirmed by FE-EPMA) meant that an X-ray energy bracket could be selected to image Fe-rich regions as a proxy for melt. Despite differences in conclusions, it is clear that combining multiple analytical methods is a powerful approach in measuring melting experiments.

4.2.3. Linking subducted oceanic crust to ultradeep diamond genesis. Until recently, there have been no samples of subducted material from the lower mantle. However, a set of rare "ultradeep" diamonds from Brazil were found to contain unusual mineral assemblages (e.g., nepheline, kalsilite, ulvöspinel, $CaSiO_3$, pyroxenes and olivine) trapped as tiny inclusions within the crystals [25, 58]. Many inclusions contain multiple phases. Compositions were characterised using EPMA and Raman spectroscopy, and it was concluded that the polyphase inclusions represent single lower mantle phases that unmixed during decompression [25]. However, the re-integrated composite inclusions often still had unusual compositions relative to known mantle phases: For example, the Na- and K-rich compositions correspond to phases known as "calcium ferrite phase" (CF) and the hexagonal phase "NAL". These phases are not present in ambient mantle peridotite, and their presence led Walter *et al.* [25] to conclude that the composite inclusions represent crystals from deeply-subducted oceanic crust that were trapped during diamond growth. They represent the first direct evidence of whole-mantle recycling of ancient material.

Walter *et al.* [25] supported their conclusions using phase relations determined by previous high PT DAC experiments performed on basaltic compositions [24] alongside lower-P experiments. The study also resulted in further experiments being performed, to better understand the inclusions' origins (e.g., [59, 60]). In one such study, we performed laser-heated DAC experiments to examine the effect of K on stability of the unusual diamond-hosted NAL and CF phases. XRD measurements identified both phases, but in some experiments, additional K-hollandite and an unknown phase were identified: This number of phases should not have been stable from thermodynamic arguments. Follow-up STEM EDS (Fig. 4) and TEM electron diffraction measurements allowed the identification and quantification of the compositions of all phases, helping to better constrain phase boundaries. The unknown phase was probably δ-AlOOH, where the unexpected presence of water had allowed the stabilisation of an additional phase. Occasional 120° triple junctions confirmed that the experiments reached equilibrium. Combining XRD measurements with follow-up microbeam work were a powerful way to interpret and confirm unexpected results.

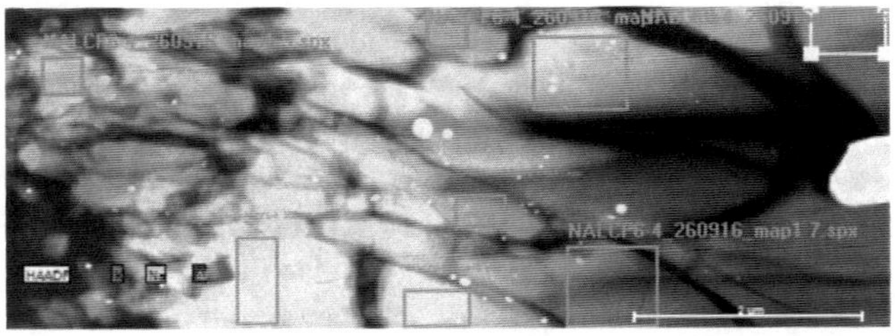

Figure 4. STEM EDS map of a sample that was equilibrated at sub-solidus conditions, containing NAL and K-hollandite (pink/Al+K rich), CF (green/Na-rich), and an aluminous phase (blue). Scale bar 2 μm.

4.2.4. Metal-silicate partitioning experiments to investigate the composition of planetary cores: characterisation of nm-scale textures. The conditions involved in the accretion of Earth and segregation of its core can be investigated by interpreting the depletions of various siderophile (iron-loving) elements in its silicate layers (e.g., [61]). For this, we must know how the various elements distribute themselves between molten metal and silicate, and how that distribution changes as a function of pressure, temperature and composition. The quest to extend partitioning experiments to lower-mantle conditions has necessitated the use of laser-heated DAC experiments. Usually, such experiments are run by sandwiching a small flake of iron metal foil between disks of silicate glass or compressed powder, where one or both phases are doped with elements of scientific interest [16, 38, 53].

Because the focus of such experiments is compositional, the samples are analysed following rapid quench, extraction and FIB preparation. The experimental run products invariably consist of a central metal sphere (~ 5 - 10 μm diameter) with a quenched silicate melt rim (~ 1 - 5 μm wide; Fig. 5). Phases frequently contain entrained or exsolved blebs (Fig. 2): whether these should be avoided during analysis depends on the interpretation of their origin. The compositions of experimental run products have typically been measured by WDS by EPMA [14, 15, 36-38, 40] or EDS by SEM [43], although other studies utilised TEM or nanoSIMS [16, 45, 53]. The small sample size and geometry presents unique analytical challenges that are especially problematic in experiments where element partitioning between coexisting chemically and physically dissimilar phases is desired [34, 35].

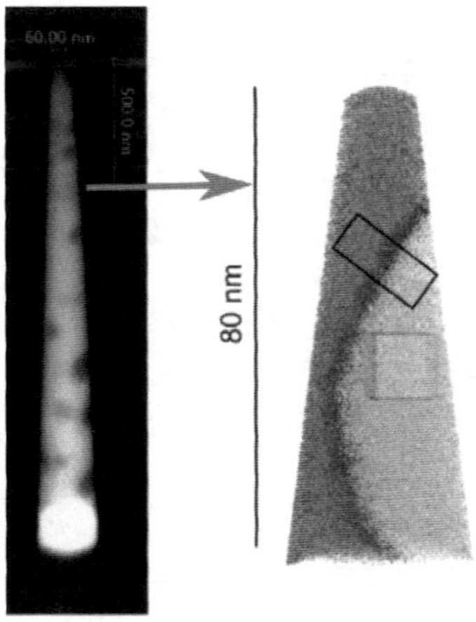

Figure 5. Left: BSE image of a needle cut by FIB for APT measurement of a DAC experimental metal containing an immiscible silicate phase. The oxide spheres (dark; blue on right) and Fe-rich matrix (light; pink on right) are clearly visible. Needle is around 2 μm long from the bright Pt base, and 60 nm wide just below the tip. Right: APT reconstruction of the positions and identities of the atoms in a needle from the same experiment. Fe - pink; C - maroon; O - light blue; Si - grey. Trace elements not shown. Black rectangle is position of the proxigram cylinder (see Fig. 6).

An interesting aspect to using multiple analytical techniques is the ability to measure at different spatial scales. The metal phase in metal–silicate partitioning experiments characteristically contains small (10's – 100's nm) spheres of a silicate material that are more clearly visible by TEM than BSE imaging. A typical approach is to interpret these spheres as quench exsolution and use a "broad" (e.g., 1 μm) electron beam to integrate these with the matrix metal for a bulk phase composition. This

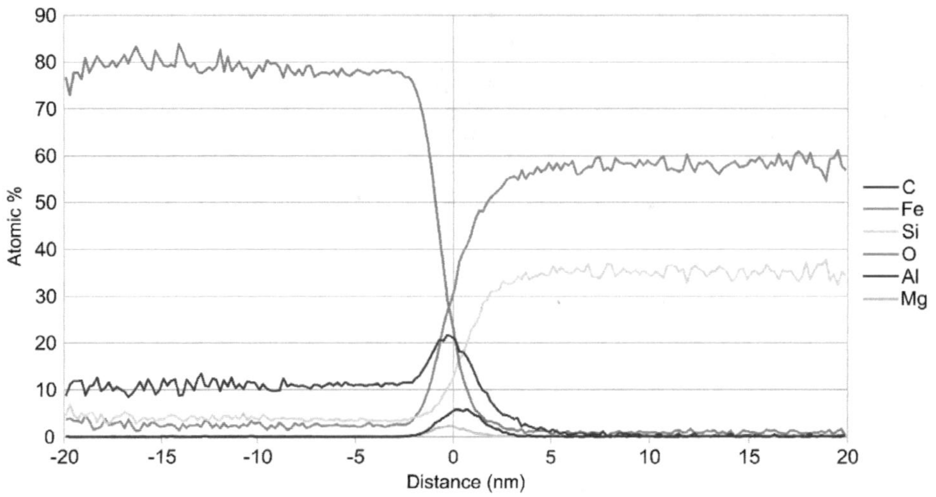

Figure 6. Proxigram (average at% concentrations) through the cylinder marked in Fig. 5, binned as a function of distance to a curved iso-surface of Si concentration. Three regions are seen: left, C-bearing Fe metal matrix; centre, interface rich in trace elements (Al, Mg) and C; Right, almost stoichiometric SiO_2 in the interior of the sphere. Matrix and sphere are homogeneous.

approach indicates a high simultaneous solubility of Si and O within Fe metal at core formation conditions (e.g., [37]). However, this conclusion relies on an assumption that the Si-O-rich spheres exsolved during quench and were soluble at high temperature. In order to characterise the spheres and any compositional gradients around their interfaces, a quenched metal from a DAC experiment was analysed by APT, giving several orders of magnitude higher spatial resolution in 3D space. It was determined that the spheres were pure stoichiometric SiO_2, that only low levels of Si and O remained in the metal, and that there were no compositional gradients in Si or O leading in either phase [62]. This has implications for the interpretation of the origin of the spheres. APT also allows the verification EPMA measurements of such metals, which tend to be plagued by low analytical totals and may contain significant inaccuracies from matrix corrections invalidated by heterogeneous analytical volumes.

4.2.5. Extreme thermal and chemical gradients in DAC experiments. A major advantage in combining ex-situ micro-analytical techniques with in-situ XRD has been in highlighting the limitations of the DAC technique, and informing the interpretation of XRD data.

It is well-known that thermal gradients in laser-heated DAC experiments are large: this is a consequence of laser heating, where radiation interacts with a laser-absorbing material. If the beam has a Gaussian profile, the centre will naturally be hottest, although beam-shaping optics are commonly used to counter this [63]. Homogeneous heating is especially difficult when working with a sample that is physically mixed with a metal to interact with the laser, or when melting occurs, allowing the re-distribution of material and resulting in heterogeneous heating.

The thermal profile can be used as an advantage. For example, Tateno *et al.* [41] qualitatively use distance from the heating spot as a proxy for temperature, and interpret their concentric pattern of phase stability as mirroring crystallisation sequence, thus maximising the information retrieved from their experiments. However, this highlights the importance of aligning the X-ray beam and the laser-heated spot precisely for in-situ measurements of heated DAC experiments, else the measured XRD pattern may be recording additional peaks from lower-temperature phases peripheral to the area of interest.

Ex-situ microanalytical work using a small spot size has highlighted a particular chemical consequence of heterogeneous heating. Thermal diffusion, known as Soret diffusion, can be driven by a temperature gradient. In silicates in DAC experiments, it appears that Fe, Al and K migrate in the direction from hot to cold; Si appears enriched in the hot spot; and Mg can migrate in either direction, dependent on material (Fig. 7; [42]). This affects solids as well as liquids.

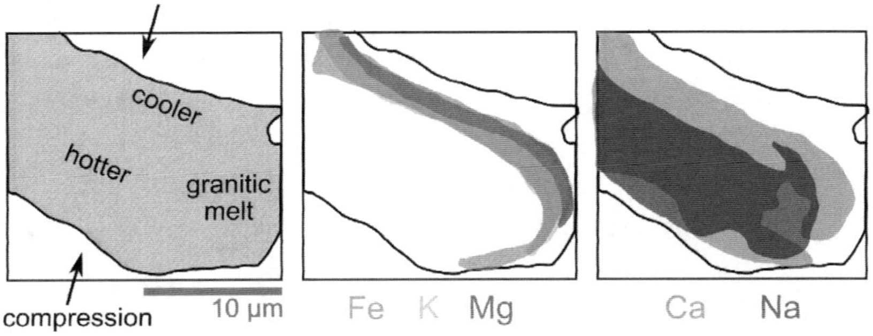

Figure 7. Extreme chemical segregation within an initially homogeneous granitic composition due to thermal diffusion. Fe, Mg and K have migrated away from the hottest region, whereas Ca has remained relatively immobile. Simplified sketch figure summarises fig. 5 of Sinmyo and Hirose [42], where colours indicate areas of peak concentration. Sample was heated at 1,900 K for 63 minutes.

Samples are most affected when a laser-absorbing powder is mixed into the sample, and the effect can be mitigated against using metal coatings or no laser absorber, insulation layers, and short experimental run times [42]. Soret diffusion may not be identified in XRD measurements and poses a major difficulty in interpreting DAC experiments, especially when melting is involved. A homogeneous starting material may no longer be homogeneous, changing the stabilities of various phases: for example, a localised concentration of MgO will increase the stability of ferropericlase [41], and a higher concentration of FeO may decrease the melting temperature. Soret diffusion provides an additional hypothesis for the presence of concentric structures around laser-heated spots, and may be responsible for some of the discrepancies between different studies [42]. Problems with diffusion and changes to the localised bulk composition can be checked by following up in-situ XRD measurements with ex-situ imaging and compositional measurements. Variations in the localised "bulk" compositions may indicate a problem.

5. Conclusion

DAC experiments allow various materials to be held at the static pressures found deep in planetary interiors, and their popularity is in part due to their ease of use. Samples can be heated to extreme temperatures, and the technique is compatible with a wide range of analytical techniques. DAC experiments have found particular popularity in the Earth and planetary sciences, as well as in materials sciences, chemistry, and biological materials. Measurements can be made through the diamond anvils, either optically or using a synchrotron X-ray source, and additional measurements can be made after the experiment by extracting the sample. The main limitations of DAC experiments are the very small sample size (e.g., 10's µm heated spot) and the high potential for heterogeneous heating and diffusion. Using just XRD for a laser-heated experiment could miss important features such as disequilibrium and diffusion. By combining in-situ and ex-situ techniques, experimental products can be fully characterised. An ideal case could consist of in-situ XRD to identify the high-PT phases, confirm PT conditions and to refine equation of state; followed by ex-situ micro- and nanoscale imaging and compositional measurements, to relate XRD measurements to experimental textures and to characterise phase compositions and elemental distribution within the DAC experiment.

Acknowledgements
I would like to acknowledge my collaborators Jon Wade, James Douglas, Nobuyoshi Miyajima and Sylvain Petitgirard for their assistance in obtaining the analytical data for figs. 4, 5 and 6 of this work, Oliver Lord for experimental assistance, and Jennifer Frost for performing the experiment of Fig. 4. I also thank David Rubie, Vera Laurenz, Ingrid Blanchard, and various others in addition to the aforementioned people, for helping me to navigate the sometimes-tricky field of DAC experiment microanalysis. Finally, I am grateful to a careful reviewer for helping to improve the clarity of this work.

References
[1] Jayaraman A 1983 *Rev. Mod. Phys.* **55** 65-108
[2] Dubrovinsky L, Dubrovinskaia N, Prakapenka V B and Abakumov A M 2012 *Nature Commun.* **3** 1163
[3] Mao H K, Xu J and Bell P M 1986 *J. Geophys. Res. Solid Earth* **91** 4673-4676
[4] Walter M J, *et al.* 2015 *Chem. Geol.* **418** 16-29
[5] Boffa Ballaran T, Kurnosov A and Trots D 2013 *High Press. Res.* **33** 453-465
[6] Shen G and Mao H K 2017 *Rep. Prog. Phys.* **80** 016101
[7] Ohishi Y, *et al.* 2008 *High Press. Res.* **28** 163-173
[8] Baron M A, *et al.* 2017 *Earth Planet. Sci. Lett.* **472** 186-196
[9] Hirose K, Fei Y, Ma Y and Mao H-K 1999 *Nature* **397** 53-56
[10] Kesson S E, Fitz Gerald J D and Shelley J M 1998 *Nature* **393** 252-255
[11] Lord O T, *et al.* 2010 *J. Geophys. Res. Solid Earth* **115** B06208
[12] Lord O T, *et al.* 2009 *Earth Planet. Sci. Lett.* **284** 157-167
[13] Morard G, *et al.* 2017 *Earth Planet. Sci. Lett.* **473** 94-103
[14] Badro J, Siebert J and Nimmo F 2016 *Nature* **536** 326-328
[15] Bouhifd M A and Jephcoat A P 2003 *Earth Planet. Sci. Lett.* **209** 245-255
[16] Fischer R A, *et al.* 2015 *Geochim. Cosmochim. Acta* **167** 177-194
[17] Boffa Ballaran T, *et al.* 2012 *Earth Planet. Sci. Lett.* **333-334** 181-190
[18] Lavina B, *et al.* 2011 *Proc. Natl. Acad. Sci.* **108** 17281-17285
[19] Mao Z, *et al.* 2012 *Earth Planet. Sci. Lett.* **331-332** 112-119
[20] Knittle E and Jeanloz R 1987 *Science* **235** 668-670
[21] Fiquet G, *et al.* 2000 *Geophys. Res. Lett.* **27** 21-24
[22] Hirose K, Takafuji N, Sata N and Ohishi Y 2005 *Earth Planet. Sci. Lett.* **237** 239-251
[23] Murakami M, *et al.* 2004 *Science* **304** 855-858
[24] Ricolleau A, *et al.* 2010 *J. Geophys. Res. Solid Earth* **115** B08202
[25] Walter M J, *et al.* 2011 *Science* **334** 54-57
[26] Sanloup C, *et al.* 2013 *Nature* **503** 104-107
[27] Morard G, *et al.* 2013 *Earth Planet. Sci. Lett.* **373** 169-178
[28] Schmidt C and Rickers K 2003 *Amer. Mineralogist* **88** 288-292
[29] Petitgirard S, *et al.* 2009 *Rev. Sci. Instrum.* **80** 033906
[30] Manning C E, Wilke M, Schmidt C and Cauzid J 2008 *Earth Planet. Sci. Lett.* **272** 730-737
[31] Andrault D, *et al.* 2012 *Nature* **487** 354-357
[32] Petitgirard S, *et al.* 2012 *Rev. Sci. Instrum.* **83** 013904
[33] Wirth R 2009 *Chem. Geol.* **261** 217-229
[34] Jennings E S, Wade J, Laurenz V and Petitgirard S 2019 *Microsc. Microanal.* **25** 1-10
[35] Wade J and Wood B J 2012 *Phys. Earth Planet. Inter.* **192-193** 54-58
[36] Bouhifd M A and Jephcoat A P 2011 *Earth Planet. Sci. Lett.* **307** 341-348
[37] Siebert J, Badro J, Antonangeli D and Ryerson F J 2012 *Earth Planet. Sci. Lett.* **321-322** 189-197
[38] Blanchard I, Siebert J, Borensztajn S and Badro J 2017 *Geochem. Perspect. Lett.* **5** 1-5
[39] Siebert J, *et al.* 2018 *Earth Planet. Sci. Lett.* **485** 130-139

IOP Conf. Series: Materials Science and Engineering **891** (2020) 012015 doi:10.1088/1757-899X/891/1/012015

[40] Mahan B, *et al.* 2018 *Geochim. Cosmochim. Acta* **235** 21-40
[41] Tateno S, Hirose K and Ohishi Y 2014 *J. Geophys. Res. Solid Earth* **119** 4684-4694
[42] Sinmyo R and Hirose K 2010 *Phys. Earth Planet. Inter.* **180** 172-178
[43] Chidester B A, Rahman Z, Righter K and Campbell A J 2017 *Geochim. Cosmochim. Acta* **199** 1-12
[44] Miyajima N, *et al.* 2010 *J. Microscopy* **238** 200-209
[45] Frost D J, *et al.* 2010 *J. Geophys. Res. Solid Earth* **115** B02202
[46] Gault B, Moody M P, Cairney J M and Ringer S P 2012 *Atom probe microscopy.* (New York, NY: Springer Science & Business Media)
[47] Valley J W, *et al.* 2014 *Nature Geosci.* **7** 219-223
[48] White L F, *et al.* 2017 *Nature Commun.* **8** 15597
[49] Lewis J B, Isheim D, Floss C and Seidman D N 2015 *Ultramicroscopy* **159** 248-254
[50] Kilburn M R and Wacey D 2014 in: *Principles and practice of analytical techniques in Geosciences.* (London, UK: Royal Society of Chemistry). 1-34
[51] Zinner E 2007 in: *Treatise on geochemistry.* (Holland H D and Turekian K K; Eds.). (Oxford, UK: Pergamon Press). 1-33
[52] Badro J, *et al.* 2007 *Earth Planet. Sci. Lett.* **262** 543-551
[53] Suer T-A, *et al.* 2017 *Earth Planet. Sci. Lett.* **469** 84-97
[54] Garnero E J and McNamara A K 2008 *Science* **320** 626-628
[55] Murakami M, Hirose K, Sata N and Ohishi Y 2005 *Geophys. Res. Lett.* **32**
[56] Andrault D 2001 *J. Geophys. Res. Solid Earth* **106** 2079-2087
[57] Nomura R, *et al.* 2014 *Science* **343** 522-525
[58] Harte B 2010 *Mineral. Mag.* **74** 189-215
[59] Armstrong L S and Walter M J 2012 *Eur. J. Mineralogy* **24** 587-597
[60] Cerantola V, *et al.* 2017 *Nature Commun.* **8** 15960
[61] Walter M J, Newsom H E, Ertel W and Holzheid A 2000 in: *Origin of the Earth and Moon.* (Tucson, AZ: University of Arizona Press). 265-289
[62] Jennings E S, *et al.* 2018 in: *EGU General Assembly 2018.* 19120
[63] Prakapenka V B, *et al.* 2008 *High Press. Res.* **28** 225-235

New horizons in quantitative compositional mapping – Analytical conditions and data reduction using XMapTools

P Lanari and F Piccoli

University of Bern, Institute of Geological Sciences, Baltzerstrasse 1+3, 3012 Bern, Switzerland

E-mail: pierre.lanari@geo.unibe.ch

Abstract. Quantitative compositional mapping is a key technique for imaging solids and, therefore, for trying to understand compositional zoning systematics in a broad range of geological materials. Compositional mapping requires an adapted analytical method for efficient data collection, a post-processing calibration stage to obtain elemental concentration values, and specific computer tools for exploring and visualising the data. In this contribution, we describe optimal analytical conditions, possible physical limitations and data reduction strategies for compositional maps of major, minor and trace elements obtained by electron probe microanalysis (EPMA) and laser ablation inductively-coupled plasma mass spectrometry (LA-ICP-MS). Data reduction is performed using the computer software XMapTools, which incorporates a large variety of options and tools for data processing, data calibration and data visualisation based on single and multi-channel maps or via binary, ternary and spider diagrams.

1. Introduction

Rock samples typically exhibit complex textures, mineral distributions and/or compositional zoning requiring chemical analysis of major, minor and trace elements over mm- to cm-sized areas. This 'natural constraint' fostered the development of analytical techniques [1-4] and software solutions [5-9] for obtaining maps of element concentration in solid geological materials. The two most popular scientific techniques for compositional mapping are electron probe microanalysis (EPMA) for major to minor elements and laser ablation inductively-coupled plasma mass spectrometry (LA-ICP-MS) for trace elements. In practice, these two mapping techniques have both advantages and disadvantages – they are complementary rather than substitutes for one another [10]. As we shall demonstrate in the following, the major advantage of EPMA lies in its high spatial resolution, in the range of 1 - 10 μm^3, whereas LA-ICP-MS analyses involve much larger analytical volumes (> 100 μm^3) but with the power of detecting small compositional variations of critical trace elements at the ppm level. However, both techniques require specific data reduction schemes for data correction, calibration and visualisation. In this study, we present recent improvements of XMapTools [6, 11] (available via https://www.xmaptools.com), an advanced software toolbox especially designed to achieve optimum workflow and feature extraction from compositional maps of rock samples. This software solution can readily handle maps obtained from a large variety of instruments including both EPMA and LA-ICP-MS (e.g., [6, 12]).

Quantitative compositional mapping by EPMA has become a routine technique in many laboratories worldwide. It requires the acquisition of X-ray maps (i.e., raw data not corrected for background or matrix effects) measured at specific X-ray emission lines and a post-collection calibration stage called analytical standardisation to derive fully quantitative data of element concentrations. The calibration can be performed either based on an empirical or semi-empirical correction schemes [1-3] or against internal standards assuming no matrix effects [4]. The advantages and limits of each strategy are discussed in Lanari *et al.* [6]. The internal standardisation procedure implemented in XMAPTOOLS [11] requires high-precision spot analyses of every mineral phase to be calibrated. This technique was improved by adding a pseudo-background correction [6].

Quantitative compositional mapping of trace elements and isotopes by LA-ICP-MS is catching up. It started with the seminal study of Woodhead *et al.* [13] and has improved at a fast rate in the past 10 years. Mapping by LA-ICP-MS involves the acquisition of successive raster images of the focussed laser beam. Each acquisition line is calibrated using a standard reference glass and a known element concentration as internal standard. The calibrated data are then stitched together to form compositional maps of trace element distributions. The calibration is usually performed using the modules available within the commercial software IOLITE [14-16], whereas subsequent image processing and data visualisation is easier and more flexible if done with XMAPTOOLS [12].

2. Analytical conditions

This section briefly outlines some important aspects of each analytical procedure along with recommended instrument operating conditions. Common artefacts affecting the spatial and chemical resolutions that were not investigated in previous studies (e.g., [4, 6, 11, 12]) are further discussed.

2.1. Quantitative compositional mapping by EPMA

A standard analytical setup for quantitative compositional mapping of silicates and oxides by EPMA is presented in Table 1. This setup is optimised for modern EPMA instruments equipped with five wavelength-dispersive X-ray spectrometers (WDS) and a single energy-dispersive X-ray spectrometer (EDS). The final image resolution is defined according to the total number of pixels of a map: low resolution for an image containing less than 150,000 pixels; medium resolution between 150,000 and 500,000 pixels; and high resolution above 500,000 pixels. The area of interest on the specimen is scanned via mechanical beam tracking mode, i.e., by moving the sample under a static electron beam using a high-precision computer-controlled x-y-z stage. In practice, the accelerating voltage of the electron beam is fixed at 15 keV, as it is for 'typical' spot analyses. The specimen current, measured at the Faraday cup, is increased up to at least 100 nA to compensate the short dwell times (Table 1). The characteristic X-ray emission caused by beam-specimen interaction is continuously analysed using both WDS and EDS spectrometers, resulting in the acquisition of 15 to 20 elemental maps in two acquisitions (as only five elements can be measured by WDS, the area is often scanned twice). The measurement time for a single area of interest takes between one night and few days, depending on the image resolution and the dwell time.

Two effects are further discussed in the following: (1) the size of the interaction volume, which can result in overlapping effects or under-sampling of a specific mineral and (2) beam damage during analysis. Other potential artefacts such as secondary fluorescence effects, thermal or beam current instability, spectrometer focus and surface irregularities are discussed in Lanari *et al.* [6].

2.1.1. Interaction volume and spatial resolution. In compositional mapping by EPMA, the spatial resolution is determined by the spacing of spot measurements and the X-ray excitation volume of the electron beam. The pixel size *dx* in the final image is equal to the distance between two spots. Any interaction volume larger than *dx* results in overlapping, increasing the number of mixed pixels, which

Table 1. Standard analytical setup for quantitative compositional mapping by EPMA.

	Setup 1 (medium resolution)	Setup 2 (high resolution)
Accelerating voltage (keV)	15	
Beam current (nA)	100 (eventually up to 500)	
Map size in pixels (# pixels)	500 × 500 (250,000)	1,000 × 1,000 (1,000,000)
Step size (μm)	1 - 40	
Map surface (mm^2)	0.25 - 400	1 - 1,600
Dwell time (ms)	100 - 300	60 - 200
Map acquisitions	2	
# elements WDS	10 (e.g., Si-Kα, Ti-Kα, Al-Kα, Fe-Kα, Mn-Kα, Mg-Kα, Ca-Kα, Na-Kα, K-Kα, Cr-Kα/Ni-Kα)	
# elements EDS	5 - 10 (e.g., P-Kα, S-Kα, Zr-Kα, Ce-Lα, La-Lα, etc.)	
Mapping time (h)	14 - 42	33 - 112

are pixels having a mixed composition between two or more phases. In contrast, any interaction volume smaller than dx results in under sampling and may cause small phases to be missed. It is, therefore, critical to quantitatively evaluate the size of the excitation volume for a large variety of minerals.

Electron-specimen interaction volumes were investigated using a Monte Carlo model and the computer programme CASINO 2.41 [17] for a set of 14 common rock-forming minerals including garnet (almandine ρ = 4.19 g.cm^{-3}; Z = 15.6), clinopyroxene (omphacite ρ = 3.34 g.cm^{-3}; Z = 14.6), white mica (muscovite ρ = 2.82 g.cm^{-3}; Z = 11.32), feldspar (albite ρ = 2.62 g.cm^{-3}; Z = 11.73), amphibole (glaucophane ρ = 3.34 g.cm^{-3}; Z = 10.6), calcite (ρ = 2.71 g.cm^{-3}; Z = 12.6), olivine (forsterite ρ = 3.27 g.cm^{-3}; Z = 10.6; fayalite ρ = 4.39 g.cm-3; Z = 18.7), epidote (zoizite ρ = 3.30 g.cm-3; Z = 12.1), chlorite (clinochlore ρ = 2.65 g.cm^{-3}; Z = 12.2), kyanite (ρ = 3.61 g.cm^{-3}; Z = 10.7), magnetite (ρ = 5.15 g.cm^{-3}; Z = 21.0), quartz (ρ = 2.62 g.cm^{-3}; Z = 10.8), and orthopyroxene (ferrosilite ρ = 3.95 g.cm^{-3}; Z = 14.2). The accelerating voltage and beam diameter were kept constant at 15 keV and 1 μm. A selection of four representative cases is presented in Fig. 1. The maximum diameter of the interaction volume d was estimated for each mineral from the trajectories of 10,000 simulations using a home-made routine defining a 95 % confidence interval of successive layers (cylinder shape) with a thickness of 0.2 μm between 0 and 2 μm depth and 0.5 μm below 2 μm depth. This maximum diameter of the interaction volume decreases with increasing mean atomic number Z and density ρ (unit g.cm^{-3}) of the target material (Fig. 2). This trend can be reproduced using the following relation:

$$d = 4.034 \exp(-0.02139\, Z\rho^2) + 1.548 \exp(-0.006457\, Z\rho^2) \quad (1)$$

This equation can be used for approximating a first order maximum diameter of interaction volume in common silicates and oxides at 15 keV and 1 μm beam diameter and may, therefore, assist in the interpretation of compositional maps. This technique was implemented in XMapTools 3.0.1; two application examples based on published datasets are presented in Fig. 3.

The purpose of the first dataset was to derive local bulk compositions to be used in thermodynamic models [18]. An area of 12.3 mm × 12.3 mm was analysed with a step size of 12 μm resulting in an image size of 1024 × 1024 pixels corresponding to 1,048,567 fully quantitative analyses. Only a small portion of this map was re-investigated here (Fig. 3a). The major drawback of this setup is that the electron-specimen interaction volume is much smaller than the pixel size (Figs. 3a to 3c). The resulting under sampling effect is strong as the analysed pseudo-surface ranges between 1 % (garnet) and 5 % (plagioclase) of the pixel surface in the final image (Fig. 3c). In this example, the average grain

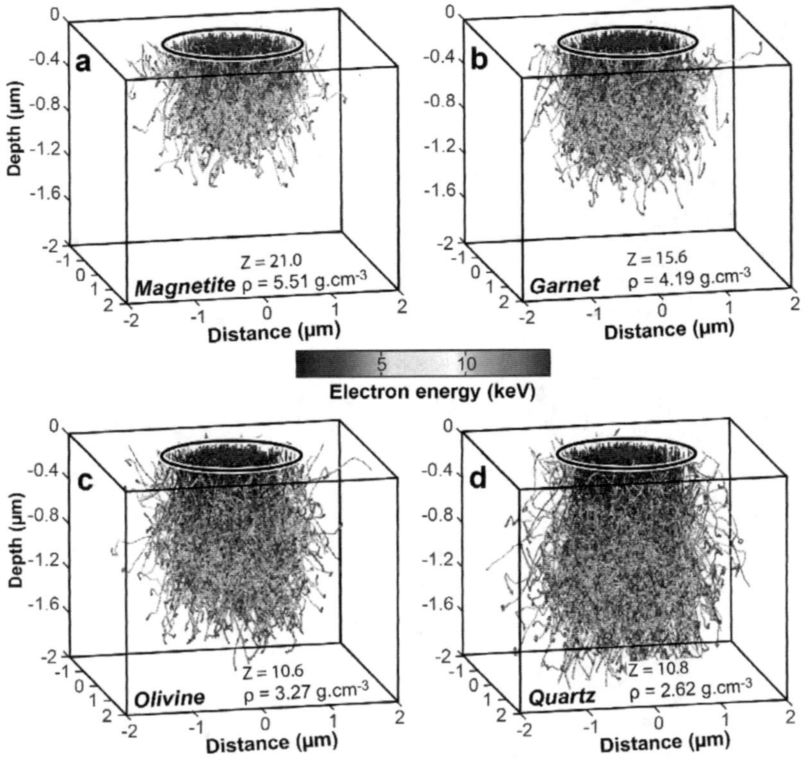

Figure 1. Illustration of possible electron-specimen interaction in four rock-forming minerals covering a broad spectrum of mean atomic number (Z) and density (ρ). A total of 10,000 simulations were performed for each mineral using the programme CASINO 2.41 (see text), but only a selection of 1,000 trajectories is shown in this figure. The size of the excitation volume is increasing with decreasing values of Z and ρ as also shown by Eq. (1). Colours show the energy of the electron (in keV). The black circle shows the beam size at the sample surface.

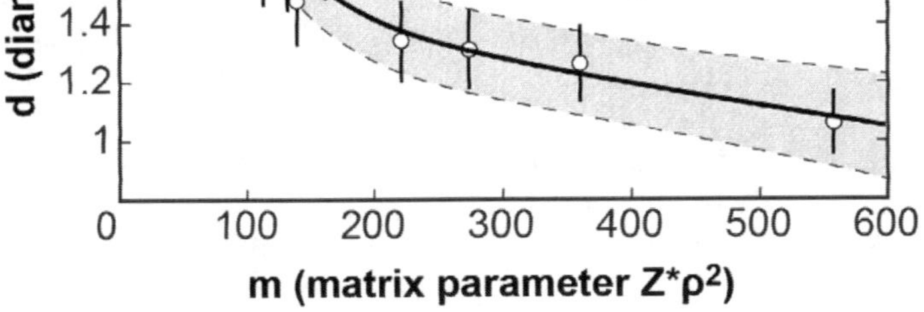

Figure 2. Diameter of electron-specimen interaction volume as a function of a matrix parameter m defined as the product of the mean atomic number (Z) and the square of the density (ρ^2) of the material. A bi-exponential model corresponding to the sum of two exponential functions was used for fitting (see Eq. (1)). The uncertainty on the value of d obtained by the Monte Carlo technique was arbitrary fixed to 10 %. Note that these data points were generated for an accelerating voltage of 15 keV and a spot size of 1 µm.

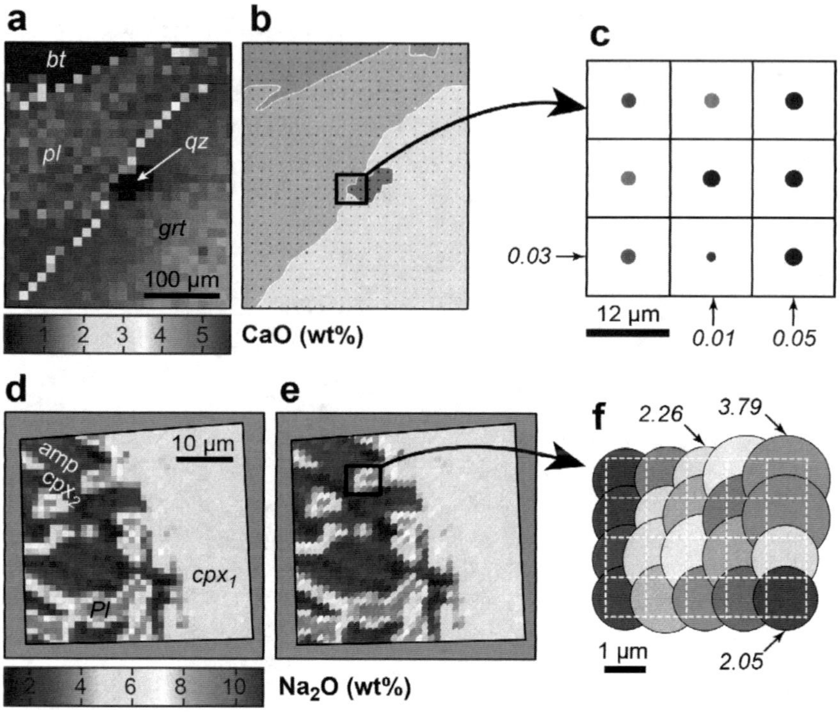

Figure 3. Possible effects of the pixel size (*dx*) on the spatial resolution of X-ray maps with a beam size of 1 µm. a) Small area of a CaO map (in wt%) with a step size of 12 µm obtained for a high-grade metamorphic rock from the Central Himalaya (modified from [18]). b) Image with circle pixels representing the electron-specimen interaction volume (see Eq. 1). Garnet has a smaller interaction volume than quartz (Fig. 1). c) Detailed view illustrating the under sampling effect. The numbers correspond to the diameter of the interaction volume (in µm) calculated for a specific pixel. d) Small area of a Na_2O map (in wt%) with a step size of 1 µm obtained for an eclogite from the NW Himalaya (modified from [21]). e) Image with circle pixels representing the electron-specimen interaction volume. f) Detailed view illustrating the overlapping effect. The numbers correspond to the diameter of the interaction volume (in µm) calculated for a specific pixel. Abbreviations: bt - biotite; pl - plagioclase; qz - quartz; grt - garnet; cpx clinopyroxene; amp - amphibole.

size is much larger than the step size and under sampling does not affect significantly the local bulk composition estimates. However, small phases (5 - 10 µm width) might not be analysed with this setup. Increasing the beam diameter to the same value of the step size improves the detection of small phases (e.g., [19, 20]), but significantly increases the number of mixed pixels and is, therefore, less suitable for quantitative analysis of zoned materials.

The second dataset shows an opposite case in which the electron-specimen interaction volume is larger than the pixel size (Figs. 3d to 3f). An area of 0.52 mm × 0.67 mm was analysed with a step size of 1 µm corresponding to 348,400 fully quantitative analyses [21]. Over sampling increases significantly the number of mixed pixels. In Fig. 3c, only one pixel shows the true composition of Na_2O in plagioclase, whereas a two-pixel rim exhibits mixed compositions towards amphibole or clinopyroxene. The degree of over sampling ranges between 200 % for clinopyroxene (cpx2) and 400 % for plagioclase. Mixed pixels are the main source of misclassification and are challenging to calibrate. These two examples show that a good compromise for the analysis of geological materials is a step size of ~2 µm.

2.1.2. Beam damage. The electron beam exposure produces heat over time and this heat can drive element diffusion in the target material. In addition, the trapping of electrons can generate an electric field within the sample and cause migration of positive ions toward a region of negative charge located outside of the analysis volume. These effects known as beam damage can significantly affect the content and distribution of mobile elements during the acquisition of spot analyses and mapping. Beam-induced changes in composition generally cause the X-ray intensity of mobile elements such as Na, Ca and K to decrease (diffusion out) and heavier element to slightly increase due to changes in density and the absorption behaviour of the specimen.

A meta-quartzite from the Glacier-Rafray Klippe in the Western Alps [22] was used to illustrate the possible effects of sample alteration during mapping under an electron beam with high-current conditions. A small area of 80 × 120 µm^2 containing albite, muscovite and quartz was mapped five times (two acquisitions each corresponding to a total of 10 scans) at the Institute of Geological Sciences of the University of Bern using a JEOL-820 Superprobe instrument, an accelerating voltage of 15 keV, a beam current of 200 nA, a beam size of 1 µm and 200 ms dwell time. The series of X-ray maps for Na-, K- and Al-Kα lines uncorrected for background are shown in Fig. 4. The mobile elements are affected by beam damage. In albite, between 5 and 35 % of the total Na-Kα counts were lost after two scans (Fig. 4b), this figure rises up to > 90 % after 9 scans (Fig. 4e). Although the original distribution of Na in albite was homogenous, this is not the case after beam damage demonstrating that diffusion was not homogenous throughout. A moderate amount of Na is preserved in bright spots 5 µm thick (Figs. 4c to 4e). In muscovite, K is more resilient to diffusion under the high-current electron beam and the K-Kα signal experienced only < 10 % count loss after two scans (Fig. 4g) and 50 - 60 % after 9 scans (Fig. 4j). In this case, the compositional zoning of muscovite is preserved (Figs. 4f and 4e). It is important to notice that in plagioclase, Al-Kα count rate increases up to 10 % in the areas with high Na diffusion (e.g., between Fig. 4k and Fig. 4o).

This example shows that beam damage can significantly affect the count rates obtained during mapping. Therefore, mobile elements (Na, K, Ca) have to be measured during the first pass of the beam over the mapped area and after the acquisition of spot analyses for calibration. It is also important to notice that beam damage caused by spot analyses can result in a significant overestimation in the concentration of the element after calibration.

2.2. *Quantitative compositional mapping by LA-ICP-MS*

A standard analytical setup for quantitative compositional mapping of silicates such as garnet by LA-ICP-MS is presented in Table 2. This setup is optimized for a 193 nm exciter laser coupled to a quadrupole ICP-MS. Mapping can be performed on standard polished thin sections or epoxy mounts. The area of interest is scanned using parallel rasters analysed across the sample surface. A single raster analysis is preceded by 20 - 30 s background analysis and a pre-ablation stage with larger laser beam size followed by 10 s pre-washout. The pre-ablation aims to minimise re-deposition effects. Standards (including standard reference materials such as glasses SRM610 and SRM612) are analysed at the start and end of each mapping run and every 0.5 to 1 h during mapping. The measurement time for a single area of interest takes between few hours and few days, depending on the size of the area (Table 2).

3. *Data reduction using XMAPTOOLS*

This section briefly summarises the main workflow procedures implemented in XMAPTOOLS (see also [6, 11]).

3.1. *Workflow for EPMA data*

Step 1 – Importing raw data and corrections. Corrections applied to X-ray maps include a dead time correction for deriving true count rates, correction for intensity deviations caused by sample surface topography and/or time-dependent drift. The corrections applied to the spot analyses (which are used as internal standard) include the elimination of mixed compositions and the adjustment of position (caused by sample displacement during the analysis) based on statistical criteria.

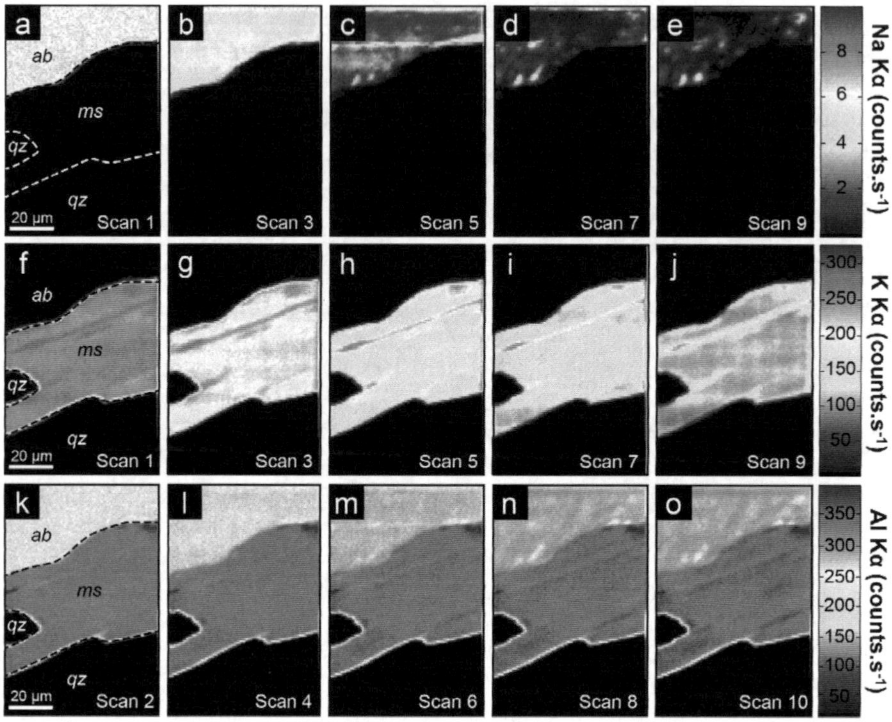

Figure 4. Beam damage effects in albite and muscovite from a meta-quartzite (Glacier-Rafray Klippe, Western Alps). The count rates (unit counts per second) of mobile elements Na and K in albite and muscovite progressively decrease with time (from left to right) because of beam damage. Note that the count rates of Al in both plagioclase and muscovite are increasing from left to right. Mineral abbreviations: ab - albite; ms - muscovite; qz - quartz.

Table 2. Standard analytical setup for quantitative compositional mapping of garnet by LA-ICP-MS.

	Setup 1 (small resolution)	**Setup 2 (medium resolution)**
Laser wavelength (nm)	193	
Fluence (energy density, J.cm^{-2})	5 - 7	
Repetition rate (Hz)	10	
Sampling mode	rastering	
Carrier and make-up gases (L.min^{-1})	He (0.5 - 0.7) and Ar (0.9 L.min^{-1})	
Map size in pixels (# pixels)	100 × 100 (10,000)	300 × 300 (90,000)
Laser spot diameter (µm)	16 - 24	
Map surface (mm^2)	1.6 × 1.6	4.8 × 4.8
Scan speed (µm.s^{-1})	22 - 30	
Dwell time (ms) (elements)	10 - 20 ms (Sr, Y, Zr, La, Ce, Pr, Nd, Sm, Eu, Gd, Tb, Dy … … Ho, Er, Tm, Lu, Th, U) 5 - 10 ms (Na, Mg, Si, P, K, Ca, Ti, Cr, Mn, Fe)	
Total sweep time (s)	0.30 - 0.5	
Mapping time (h)	5 - 7	24 - 36

Step 2 – Multi-channel classification (and manual phase identification). An automated technique based upon the K-means algorithm is implemented in XMAPTOOLS. Alternatively, a step-by-step manual classification can be performed using the modules *BINARY* or *TRIPLOT*. Individual masks are created from a range of composition defined graphically [11]. An example of manual classification is shown in Fig. 5. Four masks were generated from the binary Mg versus Al. This manual procedure allows the mixed pixels to be excluded (see grey points in Fig. 5a and the corresponding black pixels in Fig. 5c). Excluding mixed pixels is important for obtaining high quality chemical images, such as maps of structural formulas. An automated algorithm eliminating the pixels located between the different masks is also provided in XMAPTOOLS (BRC correction, see [6]).

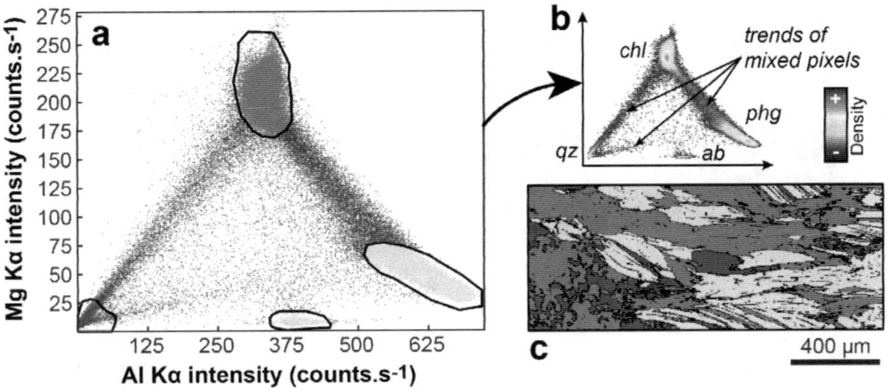

Figure 5. Manual classification in XMAPTOOLS via the BINARY module using a compositional map of a low-temperature high-pressure metasediment from the Western Alps (Briançonnais). a) Binary diagram Mg versus Al (unit: counts per second); each point corresponds to the composition of a single pixel. Selection of four groups using the option 'multi-group free shape'. The grey pixels are mixed pixels and are not classified in this example. b) Density map of the binary diagram shown in (a). c) Map showing the spatial distribution of the pixels of each group selected in (a). Abbreviations: chl - chlorite; qz - quartz; phg - phengite; ab albite.

Step 3 – Analytical standardisation. The module *STANDARDIZATION* assists the user for the calibration of the elements of each phase. An automated routine uses the variability of the mass concentration of the spot analyses to fit the slope of the calibration curve and approximate the background intensity (see Fig. 7 in [6]). The position of the calibration curve can also be manipulated manually.

Step 4 – Data exploration and visualisation. Once the maps have been calibrated, the data can be plotted as single or multi-channel maps or in binary, ternary and spider diagrams. XMAPTOOLS provides a guided environment for data exploration based on a large variety of structural formulas, diagnostic elemental ratios for major and trace elements as well as temperature and pressure conditions.

3.2 Workflow for LA-ICP-MS data

Step 1 – Importing data and corrections. The LA-ICP-MS data are imported in XMAPTOOLS as element concentration either in ppm or weight percentage of element [12]. The map size and orientation can be adjusted in the module *IMPORT TOOLS*. Holes, polish defects and all other mineral phases are masked-out using one of the classification options described above.

Step 2 – Trace element map visualisation and sampling tools. (Note that these tools are available for the analysis of any type of map, including EPMA data). XMAPTOOLS incorporates several options for data visualisation specially designed for trace element analysis. The element concentration can be displayed using either linear or logarithmic colour bar (Figs. 6a and 6b). The compositions can be extracted along any linear profile or path and plotted for all or a selection of elements (Fig. 6c). Average compositions and associated standard deviations can be extracted from spatial domains (area-of-interest) or along a smoothed profile – integrating a rectangular area or using a scanning window.

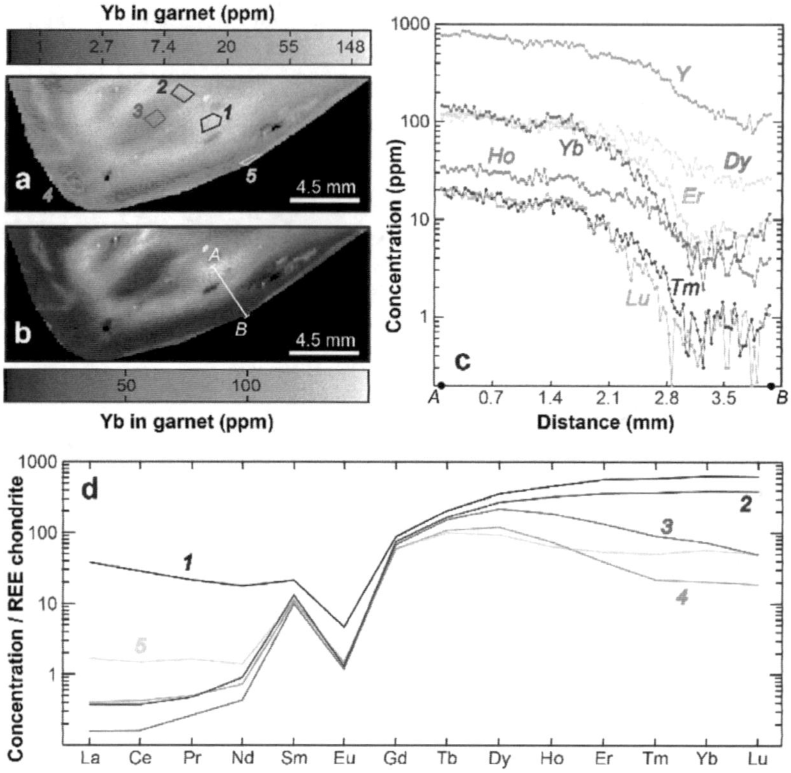

Figure 6. Garnet LA-ICP-MS map from a migmatite of the El Oro Metamorphic Complex (SW Ecuador) showing complex trace element distribution patterns. Map of Yb (in ppm) plotted using a) logarithmic, and b) linear colour scales. c) Trace element concentration profile from A to B; note that the position of the profile is shown in (b). c) Chondrite-normalised rare earth element (REE) patterns for five areas reported in (a). This diagram depicts the behaviour of REE in each zone.

Step 3 – Trace element data visualisation, spider diagrams and map transformation. Similarly to major element, trace element data can be visualised using the modules *BINARY* and *TRIPLOT*. Manual and automated clustering allow chemical groups and their spatial distribution to be characterised. Spider diagrams – plots of the abundances of a set of elements such as rare earth elements (REE) relative to their abundance in some standard – can be readily generated in XMAPTOOLS using the one of the sets of normalisation values available (Fig. 6d). Finally, the maps can be combined together within the *GENERATOR* module using any arithmetic operation generating a new map.

Acknowledgements
We thank an anonymous reviewer for providing suggestions that improved the clarity of the paper as well as Bjørn E. Sørensen and Luc Van 't dack for their efficient editorial handling.

References

[1] Cossio R and Borghi A 1998 PETROMAP: MS-DOS software package for quantitative processing of X-ray maps of zoned minerals. *Comp. Geosci.* **24** 805-814

[2] Clarke G L, Daczko N R and Nockolds C 2001 A method for applying matrix corrections to X-ray intensity maps using the Bence-Albee algorithm and Matlab. *J. Metamorphic Geol.* **19** 635-644

[3] Chouinard J and Donovan J 2015 Quantitative elemental mapping with electron microprobe and automated data analysis. *Microsc. Microanal.* **21** 2193-2194

[4] De Andrade V, Vidal O, Lewin E, O'Brien P and Agard P 2006 Quantification of electron microprobe compositional maps of rock thin sections: an optimized method and examples. *J. Metamorphic Geol.* **24** 655-668

[5] Lanari P, Wagner T and Vidal O 2014 A thermodynamic model for di-trioctahedral chlorite from experimental and natural data in the system MgO–FeO–Al2O3–SiO2–H2O: applications to P-T sections and geothermometry. *Contr. Mineral. Petrol.* **167** 968

[6] Lanari P, Vho A, Bovay T, Airaghi L and Centrella S 2019 Quantitative compositional mapping of mineral phases by electron probe micro-analyser. in: *Metamorphic geology: Microscale to mountain belts*. (Ferrero S, Lanari P, Goncalves P and Grosch E G; Eds.). *Geol. Soc. London Spec. Publ.* **478**, 39-63

[7] Ortolano G, Zappalà L and Mazzoleni P 2014 X-ray map analyser: A new ArcGIS® based tool for the quantitative statistical data handling of X-ray maps (Geo- and material-science applications). *Comp. Geosci.* **72** 49-64

[8] Ortolano G, Visalli R, Godard G and Cirrincione R 2018 Quantitative X-ray map analyser (Q-XRMA): A new GIS-based statistical approach to mineral image analysis. *Comp. Geosci.* **115** 56-65

[9] Tinkham D K and Ghent E D 2005 XRMapAnal: A program for analysis of quantitative X-ray maps. *Amer. Mineralogist* **90** 737-744

[10] Lanari P and Engi M 2017 Local bulk composition effects on mineral assemblages. *Rev. Mineral. Geochem.* **83** 55-102

[11] Lanari P, Vidal O, De Andrade V, Dubacq B, Lewin E, Grosch E G and Schwartz S 2014 XMapTools: A MATLAB©-based program for electron microprobe X-ray image processing and geothermobarometry. *Comp. Geosci.* **62** 227-240

[12] Raimondo T, Payne J, Wade B, Lanari P, Clark C and Hand M 2017 Trace element mapping by LA-ICP-MS: assessing geochemical mobility in garnet. *Contr. Mineral. Petrol.* **172** 17

[13] Woodhead J D, Hellstrom J, Hergt J M, Greig A and Maas R 2007 Isotopic and elemental imaging of geological materials by laser ablation inductively coupled plasma-mass spectrometry. *Geostand. Geoanal. Res.* **31** 331-343

[14] Paul B, Paton C, Norris A, Woodhead J, Hellstrom J, Hergt J and Greig A 2012 CellSpace: A module for creating spatially registered laser ablation images within the Iolite freeware environment. *J. Anal. Atom. Spectrom.* **27** 700-706

[15] Paul B, Woodhead J D, Paton C, Hergt J M, Hellstrom J and Norris C A 2014 Towards a method for quantitative LA-ICP-MS imaging of multi-phase assemblages: Mineral identification and analysis correction procedures. *Geostand. Geoanal. Res.* **38** 253-263

[16] Petrus J A, Chew D M, Leybourne M I and Kamber B S 2017 A new approach to laser-ablation inductively-coupled-plasma mass-spectrometry (LA-ICP-MS) using the flexible map interrogation tool 'Monocle'. *Chem. Geol.* **463** 76-93

[17] Drouin D, Couture A R, Joly D, Tastet X, Aimez V and Gauvin R 2007 CASINO V2.42 - Fast and easy-to-use modeling tool for scanning electron microscopy and microanalysis users. *Scanning* **29** 92-101

[18] Lanari P and Duesterhoeft E 2019 Modeling metamorphic rocks using equilibrium thermodynamics and internally consistent databases: Past achievements, problems and perspectives. *J. Petrology* **60** 19-56

IOP Conf. Series: Materials Science and Engineering **891** (2020) 012016 doi:10.1088/1757-899X/891/1/012016

[19] Williams M L, Jercinovic M J, Mahan K H and Dumond G 2017 Electron microprobe petrochronology. *Rev. Mineral. Geochem.* **83** 153-182

[20] Yasumoto A, Yoshida K, Kuwatani T, Nakamura D, Svojtka M and Hirajima T 2018 A rapid and precise quantitative electron probe chemical mapping technique and its application to an ultrahigh-pressure eclogite from the Moldanubian Zone of the Bohemian Massif (Nové Dvory, Czech Republic). *Amer. Mineralogist* **103** 1690-1698

[21] Lanari P, Riel N, Guillot S, Vidal O, Schwartz S, Pêcher A and Hattori K H 2013 Deciphering high-pressure metamorphism in collisional context using microprobe mapping methods: Application to the Stak eclogitic massif (northwest Himalaya). *Geology* **41** 111-114

[22] Burn M 2016 *LA-ICP-QMS Th-U/Pb allanite dating: methods and applications*. PhD-thesis. (Bern, Switzerland: University of Bern) 399

Analysis of oxide scales on oxidised EN 1.4509 ferritic stainless steel catalyst support by scanning electron microscopy

P Navarro Vicente[1], A Nuñez Galindo[1], J F Almagro Bello[1] and J A Odriozola[2]

[1] Acerinox Europa S.A.U, Departamento Técnico / Laboratorios e Investigación, Pol. Industrial Palmones s/n, 11379 Los Barrios, Spain
[2] CSIC-Universidad de Sevilla, Instituto de Ciencia de Materiales de Sevilla, Avda. Americo Vespucio 49, 41092 Seville, Spain

E-mail: pablo.navarro@acerinox.com

Abstract. This study aims to investigate the oxidation behaviour at high temperature of a commercial ferritic stainless steel EN 1.4509. Tests of high temperature oxidation in air were carried out under isothermal condition (950 ºC) with different exposure times in order to obtain the oxidation kinetics law associated to this material. The surface and cross-section were analysed by scanning electron microscopy and energy-dispersive X-ray spectrometry (SEM/EDS), where the oxide scales proved to be formed by an external Mn-Cr spinel and an inner chromia layer. These oxides were generated by the migration of metallic cations from the stainless steel to the atmosphere. The size of the generated spinel crystals increased when increasing the exposure time as well as the cross-section thickness did. Moreover, the surface modification of ferritic stainless steel EN1.4509 was studied when platinum is deposited on its surface. The feature of this surface-modified ferritic stainless steel against oxidative atmosphere was evaluated by SEM/EDS. This platinum deposition led to reach a different surface morphology in terms of crystal size and nature.

1. Background

The use of metallic structured catalytical reactors has attracted a great deal of attention as alternative to traditional catalyst [1]. These structures have been manufactured with different metals and alloys. Among the different substrate material reported, the thermal stability and the mechanical shock resistance properties is a key advantage of manufacture stainless steel structured support [2].

From a thermodynamic point of view, the oxidation reaction in metals is explained using the following reaction:

$$M\ (s) + n\frac{1}{2}O2 \rightarrow MOn\ (s) \qquad (1)$$

This reaction occurs when the free energy of the system is $\Delta G^{o} \leq 0$ [8]. The oxidation is a reaction between oxygen and metallic elements in the metal-gas interphase which produces an oxide. At a high temperature, the oxidation in metals produces a thick oxide whose growth will obey parabolic time dependence. This consideration was first described by Wagner. According to this theory, the growth takes place by the inward transport of ions [3]. The oxide scale is inhomogeneous. The two major potential phases in a ferritic stainless steel oxide scale are the chromia and Mn-Cr spinel [4].

The main aim of this work is the characterization of a commercial ferritic stainless steel EN 1.4509 used as a catalytic support under oxidative atmosphere. In the field of catalysis, when stainless steel is used as a support, it undergoes thermal treatment in order to generate an oxide scale that improves the anchoring and the deposition of the active phase [5].

The evaluation of the oxide scales formed during the thermal treatment at different exposure times provides information about the migration processes that are taking place. Surface and cross-section samples have been evaluated using scanning electron microscopy and energy-dispersive X-ray spectrometry (SEM/EDS) in order to study the changes in the topography and chemical composition of the phases during oxidation. The deposition of platinum on ferritic stainless steel is also studied for the comparison of the impact of noble metals during the oxidation reaction.

2. Experimental and discussion

2.1. Material selected

The selected material is a commercial ferritic stainless steel EN 1.4509 supplied by Acerinox Europa S.A.U. (1 mm thick). Prior to thermal treatment, samples were cut in 30 cm × 30 mm. The chemical composition of the material is shown in Table 1. All elements were analysed by X-ray fluorescence using a Panalytical PW 2606 XRF spectrometer, except for C and N, which were analysed by Leco CS 600 and TC 600 analysers, respectively.

Table 1. Elemental composition of stainless steel EN 1.4509.

Element	Al	C	Cr	Cu	Mn	Mo	N	Nb	Si	Ti
at%	0.045	0.021	17.525	0.123	0.347	0.019	0.019	0.386	0.401	0.122

2.2. Thermogravimetric experiments

The weight-gain curve, as a function of time for the ferritic material EN 1.4509 after being oxidised in air at 950 °C is shown in

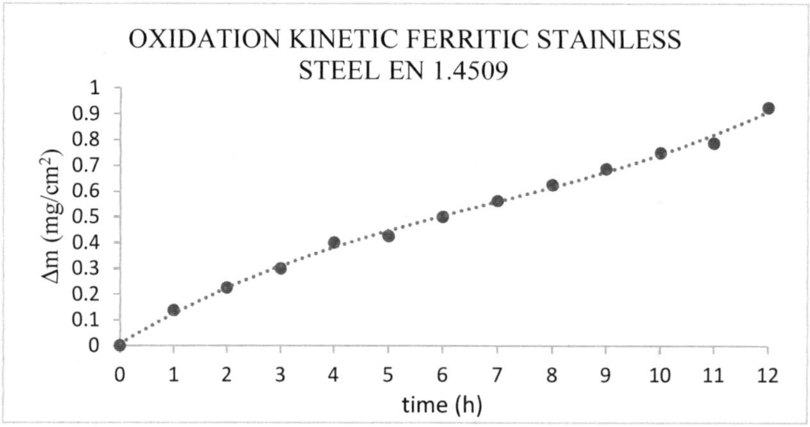

Figure 1. Mass change versus time curve for EN 1.4509 ferritic stainless steel oxidized in air for 12 hours.

The sample weight gradually increases as a function of time. The results show how the points are fitted to a parabolic curve. This kinetic behaviour indicates that the growth is controlled by the inward transport of metal cations of the substrate and/or the inward anions transport from the gas [6]. This suggest that during the early stages of the thermal treatment, the oxide layer is growing rapidly by the outward migration of the elements, but when this layer has a critical thickness, the transport is blocked, and the growing rate is reduced [7].

2.3. Microstructural evaluation

The characterisation was performed on a Zeiss ULTRA 55 FEG-SEM. Backscattered electron (BSE) images were analysed at an accelerating voltage of 10 kV and a working distance of 8.5 mm. For the purpose of investigating the morphological modifications during the oxidation process, the specimens tested under different thermal treatment are shown in Fig. 2.

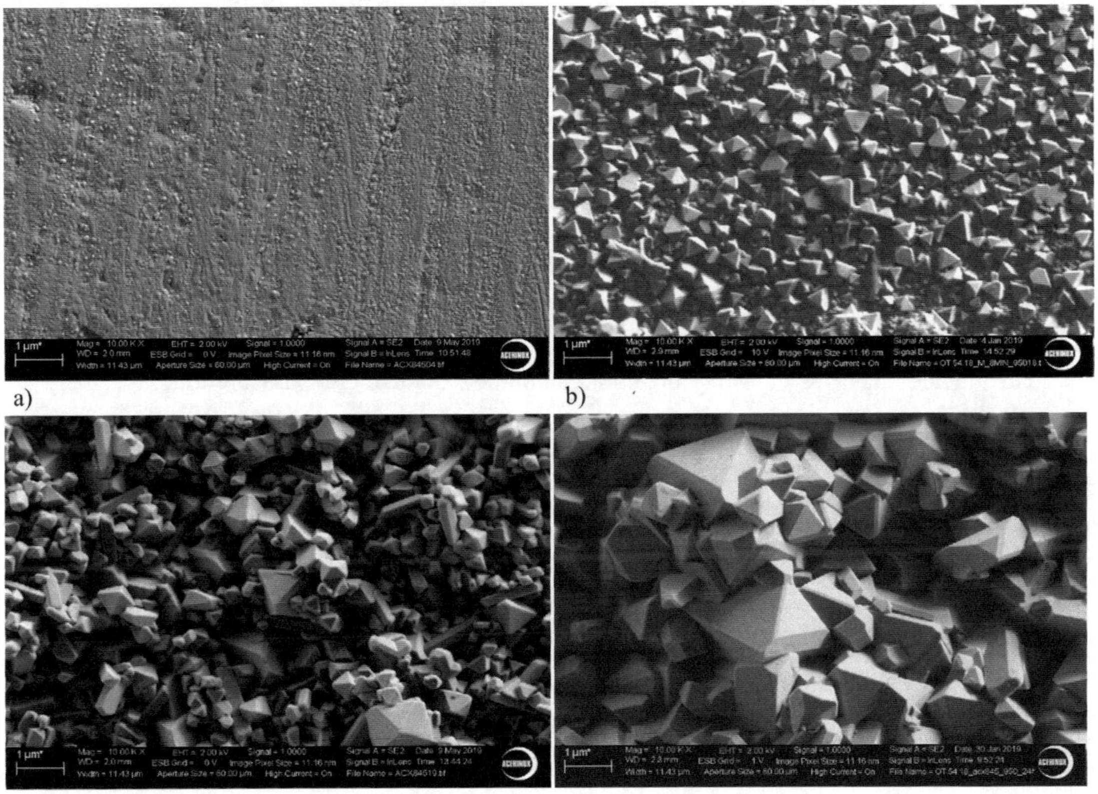

Figure 2. Surface morphology of the ferritic stainless steel EN 1.4509: a) Untreated; and thermal treated at 950 ºC in air for b) 8 min; c) 24 hours, and d) 168 hours.

Figure 2a shows thee morphology of the ferritic stainless EN 1.4509 before any thermal treatment. A modified surface was observed in Fig. 2b after a thermal treatment for 8 min at 950 ºC. This modification is attributed to the migration of metals to the surface and the formation of a crystal phase. It should be noted that surface is not completely covered by these crystals. A different topography is found after 24 hours test in for this sample, the crystal size has increased and the surface is entirely coated. A similar situation is observed in Fig. 2d. If the sample is heated in a long thermal treatment of 168 hours, an increase in the crystal size is also noted.

In all samples, regardless thermal treatment time, the oxidation products grows in an octahedral shape which is associated to stainless steel spinels in the literature [8]. Table 2 shows the EDS point analysis composition of the oxides products corresponds to a sample heated during 24 h at 950 ºC. All analyses were performed at an accelerating voltage of 10 kV.

Table 2. Chemical composition from EDS point analysis for crystals formed (in at%).

Sputtering	Phase	O	Al	Si	Ti	Cr	Mn	Fe	Nb
No	spinel	57.57	0.17	0.19	0.48	23.34	15.75	2.40	0.10
Yes	spinel	57.56	0.31	0.03	0.28	26.31	13.69	1.82	0
Yes	chromia	59.58	0.18	0.02	0.29	36.45	2.84	0.64	0

According to the chemical composition and the structure, this oxidation product is attributed to Mn-Cr spinel phase, with some contribution from other elements.

The spinel layer observed on the oxidized samples correspond to the most external one of the studied oxide scale. The spinel phase grow over a chromia protective layer. The Cr_2O_3 is not stable at high temperature and its volatilisation takes place thermodynamically [9]. However, the spinel layer formed over the chromia has a key role in the thermal behaviour of the material, inhibiting effectively the diffusion of chromium to the atmosphere [10].

Samples were prepared for FEG-SEM cross-section observation. A comparison between the cross-section of the heated samples is displayed in Figure 3. The BSE image reveals the variation of the oxide scale thickness generated during the oxidation process at different exposure time.

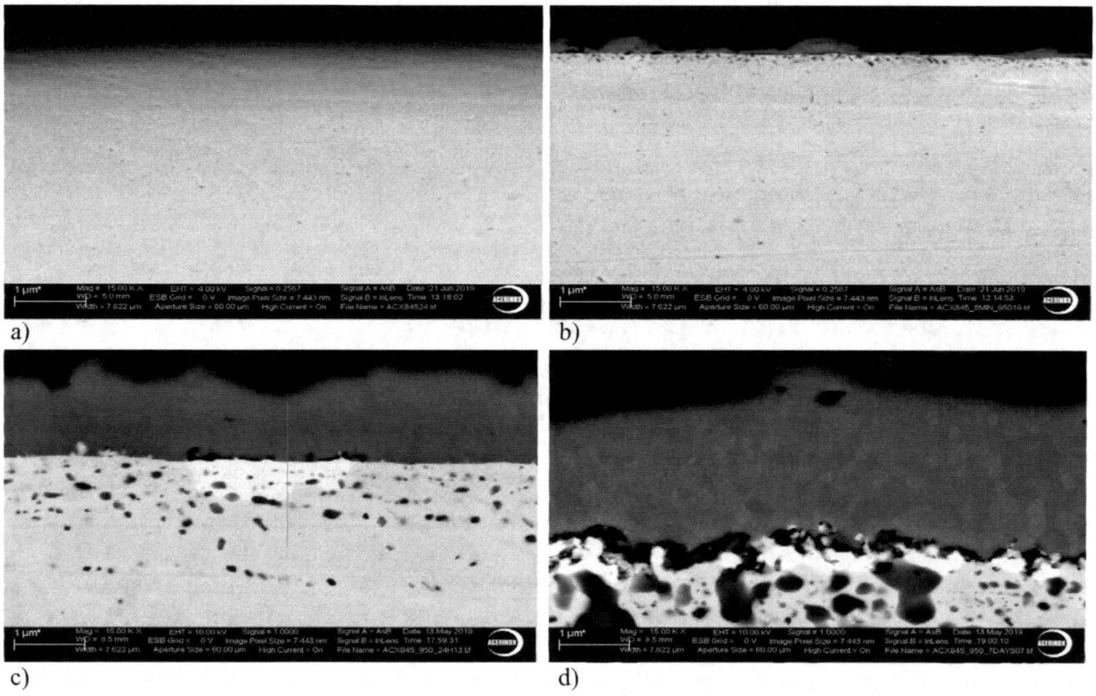

Figure 3. Cross-section view of the ferritic stainless steel EN 1.4509. a) Untreated. Thermal treated for: b) 8 min; b) 24 h; and d) 168 h.

Figure 3a is the untreated EN 1.4509 stainless steel. This sample does not exhibit any oxide scale, however, after 8 minutes at 950 °C, a thin oxide scale over the surface of the stainless steel can be appreciated in Figure 3b. When the exposure time raise 24 hours in the Figure 3c, due to the long diffusion time, the thickness of the oxide increase its size. Also is noted that the oxide scale is formed by two layers: an external spinel phase and an inner chromia protective layer [11]. As a result of the inward migration of oxygen, internal oxides were formed below the interphase oxide scale/stainless steel. According to literature, the white phase in the image is attributed to the intermetallic precipitation. A long term test of 168 hours, as shown Figure 3d, reveals that the oxide scale remains growing with the exposure time. The massive element migration has altered the interphase oxide scale/stainless steel, and due to the prolonged inward diffusion of the oxygen, the size and proximity of the internal oxides were increased. This prolonged oxidation test seems to generates an intermetallic continuous barrier.

2.4. *Chemical analysis*

The element migration during the thermal treatment was evaluated by energy-dispersive X-ray spectrometry (EDS). EDS mapping analysis in the cross-section of the oxidised specimens for 8-minute and 24-hour periods are shown in Figs. 4 and 5, respectively. The distribution and composition of the elements after testing show the influence of oxidation exposure time.

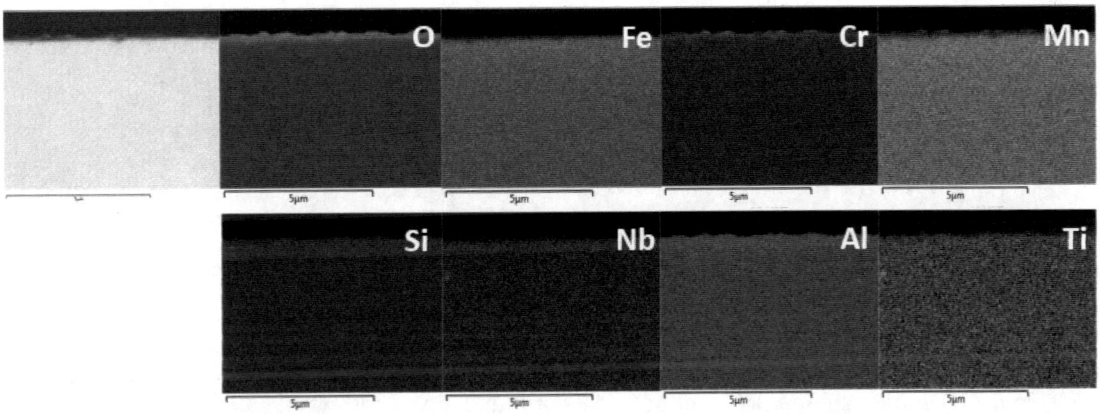

Figure 4. EDS elemental mapping of the specimen oxidised 8 min. at 950 °C.

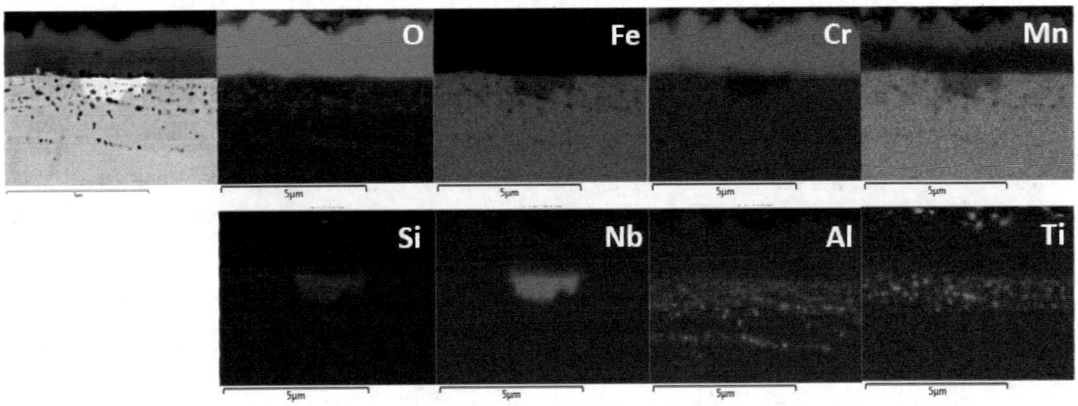

Figure 5. EDS elemental mapping of the specimen oxidised 24 h at 950 °C.

As shown in Fig. 4, oxygen is only concentrated in the most external layer of the sample, indicating the absorption in the surface, forming the protective chromia layer (with Cr coming from the stainless steel). Moreover, there is no evidence of oxygen inner migration in the matrix when forming the internal oxide products. However, Mn is slightly concentrated below the surface indicating the reaction with Cr to form spinel on the surface. This situation is quite different compared to the one after 24 h (EDS mapping in Fig. 5).

The external most layer is enriched with Mn, Cr and O, forming the spinel barrier $(Mn,Cr)_3O_4$, which is covering the Cr_2O_3 layer. Owing to the fast diffusion of Mn, this element diffuses through the chromia layer, and in response to the black band in the Mn mapping. 24 h at 950 °C is enough to promote the internal migration of oxygen, generating internal oxides such as oxides with Ti and Al, and the intermetallic precipitate, which is enriched in Nb and Si.

EDS line scanning was conducted to further investigate the distribution of elements from the external oxide layer to the stainless steel matrix. The red line in Figure 3c corresponds to the analysis extension, and the results are presented in Fig. 6.

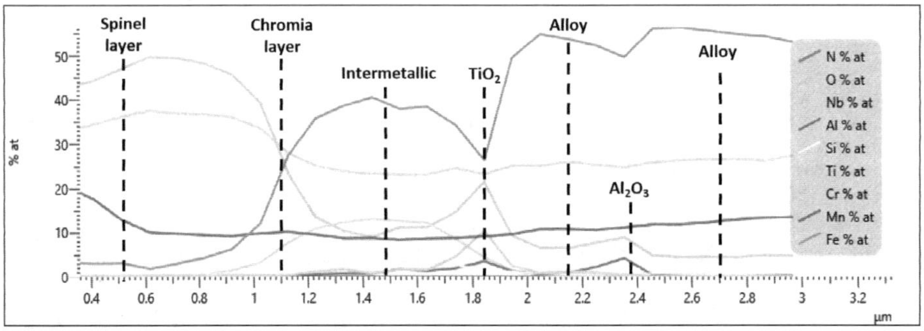

Figure 6. EDS line scanning of a ferritic stainless steel oxidised 24 h at 950 °C, see also Fig. 3c.

This analysis identifies the species present in the oxide scale, which is formed with an external Cr-Mn spinel layer over an internal layer of Cr_2O_3 corundum layer with a total thickness of 1.1 µm. The presence of internal oxides under the interphase suggest that the internal oxygen diffusion occurs, generating internal oxides rich in Ti and Al in the stainless steel matrix. According to other authors, the intermetallic phases are mainly Fe, Nb and Si, and can precipitate within the ferritic bulk [12, 13].

2.5. *Surface modification with platinum deposition*

The oxidation of the ferritic stainless steel EN 1.4509 is a surface reaction that generates spinels on the surface. The deposition of species in the stainless steel substrate modifies the oxidation products. A platinum thin film was deposited by sputtering with a plasma coating system in a stainless steel EN 1.4509, which was immediately thermally treated at 950 °C during 24 hours. The figures below present the oxides formed at 950 °C for 24 hours between a platinum sputtered and non sputtered samples.

The influence of platinum deposition can be seen by comparing Figs. 7a and 7b. After the thermal treatment, the strongly modification can be seen even at low magnification. In the micrograph, it is possible to identify the shape of the crystals formed in the modified specimen, which is in contrast with the uncoated sample. It seems that the growth of the spinel is stimulated by platinum.

The magnified images of Fig. 7a is presented in Fig. 8. Apart from the large crystal size growth of the spinels, platinum promotes the formation of another crystal phase. As is showed in the micrographs, this new crystal grow with an hexagonal appearance. The EDS point analysis as shown in Table 2 reveals the formation of chromium oxide phase Cr_2O_3 (eskolaite).

Figure 7. Surface appearance of EN 1.4509 oxidised 24 h at 950 °C. a) Pt-coating before thermal treatment. b) Uncoated.

Figure 8. SEM micrographs of the platinum deposited sample magnified at a) x5,000, and b) x10,000.

The deposition of platinum seems to modify the oxide scale growth mechanism. This alteration can also be observed in the cross-section as presented in Fig. 9. The cross-section of the Pt sputtered sample is shown in Fig. 9a, and reveals a thicker oxide scale than the unsputtered sample in Fig. 9b. Before the oxidation of the specimen, Pt was deposited as a thin film. However, after the thermal treatment, platinum is located in the oxide scale layer as a discrete metal shining inclusion (in BSE image mode). This behaviour was predicted by Stringer in 1966 [14]. The size of these particles was measured and can be found in a range between 25 nm to 170 nm.

In order to confirm that platinum is located as shining particles inside the oxide scale in Fig. 9A, an EDS mapping analysis in the cross-section of the specimen is shown in Fig. 10.

The results confirm that platinum is present inside the oxide scale forming discrete particles. The elemental distribution is similar to the uncoated samples; the signals for Cr, Mn and O are more intense. Another notable difference can be appreciated in the high concentration of internal oxides rich in Ti and Al. Furthermore, Si is present in the interphase oxide scale/metal bulk, forming a continuous barrier.

IOP Conf. Series: Materials Science and Engineering **891** (2020) 012017 doi:10.1088/1757-899X/891/1/012017

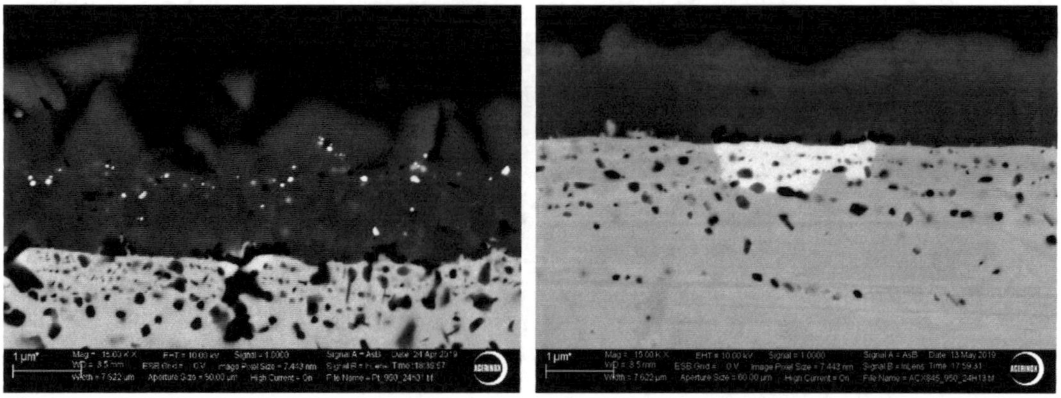

a) b)

Figure 9. Cross-section of EN 1.4509 oxidised 24 h at 950 °C. a) Sputtered Pt before thermal treatment. b) Non-sputtered.

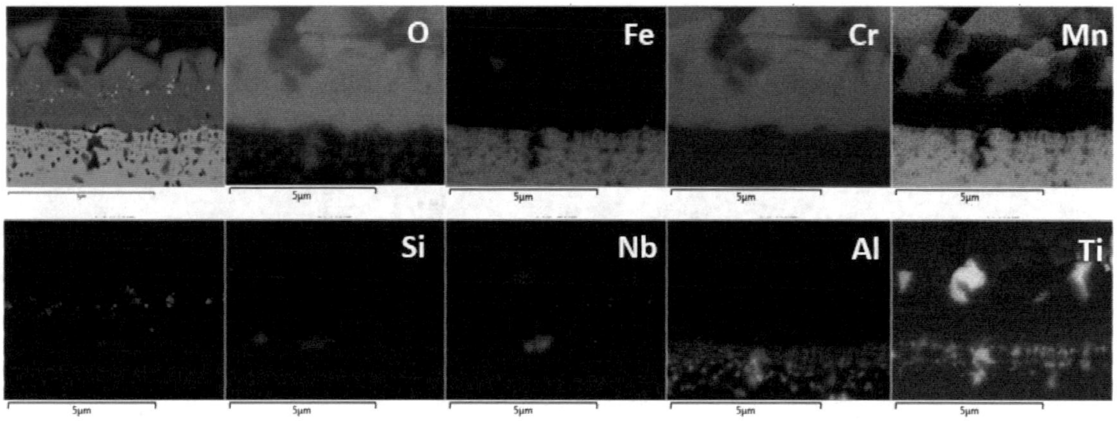

Figure 10. SEM-BSE cross-section micrographs and EDS mapping analysis over an oxidised Pt coated sample.

3. Conclusions

The oxidation of a commercial ferritic stainless steel EN 1.4509 was investigated over isothermal conditions at 950 °C in air for different exposure times. The mass gained is increased with the exposure time and the kinetic law obeys the parabolic behaviour for 12 hours. This tendency is typical from protective oxide scale growth.

All the investigated samples exhibited an external Mn-Cr spinel layer over a chromium layer. However, the thickness of the oxide scale and the spinel size increase with the exposure time.

The oxide scales are formed by the outward migration of metallic elements from the ferritic alloy and the inward migration of the oxygen from the atmosphere. The EDS mapping analysis gives the distribution and the chemical composition of the involved elements in the oxidation process.

- The external diffusion of metallics elements, such as Mn and Cr toward the surface, reveals the multilayer oxide scale nature, which is formed by the Mn-Cr spinel external layer that grows over the chromia protective layer. The Mn EDS elemental map reveals how this element reaches the surface through the chromia layer.
- The oxygen internal diffusion can be seen below the interphase oxide scale/stainless steel matrix, developing internal oxides rich in Al and Ti.

The sample modification provided by the sputtering of platinum makes a significant contribution in the oxidation process.

- The 24 hours period of thermal treatment on modified sample with platinum deposition, generate a larger crystal size compared with the obtained after 168 hours oxidation process.
- The evaluation of the cross-section shows how platinum takes part of the oxide scale, forming discrete metallic inclusions. The EDS mapping analysis confirm that the final position of the elements is inside the oxide scale, but not forming a continuous layer.

Platinum induces the growth of hexagonal crystals of Cr_2O_3 over the EN 1.4509 surface. These crystals are vertically oriented. This new phase generated is not observable in non-sputtered samples.

Acknowledgements
I would especially like to thank the European Microbeam Analysis Society for the invitation to the EMAS 2019 Workshop at the NTNU, Realfagbygget, Trondheim, Norway. This research has been done during the second year of myPhD-thesis; therefore, I would like to thank the University of Seville the thesis programme "Science and Technology of New Materials". Pablo Navarro acknowledges to Acerinox Europa S.A.U. and also to the Spanish Ministry of Economy and Business for the financial support. This work is included in FERRINOP Project, co funded by CDTI (www.cdti.es).

References

[1] Roh H S, Koo K Y, Jung U H and Yoon W L 2017 Hydrogen production from natural gas steam reforming over Ni catalysts supported on metal substrates. *Curr. Appl. Phys.* **10** S37-S39
[2] Laguna O H, Domínguez M I, Centeno M A and Odriozola J A 2016 Forced deactivation and postmortemcharacterization of a metallic microchannel reactor employed for the preferential oxidation of CO (PROX). *Chem. Engng. J.* **302** 650-662
[3] Sequeira C A C 2019 *High temperature corrosion: Fundamentals and engineering.* (Hoboken, NY: John Wiley and Sons) chapter 1, 35
[4] Wagner C 1936 *Z. Phys. Chem.* **32** 447
[5] Wei L, Zheng J, Chen L and Misra R D K 2018 High temperature oxidation behavior of ferritic stainless steel containing W and Ce. *Corrosion Sci.* **142** 79-92
[6] Khanna A 2012 High temperature oxidation. in: *Handbook of environmental degradation of materials*. (Kutz M; Ed.) (Norwich, NY: William Andrew Publishing) chapter 5, 127-194
[7] Stringer J, Wilcox B A and Jaffe R I 1972 The high-temperature oxidation of nickel-20 wt.% chromium alloys containing dispersed oxide phases. *Oxidation Metals* **1** 11-47
[8] Hosseini S N, Karimzadeh F, Enayati M H and Sammes N M 2016 Oxidation and electrical behavior of CuFe2O4 spinel coated Crofer 22 APU stainless steel for SOFC interconnect application. *Solid State Ionic* **289** 95-105
[9] Holcomb G and Alman D 2006 The effect of manganese additions on the reactive evaporation of chromium in Ni–Cr alloys. *Scripta Materialia* **54** 1821-1825
[10] Wei L, Han L, Chen L and Zhao Y 2018 Oxidation behaviour of cerium and tungsten-containing ferritic stainless steels at 1200 °C in air. *Proc. Manuf.* **15** 1588-1595
[11] Col A, Parry V and Pascal C 2017 Oxidation of a Fe–18Cr–8Ni austenitic stainless steel at 850 °C in O2: Microstructure evolution during breakaway oxidation. *Corrosion Sci.* **114** 17-27
[12] Palcut M, Mikkelsen L, Neufeld K, Chen M, Knibbe R and Hendriksen P V 2010 Corrosion stability of ferritic stainless steels for solid oxide electrolyser cell interconnects. *Corrosion Sci.* **10** 3309-3320
[13] Alnegren P, Sattari M, Froitzheim J and Svensson J E 2016 Degradation of ferritic stainless steels under conditions used for solid oxide fuel cells and electrolyzers at varying oxygen pressures. *Corrosion Sci.* **110** 200-212
[14] Stringer J 1966 The effect of alloying on oxidation: Quantitative treatments. *Metall. Rev.* **11** 113-128

IOP Conf. Series: Materials Science and Engineering **891** (2020) 012018 doi:10.1088/1757-899X/891/1/012018

About the reliability of EBSD measurements: Data enhancement

G Nolze[1] and A Winkelmann[2]

[1] Federal Institute for Materials Research and Testing (BAM), Unter den Eichen 87, 12205 Berlin, Germany
[2] AGH - University of Science and Technology, Academic Centre for Materials and Nanotechnology, Al. Mickiewicza 30, 30059 Krakow, Poland

E-mail: gert.nolze@bam.de

Abstract. An extensive set of information about the diffracting volume is carried by EBSD patterns: the crystal lattice, the reciprocal lattice, the crystal structure, the crystal symmetry, the mean periodic number of the diffracting phase, the source point from where it has been projected (projection centre), the crystal orientation, the sample topography (local tilt), the (preparation) quality or defect density of the crystal, and possible pattern overlaps. Some of this information is used regularly in conventional EBSD analyses software while others are still waiting for a more widespread application. Despite the wealth of information available, the accuracy and precision of the data that are presently extracted from conventional EBSD patterns are often well below the actual physical limits. Using a selection of example applications, we will demonstrate the gain in angular resolution possible using relatively low-resolution patterns of approximately 20k pixels in combination with pattern matching (PM) approaches. In this way, fine details in a microstructure can be revealed which would otherwise be hidden in the orientation noise.

1. Introduction

Electron backscatter diffraction (EBSD) is a technique for a quantitative microstructure characterisation of crystalline materials in the scanning electron microscope (SEM). Standard applications use EBSD patterns primarily to identify phases, phase distribution, texture description, grain size, grain shape and grain boundary analysis. New data-processing tools are continuously developed, and these can help to improve the reliable interpretation of often very complex microstructural problems in materials science.

Because small misorientations (MO) are important for understanding the mechanical behaviour of materials [1–3], it is important that the EBSD orientation measurements are precise. The computation of the MO can be performed by multiplying only two orientation matrices: $\Delta\mathbf{g}_{ij} = \mathbf{g}_i^{-1} \bullet \mathbf{g}_j$, from which the MO angle and axis can be determined.

A key factor for orientation and MO measurements is the angular resolution, which determines the precision of a crystal orientation derived from an EBSD pattern. Indirect techniques are common to determine the MO precision, e.g., the determination of MOs measured under comparable conditions at two positions in an undeformed grain. As these positions are presumed to have negligible MOs, the apparent MOs can be used to approximate an error of the experimental MO determination. From the accompanying MO histogram, the mean, the maxima, or the 95th percentile of the MO distribution can be interpreted as alternative measures for the orientation error.

Of widespread use is kernel average misorientation (KAM) angle which is the mean deviation of orientations determined around a selected pixel to the orientation of the pixel itself. Please note that this approach additionally reduces the real, or relative, orientation variation by the orientation averaging. Moreover, the definition of the kernel size for local averaging is different from software to software packages so that KAM angle distributions are only comparable if the same algorithm and parameters (maximum MO angle, number of classes, etc.) were used. In the present paper we use the KAM as calculated in MTEX [4] which is based on four surrounding pixels and a central pixel (top, bottom, right and left).

Approaches for an increase of the angular resolution in EBSD have been previously discussed, see e.g., [5]. From simple geometrical reasons it can be expected that the acquisition of higher-resolved, i.e., less-binned EBSD patterns, should be a straightforward way to increase the angular resolution of the orientation results. This more time consuming measurement effort for a higher pattern resolution is not always practical and can result in less precise results, cf. Fig. 1. In this context, it is unfortunate that the specific post-processing procedures in commercial "black-box" software are usually inaccessible. The graphs in Fig. 1 display the KAM angle distributions for maps derived from EBSD patterns with different image resolution of 160×115 and 320×230 pixels in addition to those obtained from different Hough-resolution. As anticipated, the higher the image resolution the higher the angular resolution, i.e., the blue curve shifts to smaller misorientation angles. However, to obtain this higher resolution with comparable signal quality, the dwell time must be quadrupled for 320×230 pixel patterns. Please note that the red curve in Fig. 1 is also derived from low-resolved patterns, but collected with effectively the same total exposure time because of the application of frame averaging: $t_{320\times230} = 4\text{frames} \cdot t_{160\times115}$. Taking into account that the blue and red curve were taken at different mapping positions, their similar shapes suggest an apparent correlation between pattern and angular resolution. Because of the identical measurement times but different pattern resolutions, the similarity of the KAM results indicates a software-internal, further binning of patterns to a general size and resolution. The detectable but small increase in orientation precision is very likely the result of an unintentional improvement of signal-to-noise ratio by software-internal binning and not related to the selected pattern resolution.

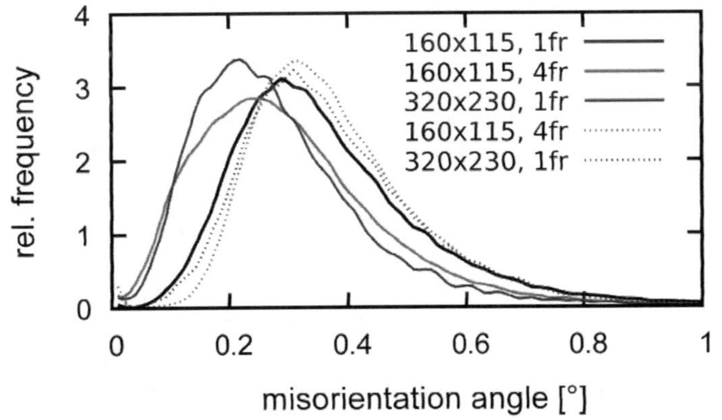

Figure 1. KAM angle distribution (polycrystalline GaP). Patterns of 320×240 pixel size deliver a slightly better angular resolution than 160×120 pixel patterns, but four frames (fr) averaged generate an equivalent distribution as the higher resolved patterns. Hough-resolution (HR) indexing using HR = 100 (dotted lines) versus the default value of HR = 60 (solid lines).

The analysis discussed above has been performed using the default Hough resolution of HR = 60 in CRYSTALIGN (Esprit 1.94, Bruker-Nano). It can be expected that an increase of the Hough space resolution HR should actually deliver a higher angular precision since it resolves the Hough space with a considerably higher resolution. As can be seen in Fig. 1, however, an increase to HR = 100 results in a histogram which is effectively identical to the one performed in a quarter of time, cf. the identically coloured, dotted lines.

While the need for higher angular resolutions in EBSD is a permanent issue, it is interesting to observe the actual improvements in orientation resolution when higher-resolved patterns are collected. In [6] patterns of 640 × 480 pixels and 128 × 96 pixels have been used for comparative measurements in deformed stainless steel. Relative to the 16-times longer acquisition time for the non-binned patterns, the achieved improvements are quite small. In [7] it is reported that the derived average geometrically necessary dislocation (GND) density seems to be unaffected by pattern resolutions higher than 160 × 128 pixels. These results are possibly related to the software-internal pattern processing discussed above. In [8, 9] the mean of the KAM distribution was two times smaller for the 640 × 480 pixel size as compared to the 128 × 96 pixel size (step size: 250 nm). This difference continues to decreases with increasing step size. For 3 μm step size, high-resolution patterns are only ≈5 % more precise than their low-resolution equivalent.

In [7] the influence of the step size (map resolution) on GND and MO fluctuation has also been investigated. In order to exclude the dependency of step size in the local deformation assessment, a parameter called the local gradient G(L) was proposed [10]. In summary, it can be stated that an increase or even an estimation of the orientation correctness without exact knowledge of the company-specific techniques used during pattern processing is not target-oriented. An independent technique is needed that re-analyses the diffraction patterns as unaffected data sources.

2. Orientation refinement by pattern matching

Pattern matching (PM) is an orientation refinement which starts with a crystal orientation taken from a conventional EBSD system, and tries to find a more precise orientation description by optimising the match between the locally acquired experimental and in real-time simulated patterns. Depending on the required resolution, this process can be much more time consuming in comparison to the Hough-based orientation determination. However, the primary goal of PM is not a further increase of measurement speed but a higher orientation precision. In this way, PM approaches can be an additional tool for applications where conventional EBSD does not have enough resolution or fails completely.

2.1. Small MOs with high precision

The first example uses a duplex steel which contains austenite precipitates in a ferrite matrix that are formed during heat-treatment. The mapped area shown in Fig. 2 is smaller than the ferrite grain so that all precipitates are orientation variants of this specific ferrite grain. For materials engineering, the most accurate description of the orientation relationship as well as the habit plane are of high interest since it is assumed that they correlate directly with the strength of the material. Therefore, a highly precise orientation measurement is desired in this case.

The KAM angle maps in Fig. 2 already indicate a clear improvement compared to the manufacturer result. The reddish coloured map in Fig. 2a changes to a much lighter coloured image in Fig. 2b. The latter also enables a clear recognition of the austenite precipitates.

The respective histograms of the KAM angle in Figs. 2c and 2d reflect quantitatively the increase in orientation precision. The according maxima, mean values and ranges which cover ≈ 95 % of all KAM angles are listed in Table 1. All values are at least four-times smaller which is interpreted as significant improvement of the orientation precision. This effect is also evident in pole figures which display the pole distribution of approximately 66,000 single austenite measurements. The scattering of the approximately displayed 400,000 dots is already small for the raw orientation data, but for the refined orientations the pole scattering is considerably reduced. Thus, the pole distribution looks sharper.

Beside the higher precision, PM has still another benefit. In about 200 out of 268,000 cases, the commercial EBSD system failed in the phase assignment, cf. the extra dots in the centre of one segment of the three-fold symmetric pole distributions in Fig. 3a. Using PM, the software realises the low correlation between the assumed phase and the locally acquired backscattered Kikuchi diffraction (BKD) pattern (EBSD patterns corrected by the background signal). A test with ferrite patterns showed they fit clearly better.

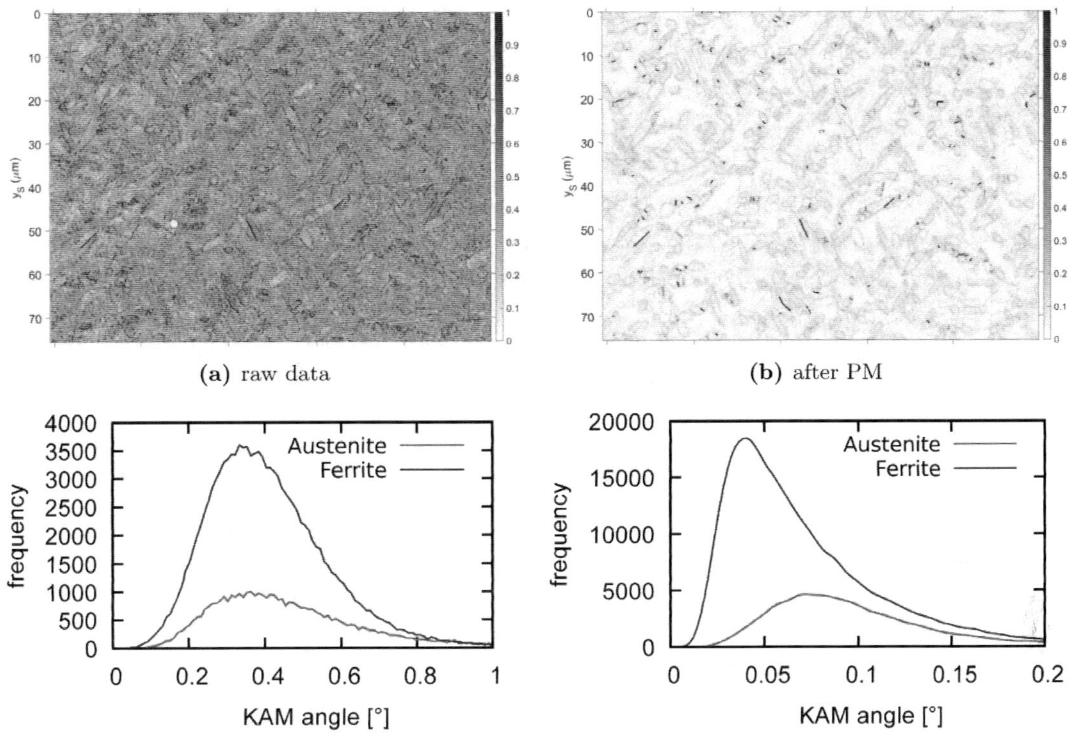

(a) raw data (b) after PM

(c) raw (d) refined

Figure 2. a) KAM angle maps derived from raw orientation data (ESPRIT 1.94, Bruker), and b) from refined orientation data after PM. The respective histogram plots are shown in c) and d). Please note the different KAM angle scales in c) and d).

Table 1. Some characteristic parameters of the KAM angle distribution from Fig. 2. The difference between raw and refined data are roughly described by a factor 4.

		maximum	mean	≈ 95 %
raw	austenite	0.36°	0.43°	0.89°
	ferrite	0.33°	0.39°	0.71°
refined	austenite	0.08°	0.09°	0.20°
	Ferrite	0.04°	0.06°	0.16°

This example demonstrates that even EBSD patterns with a comparatively low resolution of only 160 × 115 pixels possess a high orientation precision. As a matter of course, the degree of improvement depends on many factors. For ideal materials with low dislocation or defect density, like Si, the precision can reach factors of up to 10. However, as defect density increases the improvement factor in the KAM angle histogram is lowered since the histogram broadening is dominated by intrinsic misorientations and not by the orientation noise. Nevertheless, even in such cases the precision of PM is still incredible as visible on the microstructure of the next application.

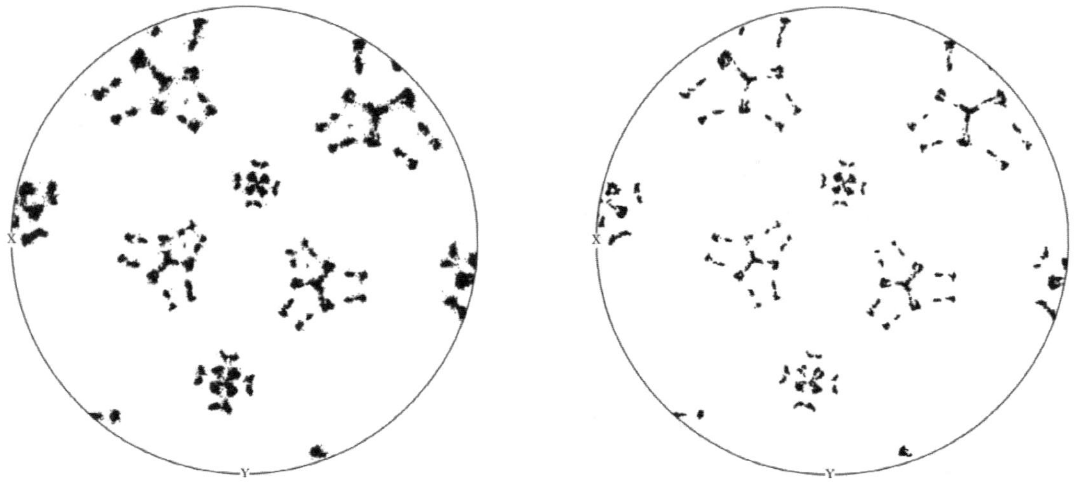

(a) 66464 raw orientations (b) 66273 orientations after PM

Figure 3. a) {011} pole figures derived from raw orientations, and b) from orientation data after PM. The reduction of scattering from a) to b) is satisfying. The number of considered measured points differs since during refinement PM realises that for ≈ 0.07 % of all measurements the phase was wrongly identified.

2.2. Large misorientations at high precision

The development of new materials or manufacturing processes often triggers a number of complications since previously used test and analytical procedures probably need to be adapted in order to discover new characteristic features and novel correlations between microstructure and observed properties. Additive manufacturing (AM) proposes a promising technology which generates totally different and unusually complex and variable microstructures.

Typical characteristics of AM are repetitive remelting, fast local cooling and an extremely heterogeneous and complicated structure derived from the applied (sequential) printing-like technique. Thus, features like a periodically varying grain size distribution, or comparatively strong lattice rotations are not surprising. The gradient defined by the microscopic, "lattice-like" periodic printing technique places higher demands on area selection (target preparation) as well as on test methods. Terms like "statistically relevant" will take on an entirely new dimension since "uniform" or "regular" are microstructure properties which no longer match to this technology.

Nevertheless, as technique with high spatial resolution, EBSD is a recommendable analytical method. It is only the angular resolution required for a precise evaluation of existing lattice rotations that needs improvement. On the other hand, the formed microstructure in AM materials is characterised by comparatively big lattice rotations. A suitable visualisation is the grain reference orientation deviation (GROD) angle which displays the angular deviation of the local orientation to the mean orientation of the grain, cf. Fig. 4. It represents a top view on the printed material. The periodicity in the checkerboard-like microstructure is consistent with the printing process but not identical since they are shifted against each other by one half. This means that the printing path is defined by the clearly visible boarder lines of the checkerboard and not by the centres of the visible squares. The colour bar in Fig. 4 suggests that 7° or more is a commonly observed misorientation angle. Thus it may be that high orientation precision is not required for AM materials.

In order to discuss this issue, Fig. 5 compares the KAM maps based on conventional and refined orientation data. Despite huge lattice rotations accumulated during fast solidification, both KAM maps display significant differences. The lower orientation noise in Fig. 5b enables a more authentic

Figure 4. GROD angle map (in degrees) reflecting considerable lattice rotations within additive manufactured material. Grain boundaries are overlaid as white lines.

impression on information which are, of course, also intrinsic in Fig. 5a. It is only because of the high orientation noise that they are not observed. This difference suggests that a high orientation precision is also beneficial in case of big lattice rotations since they enable a reliable evaluation of low-angle boundaries, although the inhomogeneity of the square-shaped regions underlines the complexity of any interpretation of such processes. They have neither the translation symmetry (homogeneous) within the observed surface plane nor in depth.

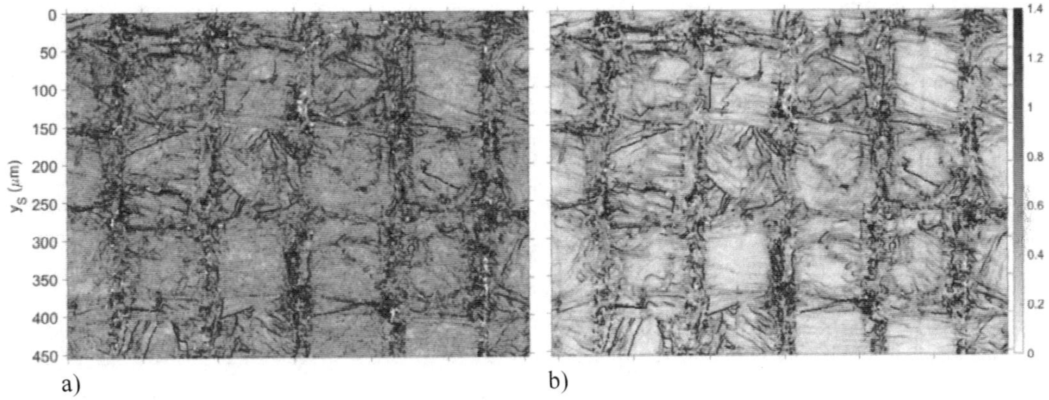

Figure 5. Comparison of KAM angle maps. The raw orientation a) enables the recognition of low and high defective regions; however, the higher precision in b) extracts much more details which cannot be discovered in a).

As mentioned above, the high precision of PM is not that evident anymore in KAM angle distributions when other effects like a high intrinsic defect density dominate the curve shape, cf. Table 2. Although the maximum and mean for the refined data differ by a factor of two, the 95 % range does not appreciably change due to the big lattice rotations within grains.

A reliable detection of small misorientations is preferably interesting for a more advanced interpretation. A typical application is the density of GNDs [11]. For the derivation of the curvature tensor a clear qualitative leap is apparent by PM, cf. Fig. 6. This gives at least some prospects regarding a more reasonable characterisation of high-gradient materials so that the data significantly contributes to a better understanding and a successful correlation between microstructure and material properties.

IOP Conf. Series: Materials Science and Engineering **891** (2020) 012018 doi:10.1088/1757-899X/891/1/012018

Table 2. Some characteristic parameters of the KAM angle distribution derived from Fig. 5

	maximum	mean	≈ 95 %
raw	0.58°	0.70°	1.64°
refined	0.28°	0.40°	1.49°

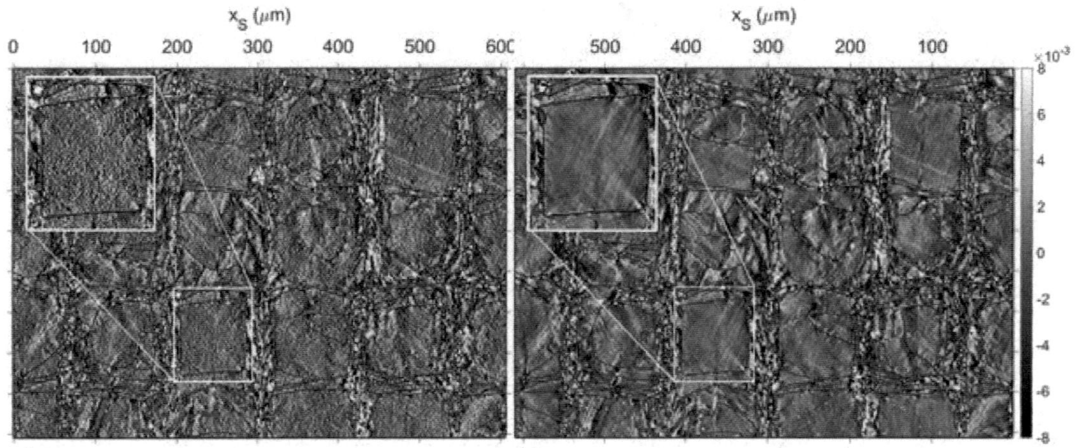

a) b)

Figure 6. One of the six measurable components of the EBSD curvature tensor $\boldsymbol{\kappa}$ – $\boldsymbol{\kappa}_{31}$ – which is used for the computation of the GND density. Both maps are derived from the identical patterns, only a) shows the result from raw orientations, whereas in b) the refined orientation data are used.

2.3. Pseudosymmetric crystal lattices

Phase transformation is a common phenomenon in material science. During this process, the atomic configuration at lower temperature tends to loose the previously higher symmetry by small shifts of atoms. Although such movements do not practically vary the intensity of bands, they are associated with a slight deformation of the crystal lattice and may require a redefinition of the basis vectors which results in clearly different lattice parameters.

Quite frequently observed is the transition from a hexagonal close-packed arrangement of statistically distributed or already ordered elements to a further ordered configuration with longer translation periodicity of orthorhombic symmetry. As one representative Ni_3Sn_2 shall be discussed. The lattice relationships are summarised in Table 3. The last row shows that during phase transformation the lattice effectively contracts in maximum only 1.7 % which is below the typical detection limit of EBSD, at least for the low-resolution patterns. To be on the safe side, for this investigation patterns of 400 × 300 pixels have been used. Later measurements demonstrated that this pseudosymmetric problem can be solved by EBSD patterns in lower resolution (160 × 115 pixels).

A single grain of Ni_3Sn_2 decomposes into numerous domains during cooling which have crystallographically three totally different orientations of pseudosymmetric character. This means that despite non-equivalent orientations the respective diffraction signals look nearly identical. Therefore, the indexing fails and randomly one of the three pseudosymmetric orientations are selected by the software. Notwithstanding that fact, the domains can be easily imaged by channelling-in contrast in SEM [12]. The question arises regarding the reliability of PM to discriminate between these three orientation variants.

Table 3. Lattice correlations of high (HT) and room temperature (RT) phase of Ni_3Sn_2. V_{uc} … unit cell volume.

	Ni_3Sn_2 (HT)	Ni_3Sn_2 (RT)
symmetry	$P6_3/mmc$	$Pnma$
lattice parameters	$a_h = 4.146$ Å $c_h = 5.253$ Å	$a_o = 7.124$ Å $b_o = 5.195$ Å $c_o = 8.152$ Å
V_{uc}	75.5 Å^3	304.9 Å$^3 \approx 4 \cdot V_{uc_h}$
relationships		$a_o \approx \sqrt{3} \cdot a_h$ (−0.8%) $b_o \approx c_h$ (−1.1%) $c_o \approx 2 \cdot a_h$ (−1.7%)

As Fig. 7 (a) demonstrates by the three different colours, a separation of the pseudosymmetric orientations is possible and also reliable. However, compared to the standard PM the applied procedure works a bit different. As usual, the refinement starts with the orientation delivered from the EBSD system. This can be either an orientation based on the orthorhombic low-temperature unit cell, or anorientation based on the high-temperature hexagonal lattice. Both have drawbacks since the $\vec{c}$-axis of the hexagonal phase is not parallel the $\vec{c}$-axis of the low-temperature phase but to the $\vec{b}$-axis, cf. Table 3. Therefore, several transformations are necessary to derive respective projections from the orthorhombic template pattern using orientations referring to a hexagonal cell. An easier method uses a refinement based on pseudosymmetric data of the orthorhombic phase. However, now $\vec{b}$ as monoclinic axis does not match the Euler-angle convention rotating around Z and X, respectively $\vec{e}_3$ and $\vec{e}_1$. Notwithstanding these complications, the subsequent refinement compares the match for all three pseudosymmetric solutions. The best is taken as most likely orientation description for this pattern. The briefly sketched procedure not only delivers a qualitatively convincing distribution of the three variants, but also satisfies in a quantitative manner. This is first demonstrated in Fig. 7b by the KAM angle distribution. For the shown map 95 % (blue line) of all derived KAM angles are smaller than 0.14° with a median (green line) of 0.04° which is certainly the main impact of higher-resolved patterns analysed.

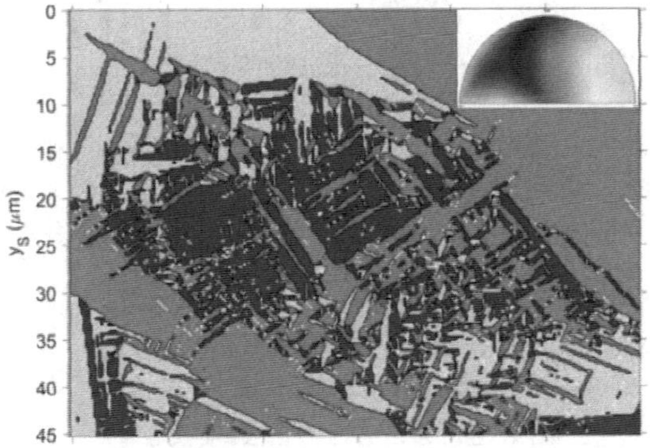

a) IPF-X map

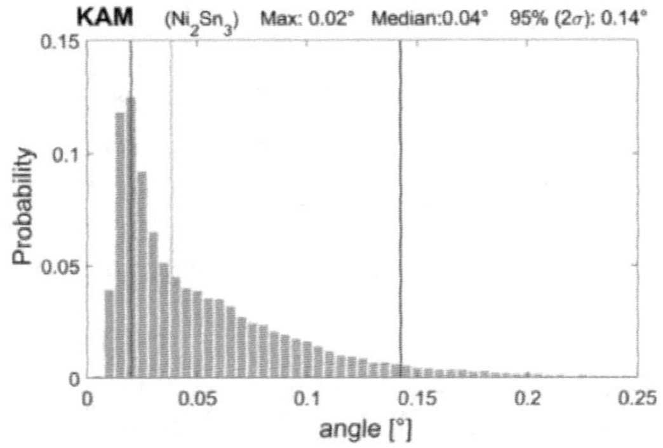

b) after PM

Figure 7. Orientation description of Ni_3Sn_2-domains covering three pseudosymmetric orientations within a former single grain. The orientation description is based on the enantiomorphic symmetry group 222 [13].

Despite these small KAM angles, MOs not only result from orientation noise but also from real existing lattice rotations. For the imaging of the MO we will now use the GROD angle, as in Fig. 4 in addition to the MO axis which transfers the local orientation in the mean orientation of the grain by a rotation around the GROD angle. The maps displaying the GROD angles and axes are pictured in Fig. 8. The colour bar of the GROD angle encodes an angle variation between 0 ... 1°. The colour distribution in Fig. 8a shows that large areas are characterized by GROD angles < 0.2°. Despite these small angles, the noise is reduced in the map which indicates a high orientation precision in agreement with the KAM angle histogram in Fig. 7. The map also proves that within comparatively big grains higher GROD angles may occur. They mainly appear in correlation to other variants which is not surprising since transformations with lattice distortions are assumed to cause stress which may result in small lattice rotations.

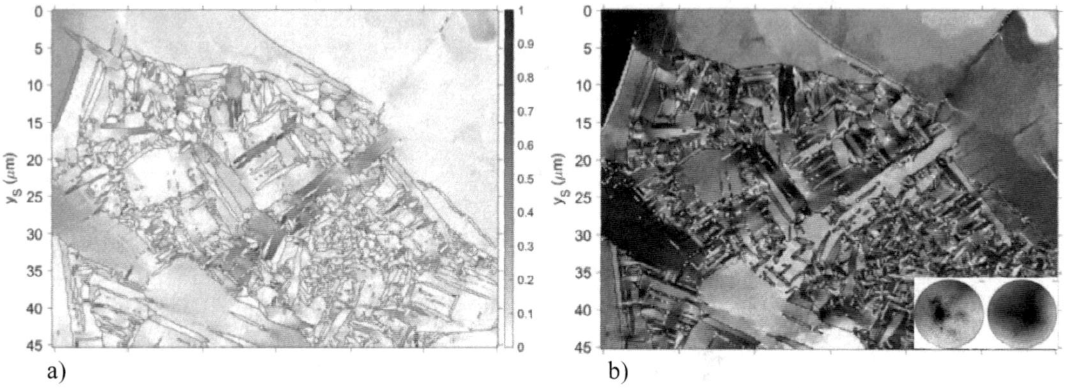

Figure 8. Grain reference orientation deviation (GROD) displayed as a) rotation angle, and b) as rotation axis. The inset image in b) displays the MO axes distribution as stereographic projection of the two hemispheres.

The MO angles help to understand and imagine the reaction of a microstructure on processes like transformations. In many cases the MO axis is of high interest, especially when the directional distribution is inhomogeneous. If no correlation to the crystal symmetry is assumed, a rotation direction colouring of the entire projection sphere needs to be considered, cf. the inset image in Fig. 8b. Many directions form a point cloud within the upper hemisphere (left) displaying as well the colour key. In contrast to the pure MO angle presentation, the colouring of the MO axis results in clearly higher colour contrasts. Please note that small MO angles commonly result in a considerable uncertainty of the MO axes. In this respect, the comparatively low colour fluctuation in Fig. 8b is surprising. In any case, it is a further indication for the high orientation precision achievable with PM.

Both the MO axis and angle, can be processed together when the used colour key is compressed so that for each specific grain the orientation variation is used to define the colour key size. Then identical MO angles are differently coloured when their MO axes varies as shown in Fig. 9.

Please note that the centre of gravity of the (compressed) colour key is defined by the mean orientation of each particular grain. The deviation of GROD angle as well as GROD axis is used to assign the colour. The disadvantage of this presentation tool is that an interpretation is only qualitatively possible since the definition of each colour depends on the compression applied.

Another domain-forming phase is perovskite. It is in fact a materials class which is characterised by numerous phase transitions describing different crystallographic subgroup relationships. Thus, in addition to the ideal cubic structure type, all other crystal systems down to triclinic phase descriptions can be found in databases. However, most of the listed phase descriptions are so similar that even X-ray

Figure 9. MO angle and axis from Fig. 8 encoded by a heavily compressed IPF-colour key. The orientation spread (polar angle) needs to be manually adapted.

powder diffraction as standard tool for phase discrimination cannot distinguish between them. Since the ability to distinguish between phases is reduced for EBSD, any attempt to try this for pseudo-cubic perovskite is doomed to failure. A safe option is to use channelling-in contrast in SEM, or FSE images to recognise non-cubic perovskites by the characteristic herringbone structure, cf. Fig. 10. This is indisputable evidence that a perovskite is not of cubic symmetry and the lattice is slightly distorted. Figure 10 displays a bimodal microstructure of coarse $BaTiO_3$ crystals embedded in a fine-grained matrix. The most often used approach in case of domains is a tetragonal distortion described by a c/a ratio very close to 1.

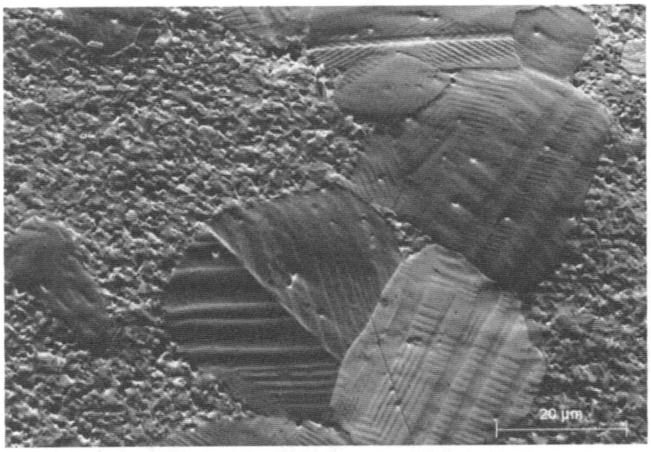

Figure 10. Forescattered electron image of a $BaTiO_3$ ceramic reflecting the characteristic herringbone structure. Please note the different colours of the grains whereas the domains do not considerably vary in colour but mainly in topography.

EBSD investigations prove that all BKD patterns can be easily indexed using a cubic crystal description. A single grain is then reflected as uniform crystal, possibly with a thin twin lamella, cf. for two of the four biggest grains in Fig. 10. In contrast, the attempt to index the patterns by a tetragonal phase description results in a more or less random selection of pseudosymmetric orientation solutions, as reported above for Ni_3Sn_2. Sometimes the indexing delivers a preferable selection of an orientation solution for a few domains, but after rotating the sample by a few degrees only this preferred orientation selection disappears so that it is assumed to be an artefact of the applied band detection.

For PM we used a similar approach as briefly sketched for Ni_3Sn_2, only that now all axes are consistent ($\vec{c}_{cub} \| \vec{c}_{tet} \| \vec{e}_3$), and only three pseudosymmetric solutions exist. The IPF-map in Fig. 11a proves that obvious misinterpretations do not occur. On the first glance, domains are uniformly indexed in a way which appears reasonable.

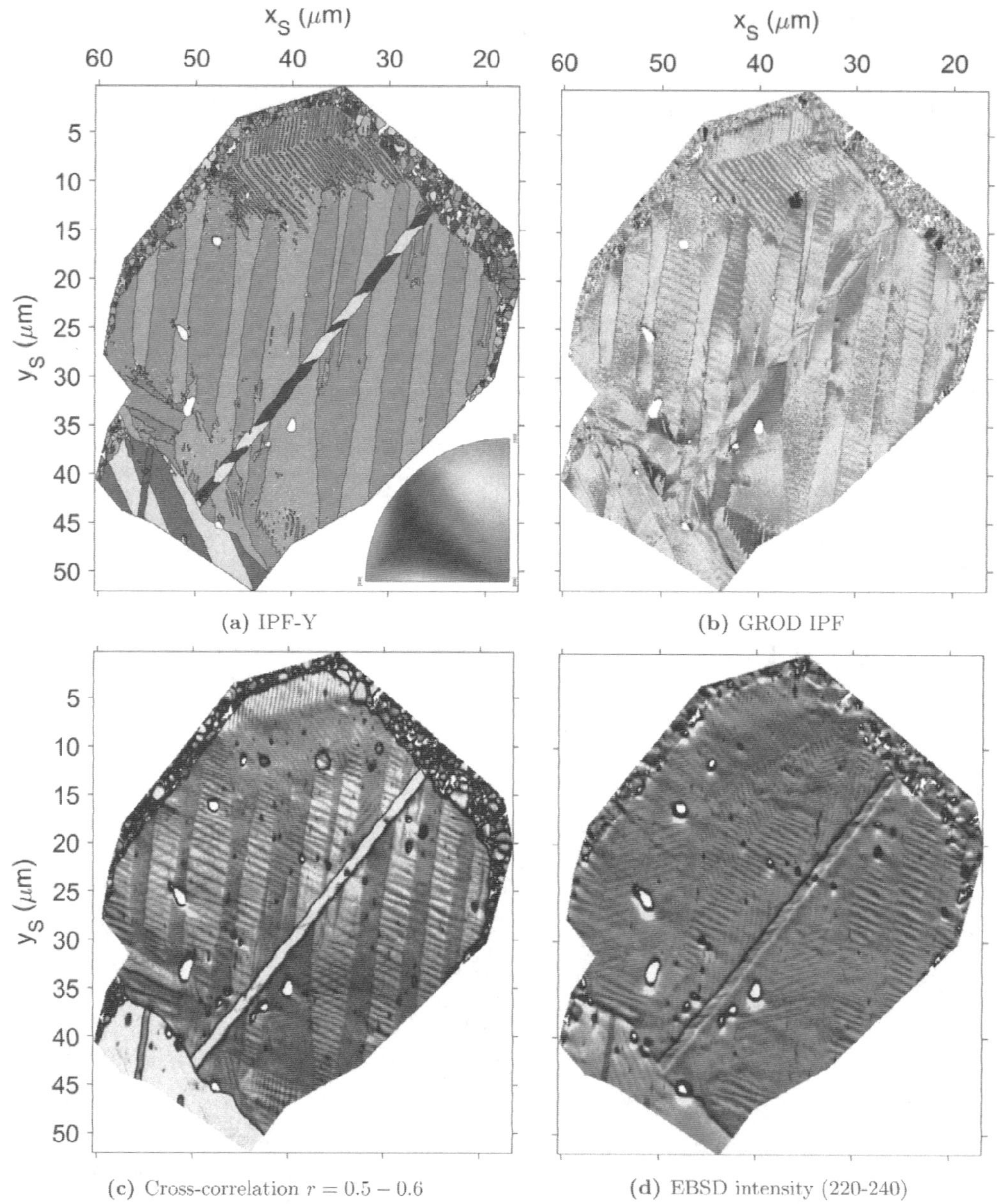

Figure 11. Orientations in a tetragonally assumed $BaTiO_3$ derived by PM and displayed as IPF-Y map. For the sake of clarity grain boundaries are displayed as black lines (a). The intragranular misorientations are imaged in b) by a compressed IPF-colour key which displays systematic orientation variations within uniformly assumed domains in (a). In order to compare the information content the cross-correlation coefficient (c) and the overall intensity derived from the EBSD patterns (d) are shown as well.

However, a comparison with FSE images raises at least tentative doubts that the displayed unambiguousness in grain recognition is totally correct. The reason for qualms is the discrepancy between the numerous domains visible in FSE images and the apparent uniformity in the IPF map. In other words, the big domains look too uniform.

According to the doubts, further investigations of possible MOs discover a systematic orientation modulation which suggests a correlation to the domains visible in FSE images like in Fig. 10. Despite the comparatively noisy signal, the modulation in form of parallel and practically equidistant stripes are well visible, see Fig. 11b. It is, however, conspicuous that there is no overall correlation. Fortunately, the differing regions accidentally correlate with the four biggest pores visible in the maps. Whereas in Figs. 11b and 11c the slope of the faintly visible, thin stripe are uniform from left top to right bottom, in Fig. 11d they are reverse. Due to this, the question is posed more and more often, whether for the different maps in Fig. 11 the identical information from EBSD patterns is used. Figs. 11a and 11b are based on the derived orientation solution only.

Figure 11c represents the map which indicates the degree of match used for Figs. 11a and 11b. Brighter colours indicate a better correlation whereas darker colours suggest that the applied PM resulted in a lower cross-correlation value r. Reasons for the signal differences in Fig. 11c may be related to the orientation description or to an unfortunately selected phase description having an imperfect c/a ratio. The map in Fig. 11d, however, is mainly triggered from the background intensity. Since the chemical composition is uniform, only channelling-in and topography are the main influencing factors. The signal in Fig. 11d is formed in higher depths so that we assume that the different slope has its reason in different information depths of BKD and background signal.

Another factor which should be considered is the influence of the angular resolution due to the sample tilt in combination with the domain alignment. For different alignments in depth the interaction volume reacts in a smearing which fits to the nearly homogeneously but darker coloured domains in Fig. 11c. In between domains show higher r-values, which suggests that they are favourably aligned and the electron probe does not average differently aligned domains as in the neighbourhood. The darker stripes are explainable by thin subdomains which produce an overlaid signal of two different but pseudosymmetric orientations. The impact of the spatial resolution at tilted sample surfaces can be also seen at the upper part of the crystal, cf. Fig. 11c. There, a herringbone structure of very fine domains can be observed with conventional EBSD. All different domains can be distinguished but the signal for the nearly vertically aligned domains at the top of the image is of much higher quality since in Fig. 11c they show higher r-values than the domains directly below in the image.

3. Summary and conclusions

In the selected examples that were presented, the potential of low-resolution (10k - 20k pixels) EBSD patterns was demonstrated to improve of orientation precision. In order to circumvent the unknown image processing of high-resolved patterns in the manufacturer software, pattern matching (PM) of stored pattern with dynamical simulations was used. The orientation data derived from the low-resolution EBSD patterns are already of higher precision than comparable investigations based on higher resolved patterns with Hough-based orientation determination. It has been shown that a higher precision can be useful for the investigation of materials with different levels of deformation. PM offers the highest precision, required in the case of discriminating pseudosymmetric orientation solutions.

While combination of higher-resolved patterns and PM can improve the orientation precision, other influences then start to become important such as: varying lattice parameter ratios, the correct projection centre position, the possible pattern distortions caused by local magnetic or electric fields, and the effective electron energy.

Acknowledgement
We are very thankful for being able to investigate the duplex steel sample provided by W. Swiatnicki (TU Warsaw). We thank K. Sommer (BAM Berlin) for the austenitic sample produced by additive

manufacturing, and A. Leineweber (TU Bergakademie Freiberg) for providing the Ni_3Sn_2 sample. In particular, we also express our gratitude to R. Saliwan Neumann and M. Buchheim for countless preparations and EBSD experiments in BAM, R. Hielscher for developing and continuous maintenance of MTEX, and J.A. Schneider (University of Alabama in Huntsville) for correcting the manuscript.

References

[1] Wilkinson A J 2001 A new method for determining small misorientations from electron backscatter diffraction patterns. *Scripta Materialia* **44** 2379-2385
[2] Bate P S, Knutsen R D, Brough I and Humphreys F J 2005 The characterization of low-angle boundaries by EBSD. *J. Microscopy* **220** 36-46
[3] Brewer L N, Othon M A, Young L M and Angeliu T M 2006 Misorientation mapping for visualization of plastic deformation via electron back-scattered diffraction. *Microsc. Microanal.* **12** 85-91
[4] Bachmann F, Hielscher R and Schaeben H 2010 Texture analysis with MTEX – Free and open source software toolbox. *Solid State Phenom.* **160** 63-68
[5] Brough I, Bate P S and Humphreys F J 2006 Optimising the angular resolution of EBSD. *Mater. Sci. Technol.* **22** 1279-1286
[6] Kamaya M 2009 Characterization of microstructural damage due to low-cycle fatigue by EBSD observation. *Mater. Charact.* **60** 1454-1462
[7] Jiang J, Britton T B and Wilkinson A J 2013 Measurement of geometrically necessary dislocation density with high resolution electron backscatter diffraction: Effects of detector binning and step size. *Ultramicroscopy* **125** 1-9
[8] Kamaya M 2009 Measurement of local plastic strain distribution of stainless steel by electron backscatter diffraction. *Mater. Charact.* **60** 125-132
[9] Kamaya M 2010 A smoothing filter for misorientation mapping obtained by EBSD. *Mater. Trans.* **51** 1516-1520
[10] Kamaya M 2011 Assessment of local deformation using EBSD: Quantification of accuracy of measurement and definition of local gradient. *Ultramicroscopy* **111** 1189-1199
[11] Pantleon W 2008 Resolving the geometrically necessary dislocation content by conventional electron backscattering diffraction. *Scripta Materialia* **58** 994-997
[12] Nolze G, Hielscher R and Winkelmann A 2017 Electron backscatter diffraction beyond the mainstream. *Cryst. Res. Technol.* **52** 1600252
[13] Nolze G and Hielscher R 2016 Orientations – perfectly colored. *J. Appl. Cryst.* **49** 1786-1802

IOP Conf. Series: Materials Science and Engineering **891** (2020) 012019 doi:10.1088/1757-899X/891/1/012019

Evolution of crystalline orientations in the production of ferritic stainless steel

A Núñez Galindo[1,3], I Collado Garcia[2], D L Sales[3] and J F Almagro Bello[1]

[1] Acerinox Europa S.A.U, Departamento Técnico / Laboratorios e Investigación, Pol. Industrial Palmones s/n, 11379 Los Barrios, Spain

[2] Universidad de Cádiz, Escuela Superior Ingeniería, Departamento de Ciencia de los Materiales e Ingeniería Metalúrgica y Química Inorgánica, LABCYP Group, Avda. Universidad de Cádiz 10, 11519 Puerto Real, Spain

[3] Universidad de Cádiz, Escuela Politécnica Superior, Departamento de Ciencia de los Materiales e Ingeniería Metalúrgica y Química Inorgánica, INNANOMAT Group, Avda. Ramón Puyol s/n, 11202 Algeciras, Spain

E-mail: andres.nunez@acerinox.com

Abstract. Ferritic stainless steel EN 1.4016 is used in a wide range of applications, the most common ones related to sheet forming. Several problems in the post-processing of these steels relates to their texture and anisotropy. Therefore, it is necessary to know the mechanisms of texture formation in the subsequent stages of metal manufacturing processes. EBSD has been demonstrated as a successful characterisation technique for this purpose.

It is known that during re-crystallisation of Fe-Cr steels, deviations from the desired γ-fibre texture promote a decrease of deep drawability. Additionally, α-fibre damages formability. Subsequent cold rolling and annealing can enhance the deep drawing properties of the steel sheet.

In this research, a standard sample and a modified one with optimised settings as regard to chemical composition and manufacturing process, to improve the formability properties, are characterised. To analyse the preferred orientation and the type of main fibre present in the material, ODF and Aztec Reclassify Phase, to calculate the content of martensite, were used.

1. Introduction

Ferritic stainless steel, in comparison to austenitic one, is low-cost, price-stable and has good engineering properties. Type EN 1.4016 (AISI 430) is used in a large range of applications, including the most common related to conformation, stretching and deep drawing. Recent research evaluates its application as a hydrogen container [1].

Several problems after each rolling and annealing processes during these steels manufacturing, lead to certain preferred crystallographic orientations (coming from their texture). Due to the impact that crystallographic texture has on the anisotropy of material properties, it is necessary to know the mechanisms that are involved in the texture of these materials during its manufacturing routing [2-10]. In this sense, deviations from the γ-fibre desired texture occur during re-crystallisation of Fe-Cr steels, and a large number of grains being oriented with their {111}-plane parallel to the sheet plane, hence promoting a decrease of deep drawability. In the same way, the α-fibre texture component transforms in such a way that <110> orientates parallel to rolling direction and {001}, damaging to formability [11-20].

The orientation measurement of polycrystalline grains (γ- and α-fibre), carried out after formability of the material, allows for evaluation of the deformation mechanism of the material. To describe crystallographic textures, a representation in a three-dimensional space is necessary. This representation is done by maps or diagrams of the orientation distribution function, ODF, defined by Euler angles φ_1, Φ, φ_2, which applied to the axes [100], [010] and [001] of the crystal structure of the grain, make it coincide with the axes RD (rolling direction), ND (normal direction) and TD (transversal direction) of the sample, respectively. In addition, the anisotropic behaviour of flat products can be characterised by the Lankford coefficients; high values of normal plastic anisotropy, r_m (greater than 1), and planar plastic anisotropy, Δr (as close to zero as possible), are good indicators of high formability [21-24].

Texture in steel depends, among other variables, on chemical composition, the remains of the primary solidification structure in the final product [25-26], and processing parameters like finishing temperature during hot rolling, percentage of cold work reduction, annealing temperature and time. Hot rolling process control can provide favourable textures so that subsequent cold rolling and annealing can enhance the deep drawing properties of the steel sheet.

In this research, two variations of the same grade of ferritic stainless steel have been analysed. One of them, type 0A, has both standard composition and processing route intended for a broad and general scope of applications. The other one, type 1C, features optimised settings in chemical composition and manufacturing process to improve the formability properties.

The technical approach, presented in a preliminary study of this research [27], includes characterisation by field emission gun – scanning electron microscopy to study texture formation, evolution and microstructural phases obtained through different steps of the processing route: casting, intermediate hot rolling, final hot rolling, annealing and pickling after hot rolling, and final annealing after cold rolling. EBSD post-processing procedures were used such as pole figure [28] to characterise the anisotropy of the material, orientation distribution function (ODF) to analyse the preferred orientation and the type of main fibre present in the material, and AZTEC RECLASSIFY PHASE [29-34], to calculate the content of martensite within the ferrite matrix.

2. Experimental procedure

Two EN 1.4016 steel samples with different initial compositions (basic: 0A, and modified: 1C) were taken from the daily production of Acerinox Europa, S.A.U. (www.acerinox.es) in different positions of the processing route: from slabs during casting, from intermediate hot rolling, from material after hot rolling with intermediate annealing and after cold rolling, and final annealing. The samples were subjected to thickness reductions between the different production stages, from slab to intermediate hot rolling by 98 % for both materials, and from this stage until final annealing by 80 % for the basic and by 87 % for the modified sample. In the modified annealing treatment, the processing speed of the materials corresponds to an increase of 20 % in the carry-in time into the furnace as compared with the basic.

The chemical composition of the samples was analysed using conventional techniques, X-ray fluorescence (XRF), spark optical emission (S-OES), and instruments available in a stainless steel factory using a Panalytical Axios Fast XRF spectrometer (www.panalytical.com), an Oblf Qsn 750 (www.oblf.de) and Leco CS 600 and TC 600 analysers (www.leco.com). Also, using the THERMOCALC software (TCW 4 using database TCFE5), the phase diagrams were calculated as a function of temperature in equilibrium conditions [35].

The materials selected for this study are summarised in Table 1, together with their global chemical composition and type of applied annealing. The composition of sample 0A is referred to as "basic" composition, while the sample 1C is a "modified" composition, with lower contents of interstitial elements such as C and N.

Table 1. Composition, annealing treatment and metallurgical properties of the as-supplied materials.

Sample identification	Composition (wt%)	Annealing time
0A	**Basic**: C: 0.050; N: 0.035; Si: 0.35; Cr: 16.3	Short
1C	**Modified**: C: 0.025; N: 0.025; Si: 0.45; Cr: 16.7	Long

The phase diagrams of the studied chemical compositions have been analysed by THERMOCALC software. Figure 1 shows the phase diagram of the basic (0A) and modified (1C) samples characterised by a dual phase ferrite-austenite structure, gamma loop and precipitation of chromium carbide type $M_{23}C_6$.

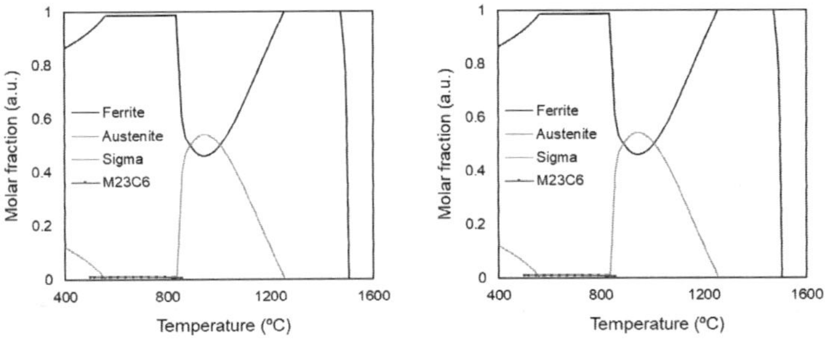

Figure 1. Phase diagrams of the material systems of sample 0A (left) and sample 1C (right).

The phase diagrams show important differences between the two materials in terms of transformation points and amount of high temperature austenite: Ac1 and Ac3 are 855 and 1,256 °C for the 0A material, and 857 and 1,180 °C for the 1C material; while the maximum austenite contents are 53.3 % (at 919 °C) for 0A and 32.8 % (at 943 °C) for 1C. This austenite is not stable so it transforms to martensite when cooling down to room temperature.

Grain size analysis was performed after polishing and etching the samples using Vilella's reagent during 30 to 75 s. The observation and measurement of grain size were performed with an Olympus GX71 light optical microscope (LOM) (www.olympus.co.uk), by Heyn lineal interception and comparison procedures. In order to assess the cooling mechanism of both ferritic stainless steel, electrochemical pickling of the cross-sectional area of the sample is carried out to reveal their macrostructure. As a result, the cooling rate from the surface to the core of each slab is observed [36].

The mechanical tests were carried out in samples with basic and modified composition after hot rolling and after final annealing using a RK100 Roell+Korthaus universal testing machine, for the study of elongation, the average r-value of Lankford and planar anisotropy and coefficient of acrimony.

EBSD characterisation was performed on a Zeiss Ultra 55 FEG-SEM (www.zeiss.es), using polished longitudinal sections of the samples. The sample sections were prepared by conventional metallographic procedures, with a final polish with colloidal silica suspension OPS (www.struers.com). The FEG-SEM used is equipped with a CHANNEL 5 EBSD system, from Oxford Instruments (www.oxinst.com). EBSD maps were acquired at 20 kV, at a working distance of 16 mm and a 0.5 μm step size. In addition, through the EBSD analysis, the software AZTEC RECLASSIFY PHASE, taken into consideration the austenite-ferrite transformation phenomenon, allows to calculate the content of martensite within the ferrite matrix, where ferrite and martensite can be distinguished from each other. The presence of martensite, was then checked by light optical microscopy [37].

3. Results and discussion

3.1. Micro- and macrostructure

Figure 2 shows light optical microscopy (LOM) images of the structure in samples from slabs for basic (0A, left) and modified (1C, right) material, respectively. Optical micrographs of both slab structure show ferritic grains with martensite needles at grain boundaries and within the matrix, as well as the distribution of carbides within the grains is noticeable. Moreover, sub-grains are revealed in the basic composition material (0A) but not in the modified one (1C). Therefore, the structure of material 1C shows a better re-crystallisation state.

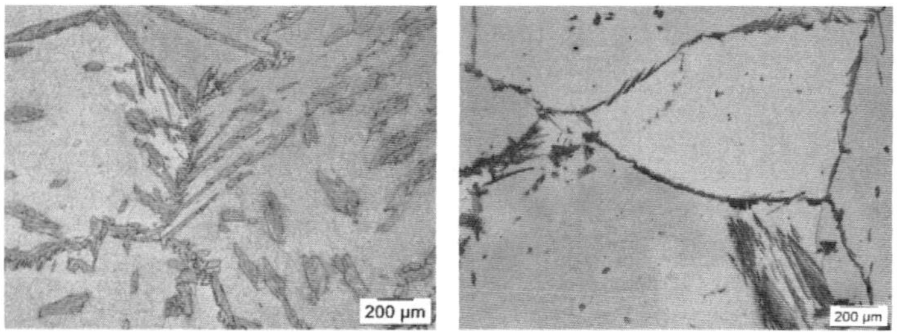

Figure 2. LOM image showing the structure of slab 0A (left) and slab 1C (right).

Figure 3 reveals a detail of the macrostructure of each sample, basic (0A) and modified (1C), in order to study the solidification mechanism of both ferritic stainless steel. The characterisation of the macrostructure indicates that the solidification mechanism is heterogeneous in the material with basic composition, whereas in the modified composition it is homogeneous.

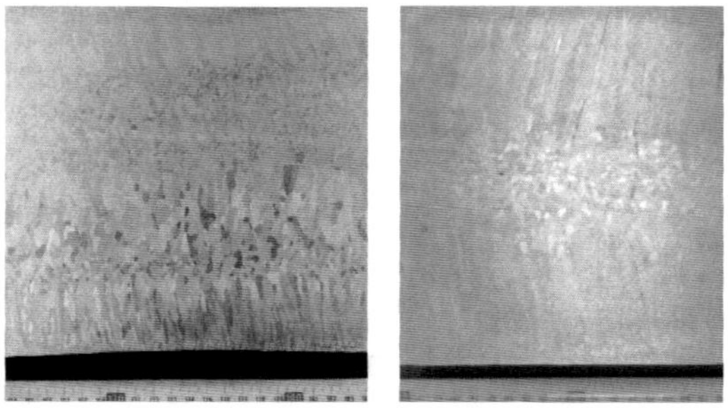

Figure 3. Macrostructure in the basic material 0A (left) and in the modified one 1C (right).

3.2. Mechanical properties

The results of elongation, anisotropy and strain hardening of both hot rolled (HR) and final annealed (FA) materials are shown in Table 2.

Table 2. Summary of mechanical tests.

Sample Id.	A (%)	r-value			r_m	Δr	n
		RD	ND	TD			
0A after HR	31	0.36	0.63	0.48	0.52	-0.21	0.18
1C after HR	28	0.34	0.69	0.54	0.57	-0.25	0.15
0A after FA	28	0.86	0.70	1.18	0.86	0.32	0.21
1C after FA	29	0.80	1.05	1.21	1.17	-0.09	0.21

The average anisotropy (r_m) and the planar (Δr) were calculated by the following equations:

$$r_m = (r_0 + r_{90} + 2r_{45}) / 4 \quad (1)$$
$$\Delta r = (r_0 + r_{90} - 2r_{45}) / 2 \quad (2)$$

The values of normal anisotropy and strain hardening in the hot rolled materials are significantly lower than those of the annealed materials, which is consistent with the corresponding microstructures: deformed and re-crystallised. On the other hand, both normal and planar anisotropy values are clearly more favourable in the 1C material, which means that it has a better performance for deep drawing operations. Therefore, the basic composition sheet exhibits more pronounced anisotropy and lower deep drawability than the modified composition sample.

3.3. Orientation distribution function, pole figures and reclassify phase

EBSD analysis allowed obtaining information of the whole sample thickness for the main phase, as the observation plane was ND-LD (ND and LD stand for normal direction and line, or rolling, direction). The FEG-SEM characterisation was carried out using a polished sample in longitudinal section (ND-LD plane).

To analyse the texture and the most important fibre present in the material, the procedure used is the orientation distribution function (ODF); it is well known that the rolling process reinforces the α-fibre (<110> parallel to RD, at $\Phi = 0°$) and re-crystallisation reinforces the γ-fibre (<111> parallel to ND, at $\Phi = 55°$). The pole figures, where random texture indicates uniform distribution in the projection, elements of symmetry for each plane can be analysed and so the anisotropy of the material.

Texture, anisotropy and content of martensite is analysed in basic (0A) and modified (1C) materials at different steps of the processing route: during hot rolling and after intermediate and final annealing.

3.3.1. Samples intermediate hot rolling

Figure 4 shows the ODF of basic and modified material in samples before final hot rolling with a thickness of 26 mm. EBSD texture results indicate the presence of γ-fibre in the modified material (Fig. 4-left) for the presence of density at $\Phi = 55°$, and the absence of γ-fibre at $\Phi = 55°$ in the basic one (Fig. 4-right) where α-fibre predominates.

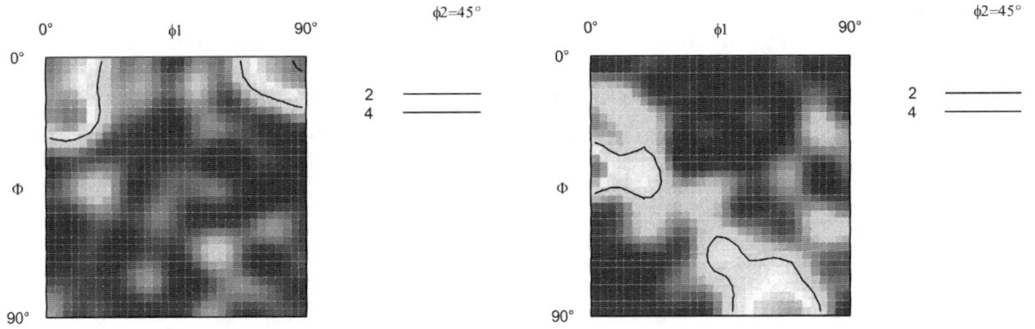

Figure 4. ODF of basic material, 0A (left) and modified material, 1C (right), before final hot rolling.

Figure 5 shows the pole figure of basic and modified material in samples during hot rolling, before finishing mill. Pole figures indicate that symmetry elements for each plane, in both materials, are not clearly defined, probably because, in this state, the materials have not developed its ideal orientations that can be found in bcc materials [28].

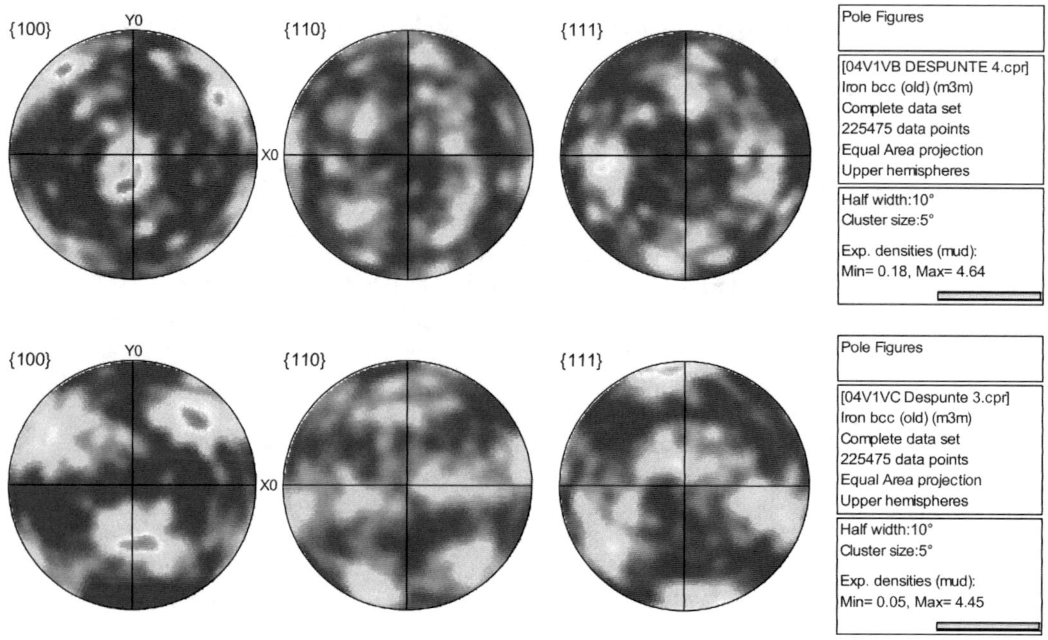

Figure 5. Above: pole figure of basic material, 0A, before final hot rolling. Below: pole figure of modified material, 1C, before final hot rolling.

Figure 6a shows the band contrast (BC), coincidence site lattice CSL) and grain boundaries (GB) map, Fig. 6b shows the CSL legend, while Fig. 6c shows the GB legend, of the basic and the modified material in samples during hot rolling before finishing mill.

The characterisation reveals the presence of a phase in the ferrite matrix (red areas in images), identified as retained martensite after quenching, in both materials [28]. Martensite and ferrite cannot be discriminated by EBSD and this software uses the density of crystalline defects to identify the martensite areas, as follows: the tetragonal elongation is greater than 2.5 - 3 % between a-, b- and c-axes, Fig. 7. This means that there is a real chance to separate martensite from Fe bcc as shown in reference [30]; in this case it would be > 3 % (3 % = 2.93138 c-axis). The specific algorithm to solve pattern is using mentioned in references [30] and [33-34]. Using martensite structure data means that it is possible recognise and index martensite in a range greater than 3 % tetragonal distortion. The content of this martensite is 27 % in basic material and 11 % in the modified one, so the content of retained martensite that corresponds to the previous austenite transformed during hot rolling is higher in basic material and, therefore, texture formation will also be higher.

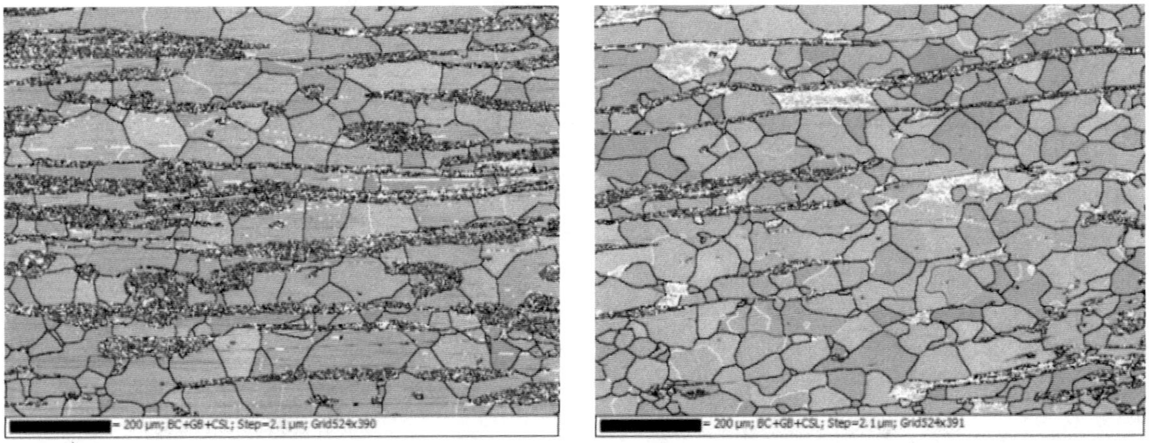

a)

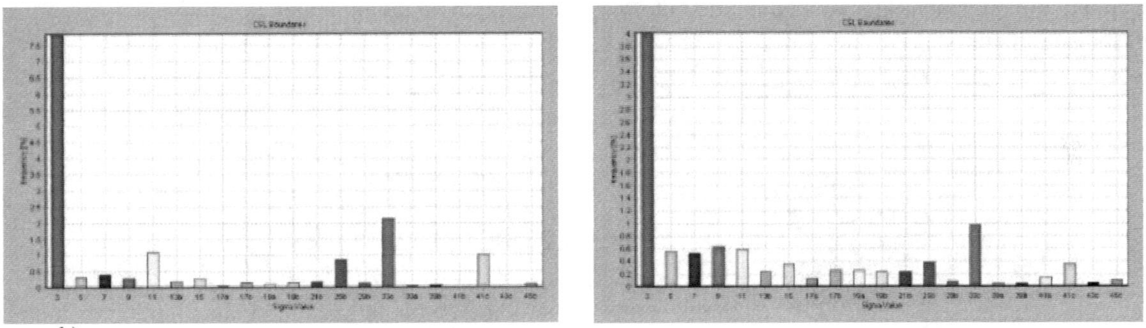

b)

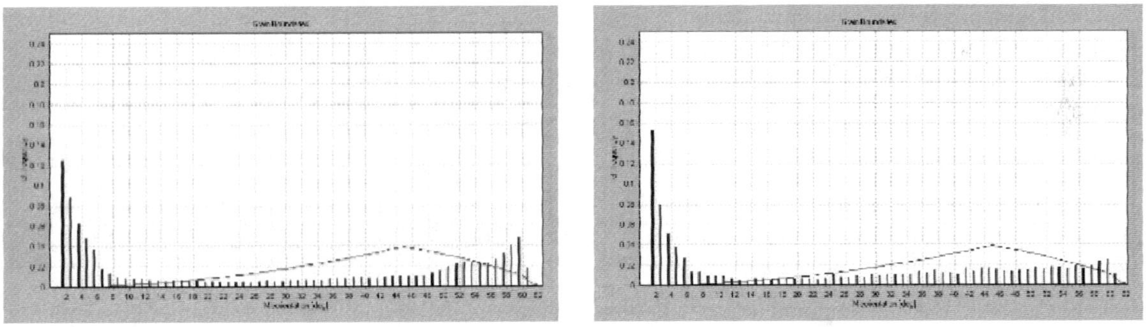

c)

Figure 6. a) Band contrast (BC), coincidence site lattice (CSL) and grain boundary (GB) map of basic material, 0A (left) and modified material, 1C (right) during hot rolling. b) CSL legend of basic (left) and modified (right) material in samples during hot rolling. c) GB legend of basic (left) and modified (right) material in samples during hot rolling.

Fe bcc

Crystal system:	Cubic
Space group:	I m -3 m
Space group number:	229
a (Å):	2.8710
b (Å):	2.8710
c (Å):	2.8710
Alpha (°):	90.0000
Beta (°):	90.0000
Gamma (°):	90.0000
Calculated density (g/cm^3):	7.83
Volume of cell (10^6 pm^3):	23.66
Z:	2.00
RIR:	11.07

Martensite

Crystal system:	Tetragonal
Space group:	I4/mmm
Space group number:	139
a (Å):	2.8460
b (Å):	2.8460
c (Å):	3.0530
Alpha (°):	90.0000
Beta (°):	90.0000
Gamma (°):	90.0000
Calculated density (g/cm^3):	7.09
Volume of cell (10^6 pm^3):	24.73
Z:	1.00
RIR:	7.36

Figure 7. Fe bcc and martensite crystallographic parameters; ICSD (Inorganic Crystal Structure Database) pattern.

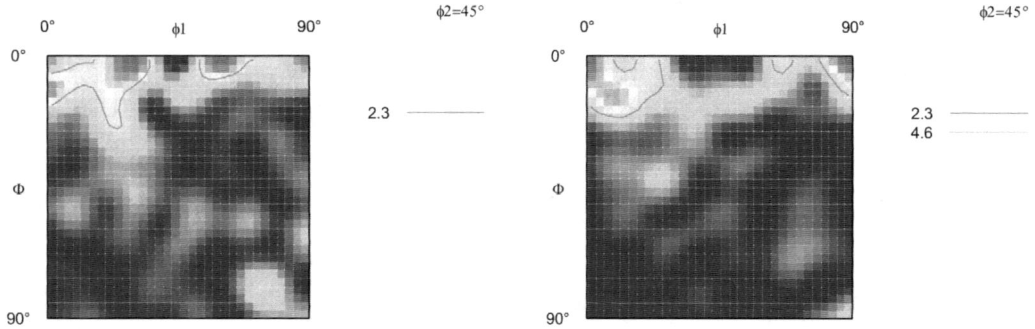

Figure 8. ODF of basic material, 0A (left), and modified material, 1C (right) with intermediate annealing.

3.3.2. Samples with intermediate annealing, after hot rolling. Figure 8 shows the ODF of basic and modified material with intermediate annealing, after hot rolling. ODFs indicate that α-fibre prevails in both materials, and that γ-fibre is not observed in either.

Figure 9 shows the pole figure of basic and modified material in samples after hot rolling. Pole figures indicate that components of re-crystallised austenite upon transformation to ferrite seem to be observed mainly in material 0A [28].

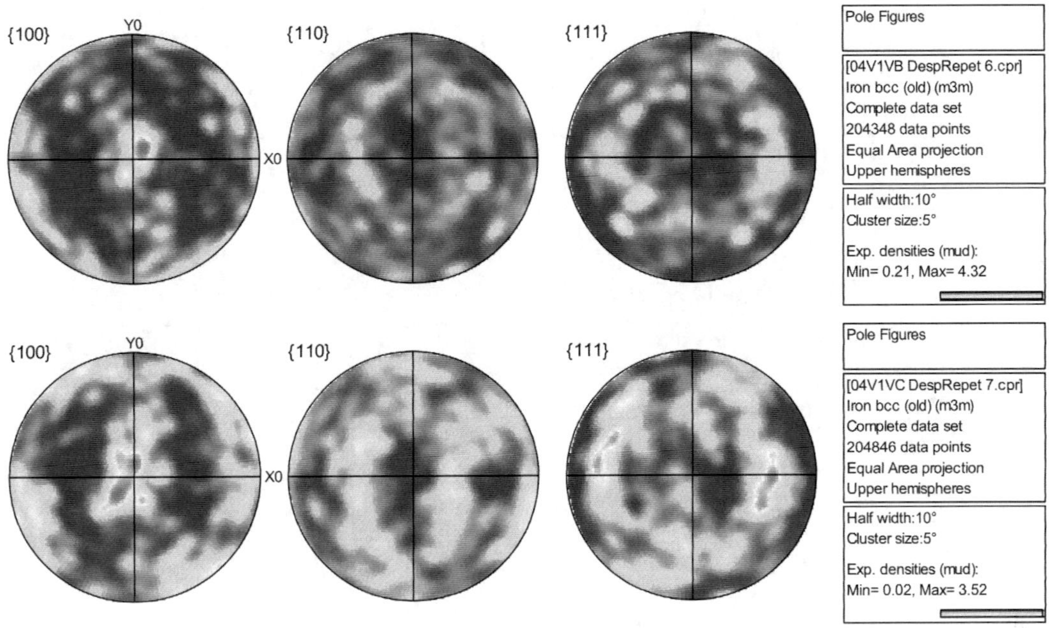

Figure 9. Above: pole figure of basic material, 0A, after hot rolling and its corresponding annealing. Below: pole figure of modified material, 1C, after hot rolling and its corresponding annealing.

Figure 10a shows the band contrast (BC), coincidence site lattice (CSL) and grain boundaries (GB) map, Fig. 10b shows the CSL legend, while Fig. 10c shows the GB legend of basic and modified material after hot rolling plus annealing. Data processing has been performed as it was done for the samples during hot rolling (Fig. 6). EBSD analysis does not show the presence of retained martensite in the ferritic matrix in any of the materials after hot rolling; also, a poor re-crystallisation in the modified material is observed.

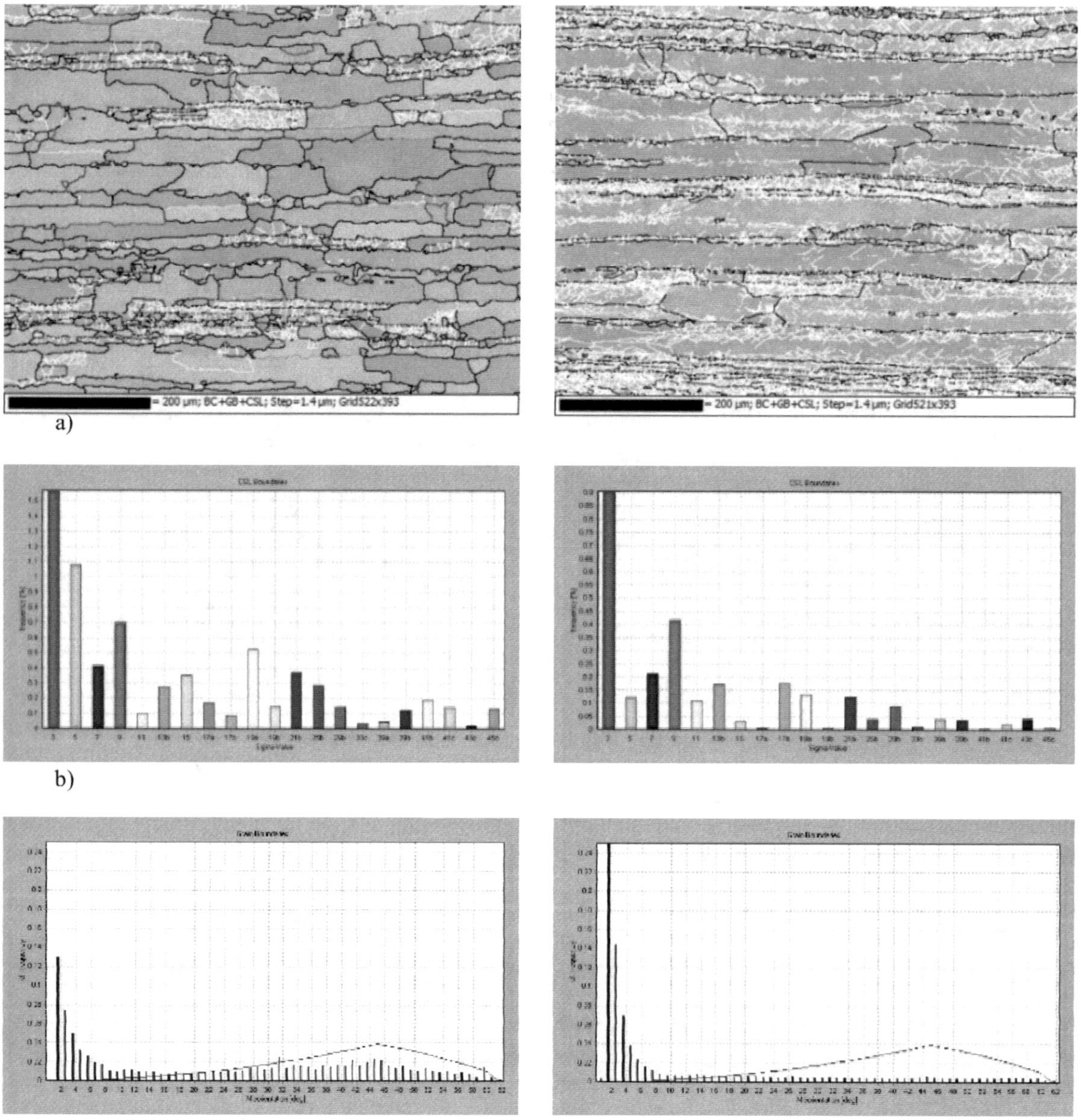

Figure 10. a) Band contrast (BC), coincidence site lattice (CSL) and grain boundary (GB) map of basic material, 0A (left) and modified material, 1C (right) after hot rolling and its corresponding annealing. b) CSL legend of basic (left) and modified (right) material in samples during hot rolling. c) GB legend of basic (left) and modified (right) material after hot rolling and its corresponding annealing.

3.3.3. Samples after cold rolling and final annealing. Figure 11 shows the ODF of basic and modified material in samples after cold rolling and final annealing. A strong γ-fibre re-crystallisation texture is developed in the modified steel after cold rolling and final annealing, while α-fibre was mostly found in the basic one.

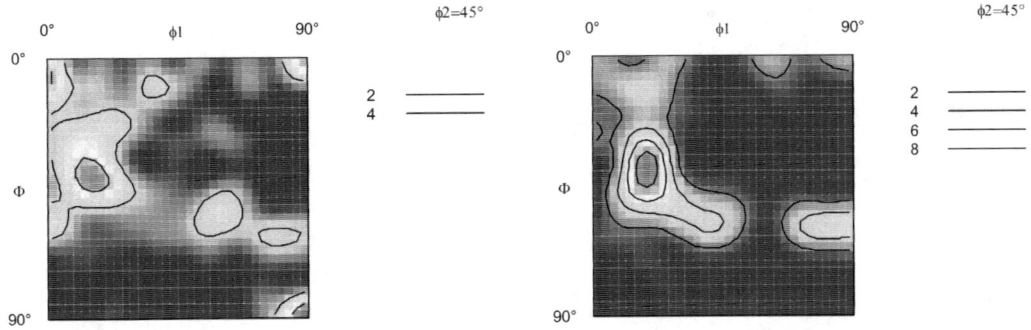

Figure 11. ODF of basic material, 0A (left), and modified material, 1C (right) after cold rolling and final annealing.

Figure 12 shows the pole figure of basic and modified material after final annealing. It is observed that in the material 0A, symmetry elements distribution in the projection is not uniform, but in the 1C material symmetry, elements are observed for each plane, texture components in cold-rolled and re-crystallised BCC steels; so the material of basic composition, presents higher anisotropy.

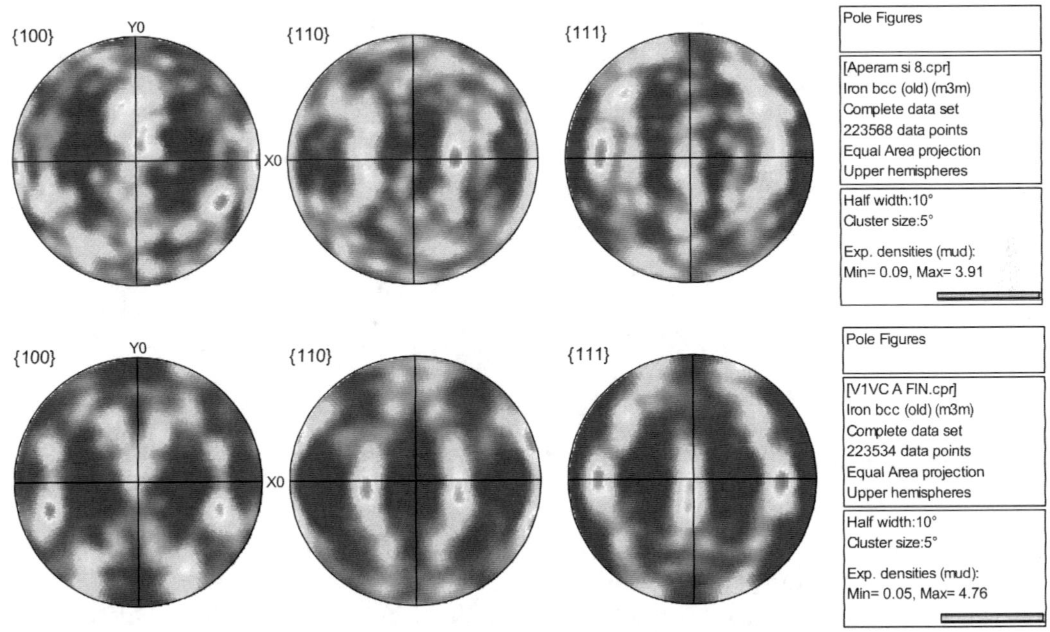

Figure 12. Above: pole figure of basic material, 0A, after cold rolling and final annealing. Below: pole figure of modified material, 1C, after cold rolling and final annealing.

Figure 13a shows the band contrast (BC), coincidence site lattice (CSL) and grain boundaries (GB) map, Fig. 13b shows the CSL legend, while Fig. 13c shows the GB legend of basic and modified material in samples after such cold rolling and final annealing. The characterisation reveals the presence of retained martensite in the basic material with a content of 29 %, which does not appear in the modified one (data processing as in Figs. 6 and 10).

IOP Conf. Series: Materials Science and Engineering **891** (2020) 012019 doi:10.1088/1757-899X/891/1/012019

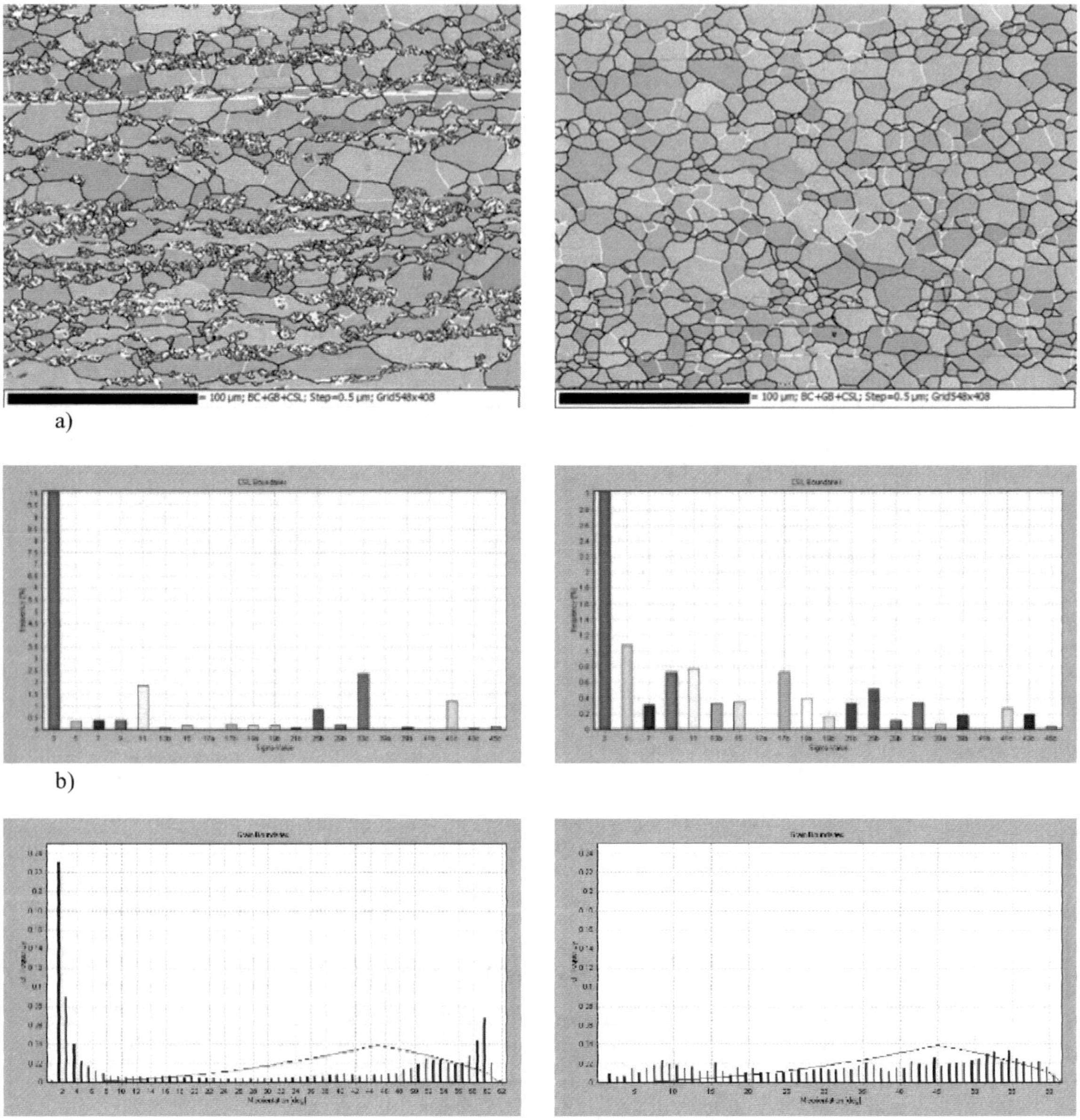

Figure 13. a) Band contrast (BC), coincidence site lattice (CSL) and grain boundary (GB) map of basic material, 0A (left) and modified material, 1C (right) after final annealing. b) CSL legend of basic (left) and modified (right) material in samples after final annealing. c) GB legend of basic (left) and modified (right) material after final annealing.

4. Conclusions

- The macrostructures indicate that the solidification mechanism is heterogeneous in the material with basic composition, while in the modified composition, it is homogeneous. As well, sub-grains are revealed in the basic composition material but not in the modified one.

- Mechanical tests show that the materials after hot rolling present small deformation resistance, but after final annealing a basic composition sheet exhibits more pronounced anisotropy and lower deep drawability than the modified composition one.
- γ-fibre increases with processing progress in both materials, from hot rolling to final annealing, although this fibre is developed strongly in the modified steel after final annealing, while the α-fibre was mostly found in the basic one.
- Lankford coefficients and EBSD results confirm the evolution of texture through the manufactured processing of ferritic stainless steel, mainly in the material with modified composition.
- In addition, the characterisation reveals the presence of a phase in the ferrite matrix, identified as martensite, higher in the basic material in the initial stages of the processing route with respect to the modified material and that are not present in the modified composition after final annealing.
- As a result, the modified composition and process delivers a better re-crystallisation status and thus the best drawability performance.

References

[1] Kim J and Tasan C C 2019 *Int. J. Hydrogen Energy* **44** 6333-6343
[2] Sornin D L, Karch A and Logé R E 2018 *J. Mater. Sci.* **53** 2965-2975
[3] Alturk R, Mates S, Xu Z and Abu-Farha F 2017 in: *TMS 2017 146th Annual Meeting & Exhibition Supplemental Proceedings*. (Springer Int. Publ.) 243-254
[4] Raabe D and Lücke K 1993 *Mater. Sci. Technol.* **9** 302-311
[5] Kestens L A I and Pirgazi H 2016 *J. Mater. Sci. Technol.* **32,** 1303-1315
[6] Gurao N P and Satyam Suwas 2016 *Metall. Mater. Trans. A* **48** 809-827
[7] Herrera C, Ponge D and Raabe D 2008 *Steel Res. Int.* **79** 482-488
[8] Gobernado P, Petrov R H, Moerman J, Barbatti C and L A I Kestens 2012 *Mater. Sci. Forum* **702-703** 790-793
[9] Aparecida K, Brandao D and Reis de Oliveira T 2018 *Mater. Res.* **21** e20170873
[10] Starink M J 2017 *Mater. Sci. Engng. A* **705C** 42-45
[11] Khatirkar R K, Nimsarkar S, Das P, Vishwakarma A and Thawale N 2010 *Trans. Indian Inst. Metals* **63** 55-62
[12] Gonçalves R J, Gomes D, Batista H J, Reis T and Mendonça B 2017 *Mater. Res.* **20** 1593-1599
[13] He Y, Jin H, Chu Y, Chen J, Zhang Z, Huang H and Wang J 2017 *Trans. Mater. Heat* **38** 83-90
[14] Zhang C, Zhang L, Shen W, Xia Y and Yan Y 2017 *Acta Metallurgica Sinica* **30** 79-88
[15] Gao F, Yu F, Misra R, Zhang X, Zhang S and Liu Z 2015 *J. Mater. Engng. Performance* **24** 3862-3880
[16] Gobernado P, Petrov R H, Ruiz D, Leunis E and Kastens I 2010 *Adv. Engng. Mater.* **12** 1077-1081
[17] De Cock T 2009 *Thesis*. (Madrid, Spain: Universidad Complutense de Madrid)
[18] Ma X, Zhao J, Du W, Zhang X and Jiang Z 2018 *Mater. Characterization* **137** 201-211
[19] Oliveira A, Magalhaes C, Silva D, Oliveira A, Dutra I and Brandao D 2016 *J. Mater. Res* **31** 2838-2849
[20] Galán R 2013 *Engineering Degree Master*. (Cádiz, Spain: Universidad de Cádiz)
[21] Serenelli M, Bertinetti M and Signorelli 2014 *Rev. Iberoamericana Ing. Mecánica* **18** 75-85
[22] Herrera C, Lima N B, Ferreira A, Plaut R L and Padilha A F 2009 *J. Mater. Processing Technol.* **209** 3518-3524
[23] Hutchinson W B 1984 *Int. Metals Rev.* **29** 25-42
[24] Gao Z, Li J and Wang Y 2018 *Steel Res. Int.* **1800397** 1-9
[25] Ennis B L, Bos C, Aarnts M P, Lee P D and Jimenez-Melero E 2018 *Mater. Sci. Engng.* **713** 278-286
[26] Presslinger H, Mayr M, Tragl E and Bemhard C 2002 *Steel Res. Int.* **77** 107-115
[27] Núñez A, Llovet X and Almagro J F 2013 *Microsc. Microanal.* **19** 959-968

[28] Suwas S and Kumar Ray R 2014 Crystallographic texture of materials. in: *Engineering Materials and Processes*. (New York, NY: Springer)
[29] Oxford Instruments 2016 *AZtec Reclassify Phase*. Application note. (High Wycombe, UK: Oxford Instruments)
[30] Oxford Instruments 2013 *Band refinement*. Technical Bulletin. (High Wycombe, UK: Oxford Instruments)
[31] Wilkinson A J, Meaden G and Dingley D J 2006 *Ultramicroscopy* **106** 303-313
[32] Maurice C and Fortunier R 2008 *J. Microscopy* **230** 520-529
[33] Oxford Instruments 2012 *AZtec Tru-I*. Application note. (High Wycombe, UK: Oxford Instruments)
[34] Oxford Instruments 2011 *AZtecHKL - In depth*. Application note. (High Wycombe, UK: Oxford Instruments)
[35] Lu H-H, Li W-Q, Du L-Y, Guo H-K, Liang W, Zhang W-G and Liu Z-G 2019 *Mater. Sci. Engng. A* **754** 502-511
[36] Mehtonen S, Karjalainen L and Porter D 2013 *Mater. Sci. Engng. A* **571** 1-12
[37] Krauss G 2017 Review. *Steel Res. Int.,* **88**

IOP Conf. Series: Materials Science and Engineering **891** (2020) 012020 doi:10.1088/1757-899X/891/1/012020

Development and validation of standardless and standards-based X-ray microanalysis

P T Pinard, A Protheroe, J Holland, S Burgess and P J Statham

Oxford Instruments NanoAnalysis Ltd., Halifax Road, HP12 3SE High Wycombe, Great Britain

E-mail: philippe.pinard@oxinst.com

Abstract. We describe one approach to standardless quantification for energy-dispersive (ED) X-ray spectrometry that can achieved unnormalised results with an accuracy better than 5 %. This approach relies on a "factory" standards database, a model of the detector efficiency, calculated peak profiles, and a multi-step process to correct spectral artefacts and extract X-ray intensities from ED spectra. These processes and components, as well as the underlying equations involved in the quantification are detailed to illustrate the similarities between this approach and the classical standard-based *k*-ratio quantification. The accuracy was assessed by measuring and quantifying standard samples at different accelerating voltages (15 and 20 kV) and input count rates (50 and 200 kcps). In all cases, the measured, unnormalised concentrations yielded a centred distribution with a relative error less than 5 %. Excluding the analysis of light elements (B, C, O, N, F) improved the relative error to around 2 %

1. Introduction

In its simplest expression, quantification in electron probe microanalysis (EPMA) is the conversion of X-ray intensities to a composition. The relationship between the measured intensity ($I_{j,k}$) of an X-ray line (k) of an element (j) and its mass fraction (c_j) is given by the fundamental X-ray intensity:

$$I_{j,k} = c_j \frac{N_A}{M_j} \frac{i\tau}{e} \frac{\Omega}{4\pi} \varepsilon_{j,k} p_{j,k} f_{j,k}(\vec{c}, E_0, \psi) \quad (1)$$

where N_A is Avogadro's number, M, the molar mass, i, the beam current, τ, the acquisition time, e, the elementary charge, Ω, the solid angle of the detector, ε, the efficiency of the detector, p, the radiative transition probability combining the fluorescence yield, Coster-Kronig transition and line weight, $f_{j,k}(\vec{c}, E_0, \psi)$, the X-ray emission calculated for a given X-ray line (j,k), composition ($\vec{c}$), accelerating voltage (E_0) and take-off angle of the detector (ψ). The X-ray emission model, also known as a matrix correction, defines the emission probability of an X-ray line under the analytical conditions used to measure the X-ray intensity. For simplicity, the subscripts j and k will be omitted in the following equations.

Castaing's k-ratio [1], Cliff-Lorimer's k-factor [2] or any other quantification approaches in X-ray microanalysis have been designed to simplify Eq. 1 by cancelling out some terms and developing

IOP Conf. Series: Materials Science and Engineering **891** (2020) 012020 doi:10.1088/1757-899X/891/1/012020

theoretical models to calculate others. Looking at the k-ratio, the normalisation of the intensity (I_{unk}) from an unknown sample by the one of a standard material (I_{std}) eliminates all terms that are not dependent on the concentration:

$$k = \frac{I_{unk}}{I_{std}} = \frac{c_{unk}}{c_{std}} \frac{\cancel{N_A/M}}{\cancel{N_A/M}} \frac{\cancel{i\tau/e}}{\cancel{i\tau/e}} \frac{\cancel{\Omega/4\pi}}{\cancel{\Omega/4\pi}} \frac{\cancel{\varepsilon p}}{\cancel{\varepsilon p}} \frac{f(\vec{c}_{unk}, E_0, \psi)}{f(\vec{c}_{std}, E_0, \psi)} = \frac{c_{unk} f'(\vec{c}_{unk}, E_0, \psi)}{c_{std} f'(\vec{c}_{std}, E_0, \psi)} \tag{2}$$

assuming that the electron dose ($i\tau$) is either the same or measured during the acquisition of both X-ray intensities. Note that the normalisation also simplifies the X-ray emission model (f → f') since only the relative emission probability must be calculated (i.e., without the ionisation cross section). Starting from Castaing, authors have developed different theoretical models in order to calculate the remaining term, $f(\vec{c}, E_0, \psi)$, including ZAF correction, $\varphi(\rho z)$ models and Monte Carlo simulation programmes. The quantification therefore combines an empirical and theoretical method, the normalisation by a standard intensity and the calculation of the X-ray emission probability. This combination is a key strength of EPMA in comparison to other analytical techniques as it provides an accurate quantification of a wide range of samples with a general and reproducible experimental method.

EPMA is not however without limitations and caveats. The measurement of X-ray intensities on standard samples entails the procurement of these standards (covering all the elements to be analysed) as well as the expertise to evaluate, store and maintain them. The quantification can only be as accurate as the accuracy of the standards. This means that the standards must be properly characterised to assess their composition, homogeneity, flatness, electrical conductivity, etc. This assessment should also be verified over time to check for any degradation (segregation, oxidation, etc.). These complications and requirements related to standards along with the desire to improve the usability and efficiency of quantitative analyses led to the development of standardless quantification approaches [3-6].

While it is often described as an inaccurate, unverifiable "black box" [7-9], standardless quantification is equally derived from Eq. 1, but uses different methods to cancel out unknown terms and calculate others in order to convert measured X-ray intensities to a composition. As selected methods can vary, this may explain the discrepancies between the different approaches. In the following paragraphs, we would like to describe the methods used in one approach to standardless quantification for energy-dispersive X-ray spectrometry (EDS) including how X-ray intensities are extracted from ED spectra. We will then present experimental work to validate the accuracy of the standardless quantification with the latest generation silicon drift detector (SDD).

2. Method

2.1. Standardless quantification

As stated above standardless quantification aims to eliminate the need of every characterisation laboratory to buy and maintain a series of standards and, therefore, lower the barrier to quantitative X-ray microanalysis. One obvious solution is to calculate X-ray intensities directly from Eq. 1 without the need of any standard. While it is an important research topic to further improve the understanding of electron-photon-matter interactions, the first-principle approach is very susceptible to inaccuracies in the physical parameters and models. A more practical approach is to still rely on the acquisitions of standard X-ray intensities (Eq. 2), but to transfer the expertise and responsibility of the standard acquisition to someone else. In other words, the standard X-ray intensities are acquired on another system than the one used to acquire the X-ray intensities from an unknown sample.

Let us define the following scenario where the unknown X-ray intensities are acquired on a system labelled "user" and the standard X-ray intensities on a system labelled "factory". Because the acquisitions on one system likely use different analytical conditions (detector, accelerating voltage, beam current, etc.) than on the other, fewer terms cancel out in the k-ratio equation (Eq. 2):

$$k = \frac{I_{unk}}{\mathbf{I_{std}}} = \frac{c_{unk}\,\cancel{N_A/M}\;\, i\tau/\cancel{e}\;\, \Omega/\cancel{4\pi}\;\, \varepsilon\, \cancel{p}\, f(\vec{c}_{unk},E_0,\psi)}{\mathbf{c_{std}}\,\cancel{N_A/M}\;\, \mathbf{i\tau}/\cancel{e}\;\, \mathbf{\Omega}/\cancel{4\pi}\;\, \boldsymbol{\varepsilon}\, \cancel{p}\, f(\vec{\mathbf{c}}_{\mathbf{std}},\mathbf{E_0},\boldsymbol{\psi})} = \frac{c_{unk} i\tau\Omega\varepsilon f(\vec{c}_{unk},E_0,\psi)}{\mathbf{c_{std}}\mathbf{i\tau\Omega\varepsilon} f(\vec{\mathbf{c}}_{\mathbf{std}},\mathbf{E_0},\boldsymbol{\psi})} \tag{3}$$

where the bold, red terms indicate those corresponding to the factory system. The acquisition time and accelerating voltage are known quantities as they are selected by the user. The port geometry of the X-ray detector on the electron microscope specifies its take-off angle. The value is normally fixed and known from the engineering drawing. The beam current, solid angle, detector efficiency, X-ray emission model and standard intensity are the remaining unknown terms in Eq. 3, obviously besides the concentration of the measured element in the unknown sample.

Use of a Faraday cup, either inserted inside the electron column or mounted on the stage, is the classical method to measure the beam current. In contrast to electron probes, scanning electron microscopes (SEM) rarely have the equipment required to measure a beam current (ammeter, stage connection, etc.). It is much harder to experimentally determine the solid angle. Even with the correct formulation [10] and all the information regarding the sensor (area, shape, entry angle), the sensor to sample distance is often unknown and may change depending on the acquisition conditions (e.g., X-ray detector may be retracted to avoid collision with other detectors inside the microscope).

Several approaches to standardless quantification require the normalisation of the quantification results. The additional equation, $\sum c_i = 1$, effectively eliminates the need to measure $i\Omega/\mathbf{\Omega}$ since it is a constant in the k-ratio equation of all measured elements for a given acquisition. Normalisation is a double-edged sword. On one hand it allows an easy assessment of the composition of a sample, on the other hand, it hides potentially serious experimental errors (charging, omission of elements, etc.).

One approach to avoid normalisation and obtain unnormalised quantification results is to calibrate both the user and factory system using an acquisition on a pure sample. This calibration resembles the k-ratio approach, except that the X-ray intensities in the numerator and denominator are not from the same X-ray line. The X-ray intensities only need to be acquired by the same X-ray detector under the same analytical conditions (i.e., accelerating voltage, beam current, insertion distance). For example, Cu-K and Zn-K X-ray intensities from an unknown brass sample can be calibrated using the Co-K X-ray intensity from a pure Co sample. The calibration acts exactly as a beam current measurement, meaning that it can be repeated to check the beam current stability or if the beam current was changed. The calibration also removes out the solid angle term from Eq. 3:

$$\frac{I_{unk}}{I_{cal}} = \frac{c_{unk}\,\cancel{N_A}/M_{unk}\;\, \cancel{i\tau/e}\;\, \cancel{\Omega/4\pi}\;\, \varepsilon_{unk}\, p_{unk}\, f(\vec{c}_{unk},E_0,\psi)}{\cancel{c_{cal}}\,\cancel{N_A}/M_{cal}\;\, \cancel{i\tau/e}\;\, \cancel{\Omega/4\pi}\;\, \varepsilon_{cal}\, p_{cal}\, f(\vec{c}_{cal},E_0,\psi)} \tag{4}$$

The calibration is performed on a pure standard, so c_{cal} is unity, and the molar masses (M_{unk} and M_{cal}) are known. In contrast to the k-ratio (Eq. 2), the radiative transition probabilities (p and p_{cal}) do not cancel out because two different X-ray lines are used. If the same calibration is performed on the factory system using the same element, the radiative transition probabilities are removed, and Eq. 3 becomes:

$$k = \frac{I_{unk}\mathbf{I_{cal}}}{\mathbf{I_{std}}I_{cal}} = \frac{c_{unk}\varepsilon_{unk}\boldsymbol{\varepsilon}_{\mathbf{cal}}\cancel{p_{unk}\mathbf{p_{cal}}}f(\vec{c}_{unk},E_0,\psi)f(\vec{\mathbf{c}}_{\mathbf{cal}},\mathbf{E_0},\boldsymbol{\psi})}{\mathbf{c_{std}}\boldsymbol{\varepsilon}_{\mathbf{std}}\varepsilon_{cal}\cancel{\mathbf{p_{std}}p_{cal}}f(\vec{\mathbf{c}}_{\mathbf{std}},\mathbf{E_0},\boldsymbol{\psi})f(\vec{c}_{cal},E_0,\psi)} \tag{5}$$

It is, however, not a requirement that the same calibration sample is used on the user and factory system. In that case, the X-ray intensities of the two calibration samples measured on the factory system are used to construct an additional ratio based on rearranging the terms of Eq. 4. For this ratio, all terms are known and multiplying it to Eq. 5 results in the cancellation of the radiative transition probability terms since they are independent of the system. Equation 5, therefore, demonstrates that it is possible to indirectly measure the beam current without a Faraday cup and cancel out the solid angle regardless of the geometry of the X-ray detector and its position inside the microscope.

It is likely that the detector efficiency will be different between the user and factory system if X-ray detectors from different generations are used. For a given system, the efficiency between the X-ray line of the unknown and calibration element could also be very different. In the results presented below, the detector efficiency is derived from a direct measurement of the efficiency of a "gold standard" Si(Li)-detector using the Bessy II synchrotron in Berlin [11]. Before and after the synchrotron measurements, X-ray intensities from standard samples were acquired using the same detector on a SEM where the take-off angle of the detector was carefully measured by tilting a sample surface away from the detector. Microscope conditions (beam current, accelerating voltage) were carefully monitored during these acquisitions. For any new type of X-ray detector (e.g., new SDD), a new model for the efficiency curve is constructed based on the synchrotron and SEM measurements of the gold standard detector, as well as an intimate knowledge of the manufacturing processes, window characteristics and detector parameters. The model is adjusted using X-ray intensities measured using the new X-ray detector, where its efficiency (ε) at the energy of the measured X-ray line is given by:

$$\varepsilon = \frac{I}{I_{cal}} \frac{I_{cal}^{GS}}{I^{GS}} \frac{\varepsilon_{cal}}{\varepsilon_{cal}^{GS}} \varepsilon^{GS} \quad (6)$$

where the subscript *GS* indicates X-ray intensities measured by the "gold standard" detector. As shown in Statham 2009 [12] and 2010 [13], the experimental efficiency determined by the synchrotron significantly improves the standardless quantification results, especially for the quantification using X-ray lines below 1 keV (e.g., O-K and F-K).

As in the standard-based k-ratio equation (Eq. 2), a model must be used to calculate the X-ray emission model and correct for matrix effects in the X-ray emission between the unknown and standard samples. In the standardless *k*-ratio equation (Eq. 5) this model is also used to correct for differences in the calibration measurement between the user and factory systems. In all the following results, the X-ray emission is calculated using the $\varphi(\rho z)$ model, XPP, as described in Pouchou and Pichoir (1991) [14].

Finally, the remaining undetermined quantity is the standard X-ray intensity. Using the factory system, X-ray intensities from a series of standards, consisting of pure elements and simple binaries (e.g., SiO_2, FeS_2), were acquired at two accelerating voltages: 5 and 20 kV. The take-off angle of the X-ray detector, thickness of carbon coating on standard block, homogeneity of standards, etc., were carefully measured prior to the measurements. As described above, a calibration standard was periodically measured, Co for the acquisitions at 20 kV and Si for those at 5 kV, instead of a direct measurement of the beam current. The X-ray intensities from the standards were then stored as ratios to the Co-K or Si-K X-ray intensity to produce a "factory" standards database.

This completes the description of the standardless quantification approach. Comparing Eqs. 2 and 5 illustrates the similitudes between standard-based quantification and this approach. Both calculate a *k*-ratio of each analysed element, both correct for matrix effects using a $\varphi(\rho z)$-model, both use the same iterative procedure to solve for the unknown concentrations and both output unnormalised quantification results. Both approaches can also be combined if needed. The trade-off between the two approaches could be summarised as the requirement for well-characterised standards versus a well-characterised X-ray detector.

2.2. Extraction of X-ray intensity

Besides the quantification method, one important consideration is how to accurately extract X-ray intensities from an ED spectrum in order to calculate the *k*-ratios (Eq. 2 or 5). This applies to both standard-based or standardless quantification. Extracting X-ray intensities is a multi-step process which typically involves:

a) Correction of the energy scale;
b) Correction of artefact peaks;
c) Removal of the background;

d) Fit of modelled or experimental peak profiles over the X-ray peaks of interest;
e) Extraction of the X-ray intensity from the fitted X-ray peak.

In some pulse processing electronics, a "strobe zero" peak is obtained by accumulating a histogram of measurements that would be obtained from zero-energy photons. The measurement is performed in time gaps between real X-ray events using exactly the same pulse processing as for a real event. By fitting a Gaussian to this histogram, the channel position for zero energy photons is determined without needing a special standard. This position can be used first to calibrate the energy scale using a single X-ray peak from an element standard and is subsequently used on every spectrum to correct for any shift in zero position. With the proper gain calibration of the electronics, this ensures that all X-ray peaks appear at the right energy in the ED spectrum.

Silicon escape and pulse pile-up peaks are artefacts from the X-ray detection [9]. While different phenomena cause these artefact peaks, both results in the reduction of the X-ray intensity of the parent peak. They may also indirectly affect other X-ray peaks as they can coincide with other "real" X-ray peaks and artificially increase their X-ray intensity. Besides the X-ray intensity, these artefact peaks may lead to the misidentification of elements in the spectrum, which would inevitably influence the accuracy of the quantification results. For these three reasons, it is important to correct these artefacts before extracting the X-ray intensities of any peak. This especially applies to acquisitions at high count rates where the influence of pulse pile-up peaks increases. Silicon escape peaks can be corrected using algorithms developed by Reed and Ware [15] and improved by Statham [16] to consider the enhanced resolution of the escape peak, whereas Statham [17, 18] and Eggert *et al.* [19, 20] describe the principles of pulse pile-up correction.

The X-ray intensity terms in all the previous equations more precisely refer to the intensity from the emission of characteristic X-ray lines. This implies stripping the background from the ED spectrum, i.e., removing the contribution of bremsstrahlung X-rays from the spectrum. From the literature, two methods have been proposed for this operation: (1) fitting a calculated background to the experimental and (2) filtering out the background using a convolution filter (also known as top-hat filter) [21-24]. The first method requires the composition of the analysed area to calculate the background as well as peak-free background points to adjust the calculated background. While the composition can iteratively be determined, any misidentification of the X-ray peaks (e.g., Si instead of Ta) would result in an incorrect background shape due to the different absorption behaviour. The second method, the convolution filter, was thus selected as it offers a consistent and operator-independent way to remove the background for any spectrum, including spectra with no clear peak-free background region, for example the 100 - 1,000 eV region in Fig. 1. The convolution filter is automatically adjusted to match the X-ray detector resolution to ensure the accurate filtering of the background for all samples at any X-ray energy. Nonetheless, the calculated background can be a useful tool to validate the element identification and the quantification results [25].

The resolution of EDS entails that X-ray lines from different elements may overlap. The X-ray peaks must be separated before extracting their respective X-ray intensities. This is normally performed using least-squares fitting of the experimental spectrum (after background removal) with either experimental [23] or calculated peak profiles [22, 24] (Fig. 2). If a convolution filter is used to remove the background, this means that the least-squares fitting is performed between the filtered experimental spectrum and the filtered peak profiles. In practice, calculated peak profiles offer more flexibility as they implicitly take into account the influence of the acquisition conditions (e.g., accelerating voltage, X-ray detector resolution and pulse processor settings). The model must however properly describe the impulse response function of the X-ray detector (i.e., charge collection, efficiency, etc.) and construct peak profiles based on accurate fundamental data of relative line weights, mass absorption coefficients and Coster-Kronig transitions [16, 25]. This is particularly true for L and M X-ray line series with large absorption edges within the line series.

IOP Conf. Series: Materials Science and Engineering **891** (2020) 012020 doi:10.1088/1757-899X/891/1/012020

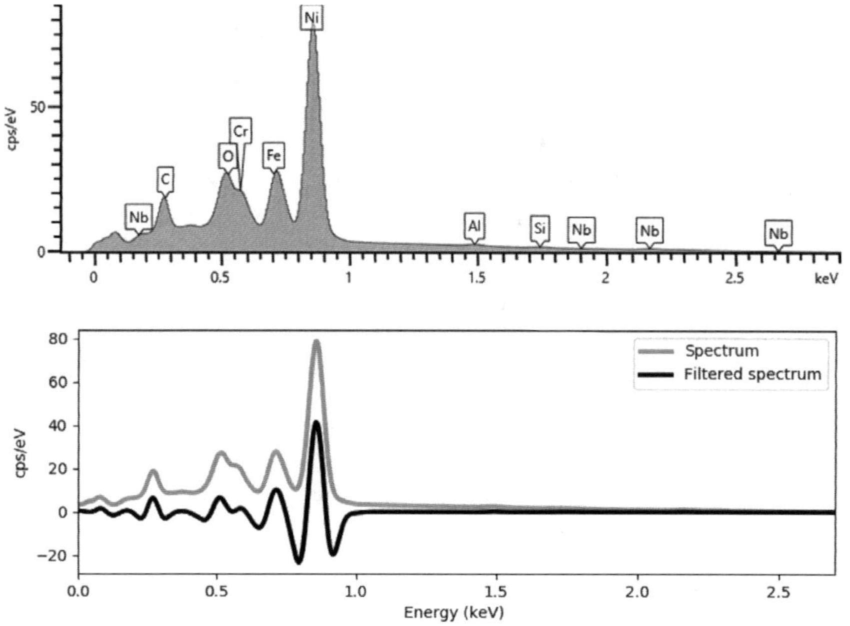

a)

b)

Figure 1. a) Spectrum of a 718-steel alloy acquired at 3 kV (Zeiss Merlin, X-Max Extreme detector, 1.3 nA, PT4, 600 s live time), and b) after applying the convolution filter. Sample is courtesy of Manchester University.

The least-squares fitting is always performed over the whole X-ray line series. In other words, all X-ray lines emitted by the transition to a specific shell (i.e., K, L, M, etc.) are fitted together. This is because X-ray lines from the same family often overlap at the EDS resolution. While several parameters could be included in the least-square fitting (e.g., resolution, relative line weights, peak positions), this would quickly result in ambiguous solutions. A more robust strategy is to only fit the relative area ratio of the overlapping X-ray line series. For example, if two X-ray line series overlap in a region of the spectrum, the following equation can be constructed and solve by least-squares fitting:

$$\sum_n I_n = \beta_1 \sum_n \hat{I}_{n,1} + \beta_2 \sum_n \hat{I}_{n,2} \quad (7)$$

where I_n is the experimental intensity in channel n of the spectrum, $\hat{I}_{n,k}$, the calculated intensity in channel n of the profile of the X-ray line series k and β_k, its relative area ratio. The experimental area for the X-ray line series k (A_k), is therefore equal to the product of the line series profile area ($\hat{A}_k$) and the fitted relative area ratio (β_k):

$$A_k = \beta_k \sum_n \hat{I}_{n,k} = \beta_k \hat{A}_k \quad (8)$$

In Eqs. 2 and 5, the detector efficiency and X-ray emission model are normally defined either at a specific X-ray energy or for a specific X-ray line. In other words, the value of these quantities may differ between the X-ray lines within a same family. In contrast to wavelength-dispersive X-ray spectrometry (WDS), the line series area cannot directly be used as the X-ray intensity in the k-ratio equation [23]. One method to convert line the experimental area of the X-ray line series k (A_k) to the experimental intensity of the X-ray line j (I_j) is to determine the most intense X-ray line in the line series and calculate its contribution to the line series area:

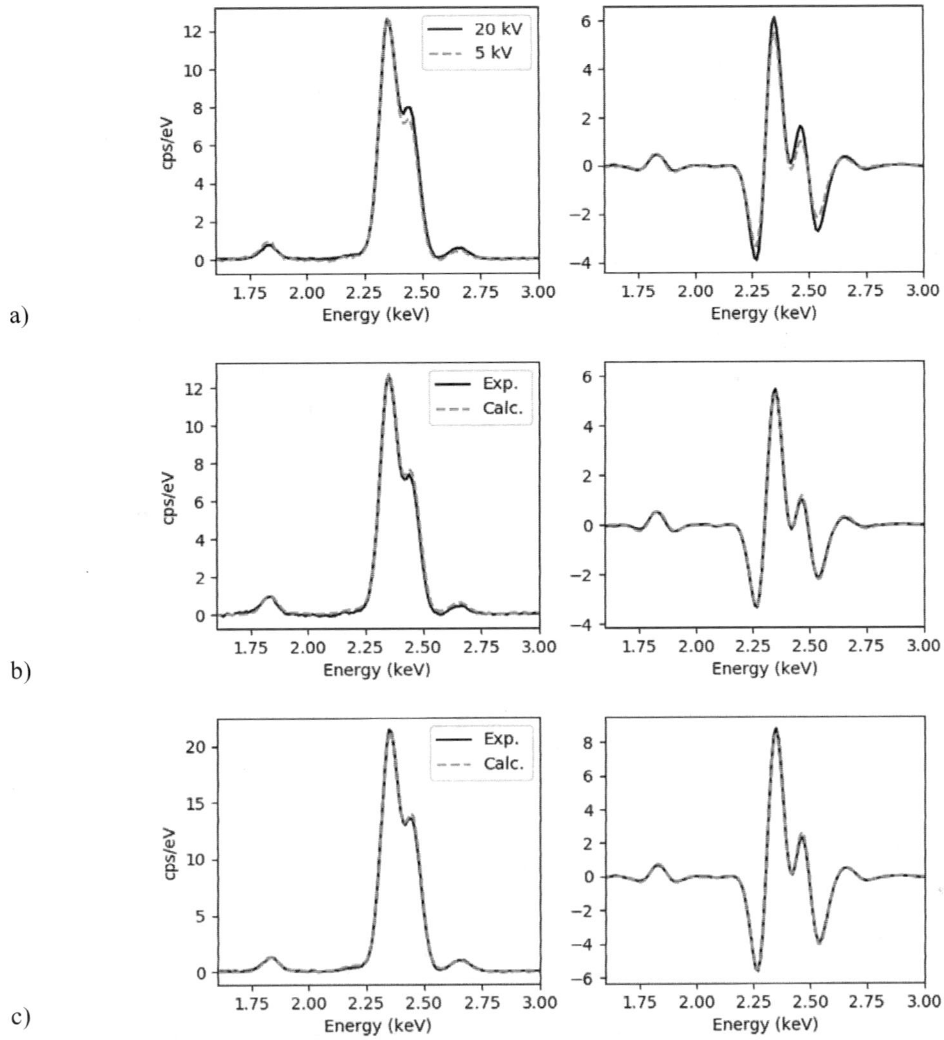

Figure 2. a) Comparison between experimental spectra of Pb-M X-ray line series acquired at 5 and 20 kV (X-MaxN 80, PT5, 100 s live time ,10 eV/ch). Comparison of these experimental spectra with calculated peak profile at b) 5 kV, and c) 20 kV. The background was removed from the left spectra, whereas the right spectra correspond to the raw experimental spectra after the convolution filter.

$$I_j = A_k \frac{\hat{I}_j}{\hat{A}_k} \quad (9)$$

where $\hat{I}_j$ and $\hat{A}_k$ are respectively the calculated intensity for the most intense X-ray line and the calculated area of its corresponding X-ray line series. Both quantities are calculated from the peak profile generated for this X-ray line series.

This concludes the description of how X-ray intensities are extracted from ED spectra, regardless whether standard-based or standardless quantification is used. It is worth pointing out that regardless of strategies used to extract X-ray intensities, theoretical models for X-ray emission and detection are required (e.g., background model, peak profile model, pulse pile-up correction). It is not possible to purely rely on empirical methods to extract X-ray intensities from ED spectra. This contrasts with WDS where X-ray intensities correspond to the peak height subtracted by a background intensity interpolated from two off-peak measurements.

3. Experimental procedure

To assess the accuracy of the standardless quantification approach, experimental ED spectra were acquired on 53 standard samples, part of a single standard block (#6260, Micro-Analysis, Consultants Ltd., St Ives, United Kingdom). The standard samples include pure elements, simple binaries (SiO_2, HgTe, InAs, Fe_3C, GaP, FeS_2, NaCl, etc.) and minerals (albite, orthoclase, wollastonite). In contrast to standard-based quantification, the standardless quantification of pure element standards is a good indicator of the accuracy of the detector efficiency and the matrix correction if the factory standards were acquired using a different X-ray detector and/or under different analytical conditions. The standard block was carbon-coated by the manufacturer. Using the thin-film measurement tool, LayerProbe (Oxford Instruments, High Wycombe, United Kingdom), the carbon thickness was measured on some of the standards and found to be uniform and equal to 13.5 ± 3 nm. The oxide thickness was less than 5 nm on most of the metallic standards, except on Ti and V where the oxide thickness was 13 and 30 nm, respectively. At the accelerating voltage used, this level of oxidation should have minimal influence on the quantification results.

ED spectra were acquired using a single Oxford Instruments Ultim Max 100 SDD with a Mn resolution of 127 eV at an input count rate of 130 kcps (in line with ISO 15632-2012). The detector was installed on a Leo 1450VP scanning electron microscope (Zeiss, Oberkochen, Germany). The elevation is 35° at a working distance of 20 mm. The channel width was fixed at 10 eV/channel (2,048 channels and 20 keV energy range). Three sets of acquisitions were performed at different accelerating voltages (15 and 20 kV) and input count rates (50 and 200 kcps). Table 1 summarises the acquisition conditions. For each acquisition condition, one spectrum was acquired on each standard. Since the X-ray emission is sample dependent, the input count rate (ICR) was adjusted on pure Co where the dead time was kept at 30 % by changing the process time. The live time was similarly adjusted to approximately acquire 1 million counts on Co irrespectively of the input count rate (ICR).

Table 1. Analytical conditions for accuracy tests.

	Test A	**Test B**	**Test C**
Accelerating voltage (kV)	20	15	20
Input count rate (kcps) on Co	50	50	200
Approx. beam current on Co (nA)	1.25	1.25	4.98
Live time (s)	20	20	5
Process time	PT5	PT5	PT3
Dead time on Co (%)	~30	~30	~30

Cobalt was used as the calibration element. The measurement on Co was repeated after every 5 – 7 acquisitions on the standard samples to minimise the effect of any beam current instability. The same acquisition conditions were used, except that the total number of counts in the spectra was set to 600,000.

The spectrum processing to extract X-ray intensities and the quantification were performed using AZTEC v4.1. All spectra were escape peak and pulse pile-up corrected. Elements in each spectrum were automatically identified. Any trace elements (< 1 wt%) [26] were excluded from the analysis as the counting time was not sufficient to properly measure these elements. Unless otherwise stated, all elements were quantified by direct measurement, including oxygen of the measured oxides and minerals. The X-ray line used for quantification was automatically selected by the software based on the overvoltage. All results use standardless quantification method and are unnormalised.

The experimental procedure described above follows the one previously performed by Burgess *et al.* [27] and Statham [12] with an input count rate of 10 kcps.

4. Results and discussion

In the following figures, the standardless "unnormalised" quantification results are compared to the nominal compositions specified by the manufacturer of the standard block. The comparison is performed by taking the ratio between the measured and nominal concentration. A value close to unity indicates a good agreement, whereas a value below or above respectively implies an underestimation or overestimation of the concentration by the standardless quantification. The results are plotted as a histogram centred at unity with a fixed scale between 0.8 and 1.2 (i.e., ± 20 % relative deviation). Any ratio falling outside this range is separately reported and excluded from the overall average and standard deviation.

4.1. Test A (20 kV, 50 kcps)

Figure 3a shows the results for the acquisitions with an accelerating voltage of 20 kV and an input count rate on Co of 50 kcps. There is one outlier, B in LaB6, with a ratio of 1.2066. All other 75 measured concentrations are within 20 % of the nominal concentrations, including 68 within 5%. This yields a well-centred distribution with a relative error of 3.3 %. The concentrations with a relative deviation greater than 5 % generally correspond to those of light elements (B, O and N).

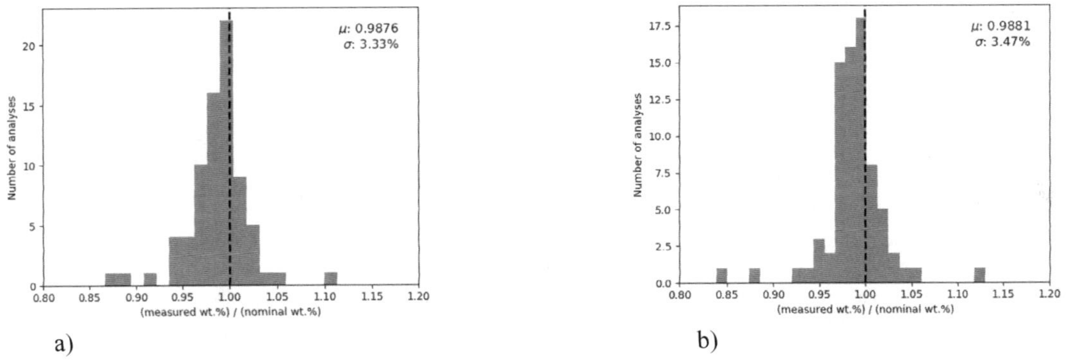

Figure 3. Comparison of measured and nominal quantification results acquired with an accelerating voltage of 20 kV, an input count rate on Co of 50 kcps, a live time of 20 s and PT5 process time. The standard block was rotated by 180° in (b).

The overall relative error drops to 2.1 % if the analysis of light elements is excluded. This relative error approaches the rule of thumb value of 2 %, usually quoted when describing the accuracy of EPMA performed by WDS, as well as the one of 1.8 % obtained by Pouchou and Pichoir [14], when developing the XPP $\varphi(\rho z)$ models. While more exhaustive than this work, their analysis of 826 *k*-ratios also excluded those of light elements. Inaccuracies in the X-ray emission model, or more precisely the mass absorption coefficients, and in the peak profile model might explain the larger errors in the quantification of light elements.

Overall, the results demonstrate that standardless quantification can achieve an accuracy of 5 % or better without any special settings or adjustment to the "user" system. It likewise confirms the accuracy of the strategy used to extract X-ray intensities from the spectra.

The measurements were repeated twice (Figs. 3a and 3b), with the stage rotated by 180° for the second set. Besides confirming the results, this technique provides a way to verify that the standard block is flat and therefore, that the take-off angle is roughly equal to the elevation.

4.2. Test B (15 kV, 50 kcps)

The experiment was repeated at an accelerating voltage of 15 kV. As shown in Fig. 4 the relative standard deviation increases from 3.3 to 4.2 %. B in LaB_6 is no longer an outlier with a ratio of 1.15, which reinforces the aforementioned hypothesis that the larger error might be attributed to inaccuracies in the mass absorption coefficients of light elements (lower absorption correction at 15 kV). Two reasons might explain the increase of the standard deviation between the experiments of Test A and B. Firstly, the overvoltage on the X-ray line used for quantification decreases at 15 kV and thus the measured X-ray intensity. Taking only the analyses with an overvoltage less than 5 at 20 kV (26 concentrations), the relative error increases from 1.9 % at 20 kV to 2.5 % at 15 kV. On average the measured X-ray intensity at 15 kV is approximately 10 % lower than at 20 kV, including more than 30 % drop for X-ray lines of the first-row transition metals. A longer acquisition time for some of these measurements might help improve the results. A second explanation might be due to the inaccuracies in the fundamental parameters used in the X-ray emission model, such as the X-ray ionisation probability or the mass absorption coefficients. In Test A, some of these inaccuracies cancel out since the unknown and standard samples were measured at the same accelerating voltage. Nonetheless, this experiment again confirms that standardless quantification can determine the composition with an accuracy better than 5 %, even if different analytical conditions are used for the "user" and "factory" system.

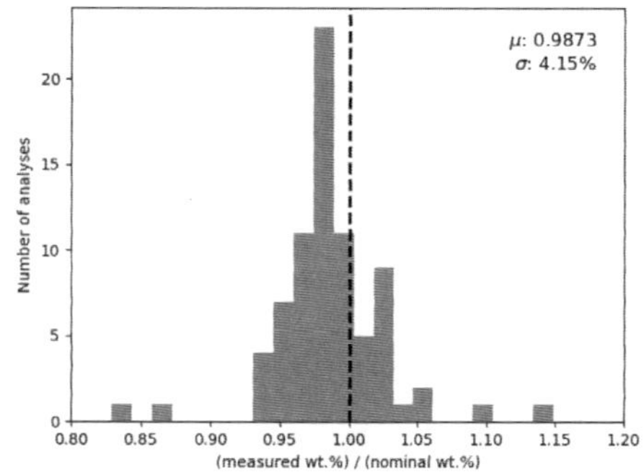

Figure 4. Comparison of measured and nominal quantification results acquired with an accelerating voltage of 15 kV, an input count rate on Co of 50 kcps, a live time of 20 s and PT5 process time.

4.3. Test C (20 kV, 200 kcps)

The results of final test are shown in Fig. 5. Despite a four-fold increase in the input count rate, Test C has a similar relative error than Test A (3.2 versus 3.3 %). There are, however, two additional outliers, N in BN (0.773) and Be in pure Be (1.234). The distribution is also slightly shifted to the left with an average ratio of 0.976 instead of 0.988. This indicates the measured concentrations are on average slightly lower than the nominal concentrations. As the input count rate on some samples can be as high as 330 kcps, an incomplete correction of some pulse pile up peaks might explain the decrease of the average ratio. The change of process time from Test A did not have an influence on the results, confirming that the associated increase of the Mn resolution to 133 eV was correctly modelled in the calculated peak profile. The similitudes between the results of Test A and C confirm the ability to increase the productivity of quantitative analyses by using standardless quantification at high count rates without sacrificing the accuracy [28].

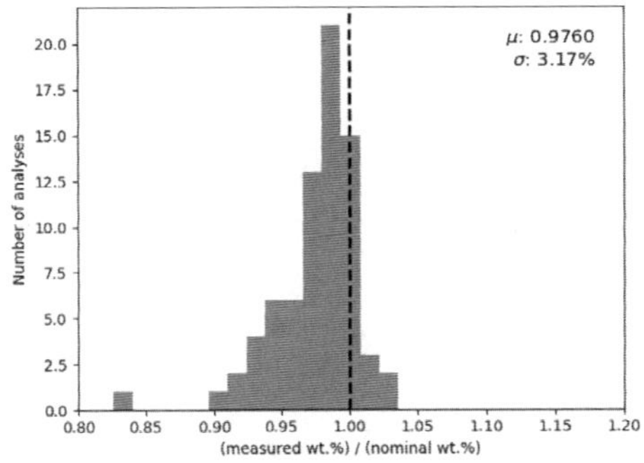

Figure 5. Comparison of measured and nominal quantification results acquired with an accelerating voltage of 20 kV, an input count rate on Co of 200 kcps, a live time of 5 s and PT3 process time.

5. Conclusion

Many factors influence the analytical techniques a scientist chooses to characterise his/her samples. One of the factors is certainly the easy of use of the technique, or in more measurable terms, the "time to result". Castaing's PhD-thesis was innovative in many ways, but one crucial realisation is the need of a theoretical model for the X-ray emission. He could have resorted to make different brass standards to quantify his copper-zinc diffusion couple, but he opted to use pure standards and develop the first matrix correction model. The X-ray emission model is key to remove the requirement to find, prepare and measure standards of similar composition to the unknowns. The ability to use pure or simple compounds as standards is a clear advantage of EPMA over other analytical techniques as it generalises and shortens the experimental procedure.

Standardless quantification further simplifies quantitative analyses by removing the need of the user to measure X-ray intensities on standards. The approach used in this work combines a detailed characterisation of X-ray detectors and the production of a factory standards database. Only a measurement on a single standard is required to obtain unnormalised results. The results presented in this and previous works confirm that this approach can achieve an accuracy better than 5 % under different analytical conditions (accelerating voltage, process time and input count rate).

References

[1] Castaing R 1951 *Application of electron probes to local chemical and crystallographic analysis.* PhD-thesis. (Paris, France: Université de Paris)
[2] Cliff G and Lorimer G W 1975 *J. Microsc.* **103** 203-207
[3] Statham P J 1981 *J. Microsc.* **123** 1-23
[4] Eggert F 1994 *Standardless electron probe microanalysis.* in: EDO conference
[5] Scheller S, Salge T and Terborg R 2010 *Microsc. Microanal.* **16** 1314-1315
[6] Trincavelli J, Limandri S and Bonetto R 2014 *Spectrochim. Acta B* **101** 76-85
[7] Newbury D E 1998 *Microsc. Microanal.* **4** 585-597
[8] Newbury D E 2002 *Microsc. Microanal.* **8** (Suppl. 2) 1464-1465
[9] Goldstein J I, *et al.* 2018 *Scanning electron microscopy and X-ray microanalysis.* (Berlin-Heidelberg, Germany: Springer)
[10] Zaluzec N J 2009 *Microsc. Microanal.* **15** 93-98
[11] Alvisi M, *et al.* 2006 *Microsc. Microanal.* **12** 406-415
[12] Statham P 2009 *Microsc. Microanal.* **15** 528-529
[13] Statham P 2010 *Microsc. Microanal.* **16** 1304-1305
[14] Pouchou J-L and Pouchou F 1991 Quantitative analysis of homogeneous or stratified microvolumes applying the model "PAP". in: *Electron probe quantitation.* (Heinrich K F J and Newbury D E; Eds.) (New York, NY: Plenum Press) 31-75

[15] Reed S J B and Ware N G 1972 *J. Phys. E: Sci. Instrum.* **5** 582-583
[16] Statham P J 2014 *Microsc. Microanal.* **20** 690-691
[17] Statham P J 1977 *X-ray Spectrom.* **6** 94-103
[18] Statham P J 2006 *Microchim. Acta* **155** 289-294
[19] Eggert F, Elam T, Anderhalt R and Nicolosi J 2011 *Microsc. Microanal.* **17** 1192-1193
[20] Eggert F, Elam T, Anderhalt R and Nicolosi J 2012 *IOP Conf. Ser.: Mater. Sci. Engng.* **32** 12008
[21] Statham P J 1977 *Anal. Chem.* **49** 2149-2154
[22] Statham P J 2002 *J. Res. Natl. Inst. Stand. Technol.* **107** 531-546
[23] Ritchie N W M, Newbury D E and Davis J M 2012 *Microsc. Microanal.* **18** 892-904
[24] Terborg R, Salge T, Pinard P T and Richter S 2016 *Microsc. Microanal.* **22** 404-405
[25] Statham P, Penman C and Duncumb P 2016 *IOP Conf. Ser.: Mater. Sci. Engng.* **109** 12016
[26] Newbury D E and Ritchie N W M 2016 *Microsc. Microanal.* **22** 520-535
[27] Burgess S, Statham P, Holland J and Chou Y-H 2007 *Microsc. Microanal.* **13** 1432-1433
[28] Pinard P T, Burgess S, Zhang J Q, Holland J and Statham P 2018 *Microsc. Microanal.* **24** 724-725

IOP Conf. Series: Materials Science and Engineering **891** (2020) 012021 doi:10.1088/1757-899X/891/1/012021

Recent features in EBSD, including new trapezoidal correction for multi-mapping

B E Sørensen[1], J Hjelen[2], H W Ånes[2] and T Breivik[1]

1 Norwegian University of Science and Technology, Department of Geoscience and Petroleum, 7491 Trondheim, Norway
2 Norwegian University of Science and Technology, Department of Materials Science and Engineering, 7491 Trondheim, Norway

E-mail: bjorn.sorensen@ntnu.no

Abstract. EBSD (electron backscatter diffraction) of geological samples can be a challenge due to the complex crystal structures of minerals and commonly 5 or more different minerals in the samples. This poses a big challenge as the EBSP (electron backscatter diffraction patterns) of the minerals like feldspar and quartz commonly are much noisier than metallurgical materials that have simpler structures. Therefore, the indexing of geological samples can commonly be quite difficult and to correct for this in the online procedure means that the data acquisition needs to be redone on the SEM. This is a costly procedure, also because the indexing can take longer than the pattern acquisition itself. We, therefore, suggest that offline techniques are better suited for geological materials because the patterns can be re-indexed using different settings and further can be processed using different software both for indexing and pattern improvement. EBSD commonly have geometric scan distortion which increases with decreasing magnification. This problem is solved by acquiring using a geometric online scan correction during data collection in the NORDIF software using calibration grid. This overcomes stitching problems and additionally removes hidden problems such as geometrically distorted grain boundary segments and grain shape distortions.

1. Introduction

Electron backscatter diffraction (EBSD) has been a technique used in metallurgy for several decades and has proven to be very useful for describing microstructures and textures in metals, ceramics and rocks.

One of the focuses in EBSD development has been speed, as one of the main applications is quality control in metal production. Offline EBSD brought even faster data collection because a part of the collection time during online EBSD acquisition is due to patterns indexing, which is slower than the acquisition time of the patterns, especially if there are several phases to index in the sample. This means that the indexing sets the time-limit for data acquisition, something that is not ideal for product control, in-situ experiments and multiphase samples such as geological samples or ceramics. In addition, offline data gives the possibility to check patterns of noisy areas to tell whether noise removal procedures remove noise and not correct observations, by directly comparing patterns in noisy regions. These observations are essential in judging if procedures such as grain reconstruction are correct. Patterns in unindexed regions are in some cases weak versions of patterns from the neighbouring indexed areas.

Hence, it would be practical if these unindexed areas could be included in the indexed points, as normal grain determination does not have any guidance whether to include these areas in the grains or leave them as unknown areas in the final EBSD map. One promising path towards overcoming this is to improve the pattern quality prior to indexing as traditional pattern treatment does not always handle static and dynamic background subtractions well. This causes to artefacts, such as rings originating from different position of the diffuse scattering in the background averaged EBSP and the pattern in each pixel of the EBSD dataset. The positions of the diffuse scattering part of the EBSP varies as a function of several parameters, including the x-y of the pixel in the map, the Z number of phases, or artefacts due to charging and topography. This is to some extent handled by dynamic background subtraction procedures, but the dynamic background subtraction cannot handle the detector noise so this has to be corrected without at the same time applying a wrong diffuse scattering noise.

Because EBSD on geological samples requires mapping of large areas, stitching of EBSD datasets is necessary. However, geometric trapezoidal distortions are common due to the tilting of the sample. This means that the maps do not fit completely together. The traditional solution has been to map small areas at high magnification to reduce the geometric distortion, but for coarse-grained samples, this is a bad approach as this gives too few different orientations, which can be used to match the maps together. Therefore, maps needs to be collected at lower magnification, giving larger amount of geometric distortion. For this purpose, an online correction procedure has been made such that the original maps have very little geometric distortion even at magnifications below 30x. This solves the problem of stitching when working with large samples, however there are still minor issues with accuracy of orientation measurements ($< 0.5°$ across map) and uneven illumination which has to be considered when collecting EBSD at low magnifications.

2. Results and discussion

2.1. *Offline versus online EBSD, the main differences*

During the early period of EBSD, data acquisition was mainly done in the so-called online mode, meaning that the electron backscatter pattern (EBSP) were collected, indexed and then deleted continuously during data acquisition. A successful online EBSD data acquisition experiment depends on a good setup prior to data collection. This setup starts with adjustment of the electron microscope to obtain good diffraction patterns, acquisition and calibration of geometry, definition of crystal structures for the phases in the sample and the settings for background subtraction and pattern determination scheme. If everything is set up correctly then indexing will work and correct phase and orientation determination will reach 90 % or higher. However, this is with the 'ideal' setup, something that is not always the case. If the setup is wrong and results turns out bad then the analysis in the SEM has to be redone. In contrast, the offline EBSD relies in the first instance on collecting good patterns in the SEM. Patterns are streamed to a disc and then indexed on a separate computer afterwards, this means first of all for a geological sample that data collection time is reduced [1-3], at least by one to two orders of magnitude for high Hough and pattern resolution. Secondly indexing can be redone without running data acquisition again. NTNU implemented offline EBSD in 2007 [1]. Figure 1 summarises the work flow in off- and online EBSD.

To obtain data fast in the SEM using online EBSD a compromise has to be made with the quality in the indexing, as the indexing is the slow part of the data collection process if high resolution of the patterns ($> 200 \times 200$ pixels) and high resolutions in the Hough is used. This means that one has to be a little less precise and accurate when it comes for example to Hough transform setup. This can work quite well for simple materials such as metal samples, though the precision of the orientation determination is reduced from 0.1° to about 0.3°. In complex materials like geological samples the fast approach is possible, however, this will be likely to introduce artefacts in the dataset that originate from pattern binning, low angle Hough resolution and the number of bands used for indexing. This relates to the complexity of geological sample both in terms of the number of phases and their complex crystal

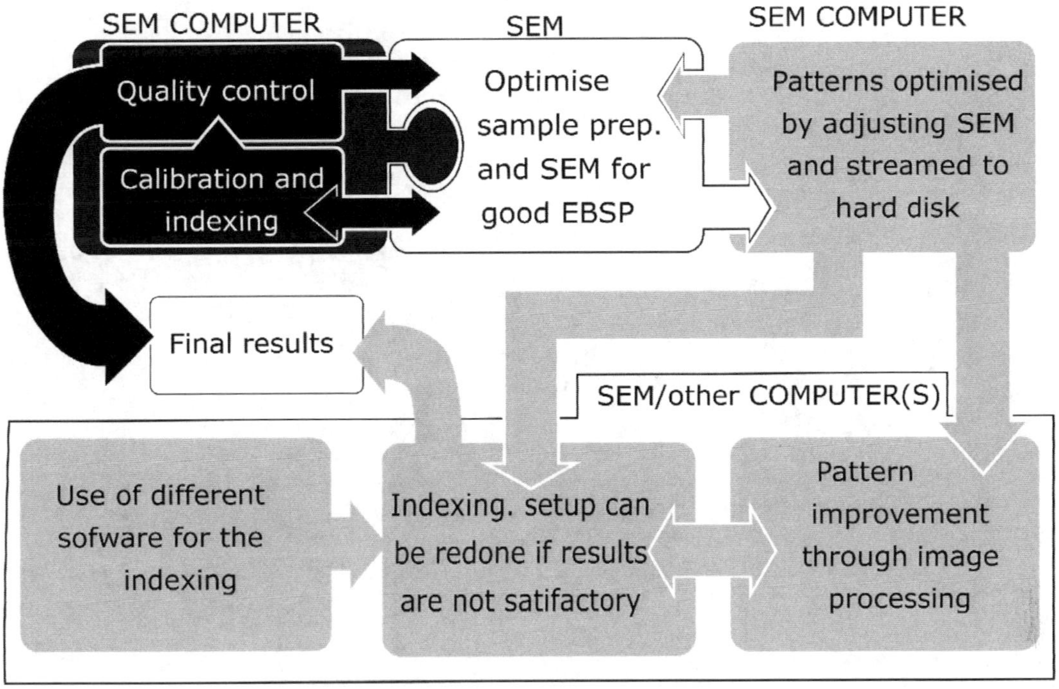

Figure 1. Comparison between on- and offline EBSD procedure. Online procedure in black offers a direct solution, however it is tied with the SEM. Offline procedure in grey opens a range of options after data acquisition on the SEM.

structures. As a result of the low symmetry of minerals, compared to the cubic metals, more bands are needed to determine the orientation and also to deal with pseudo-symmetry. Pseudo-symmetry occurs when the symmetry of a phase is close to a higher symmetry (pseudo symmetry). In the EBSP this is reflected by the majority of the bands are covered by the higher pseudo symmetry, while weaker bands reflect the lower symmetry (Fig. 2). This can cause misindexing because orientations that are symmetrically equivalent in the higher symmetry are misindexed. A well-known example of this is Dauphiné twins in quartz, which represents a 120° degree rotation about the c-axis. Quartz is trigonal and pseudo-hexagonal, meaning that if the right bands are not detected then the Dauphine twin orientations cannot be distinguished and misindexings occurs in the form of false twins (Fig. 3). By selecting the correct reflectors in the crystal structure input the false twins can commonly be avoided as long as the pattern quality is good enough.

Single/few pixel twins are commonly removed as they are assumed misindexed where patterns are noisy for example due to dirt particles or holes in the sample surface. With online EBSD there is no way of checking whether this assumption is correct. However, with off-line EBSD it is possible to check if such assumptions are correct by inspecting the EBSP of the suspected misindexing (Fig. 3). Contrastingly, areas larger than 2 - 3 pixels are true twins and should not be changed during post-processing. EBSD maps commonly also contain unindexed points in the map. Single pixel missing are normally not a problem as they can easily be interpolated from the surroundings. However commonly there are also larger areas that need to be either interpolated from neighbouring grains or maintained as unindexed areas. Off-line EBSD offers the opportunity to investigate this (Fig. 4). Here there is a five pixel wide unindexed area between two calcite grains. By EBSP inspection, it is revealed that the quality of the EBSPs gradually worsen toward the unindexed region, however weak versions of the patterns of the grain are observed in the unindexed region (Fig. 4). In this case, it is, therefore,

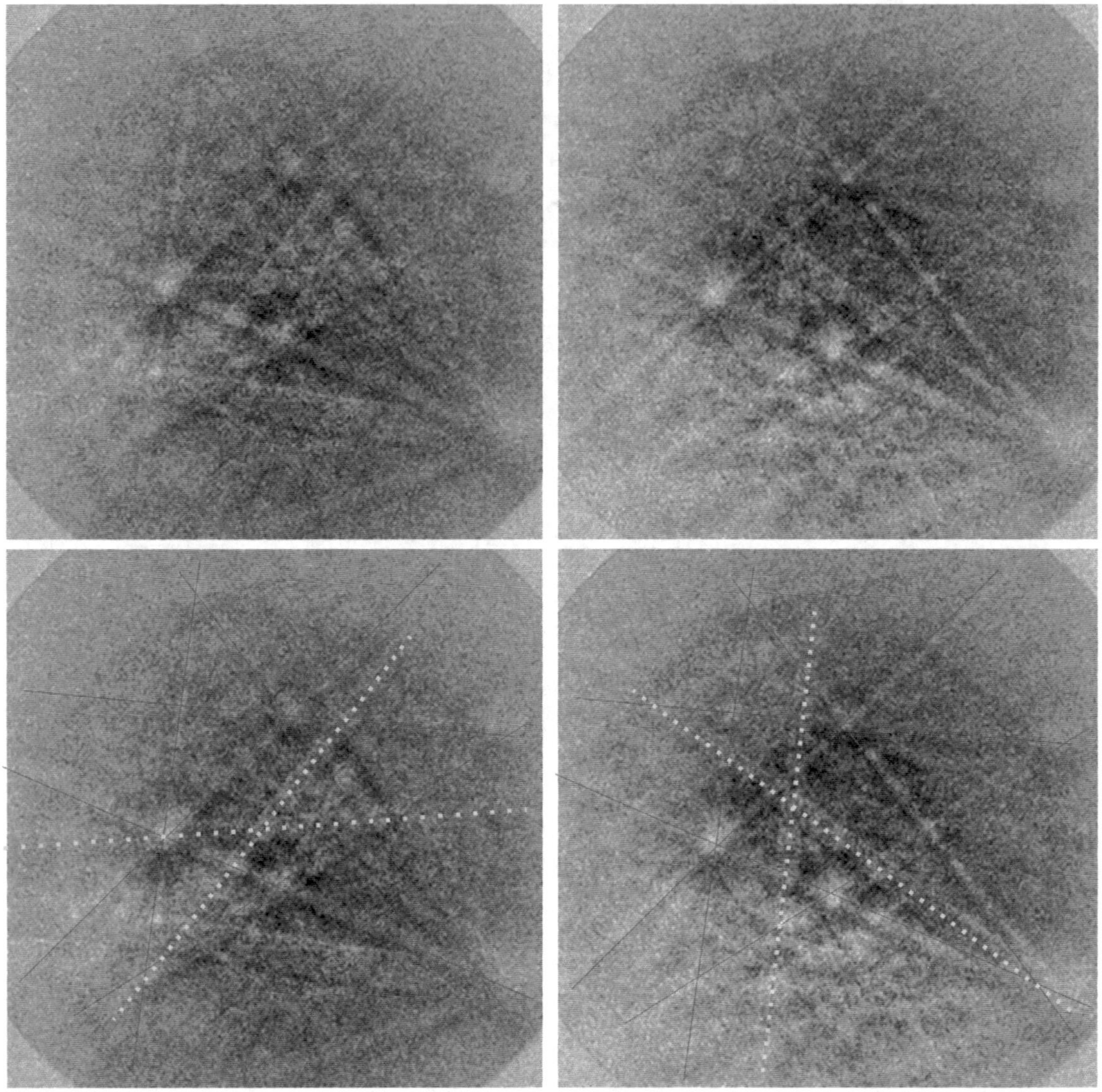

Figure 2. EBSP of two orientation of quartz across a Dauphiné twin boundary. Many bands are common, except broad bands in the background that reflect the true trigonal symmetry of the quartz Laue group. Upper figures are the raw patterns, lower the same pattern with bands that are the same marked by red and bands that differ the two orientations are marked in green. Note that the main bands are the same in the orientations and therefore misindexings are likely if the Hough indexing procedure does not account for this.

a reasonable solution to interpolate the grain boundary to the middle of the unindexed region (Fig. 4). Since the unindexed regions have weak versions of the indexed areas, it might be an even better solution to do pattern improvement and then re-index the dataset (Fig. 4). Another example is misindexed phases along grain boundaries, due to mixing of pattern from either side of the grain-boundary. As an example false dolomite occurs along calcite-calcite grain boundaries, due to the similarity of the EBSP of calcite and dolomite (Fig. 5). By looking at the EBSP it is revealed that the dolomite is indeed misindexed as the EBSP are just mixtures of the patterns of the neighbouring calcite grains.

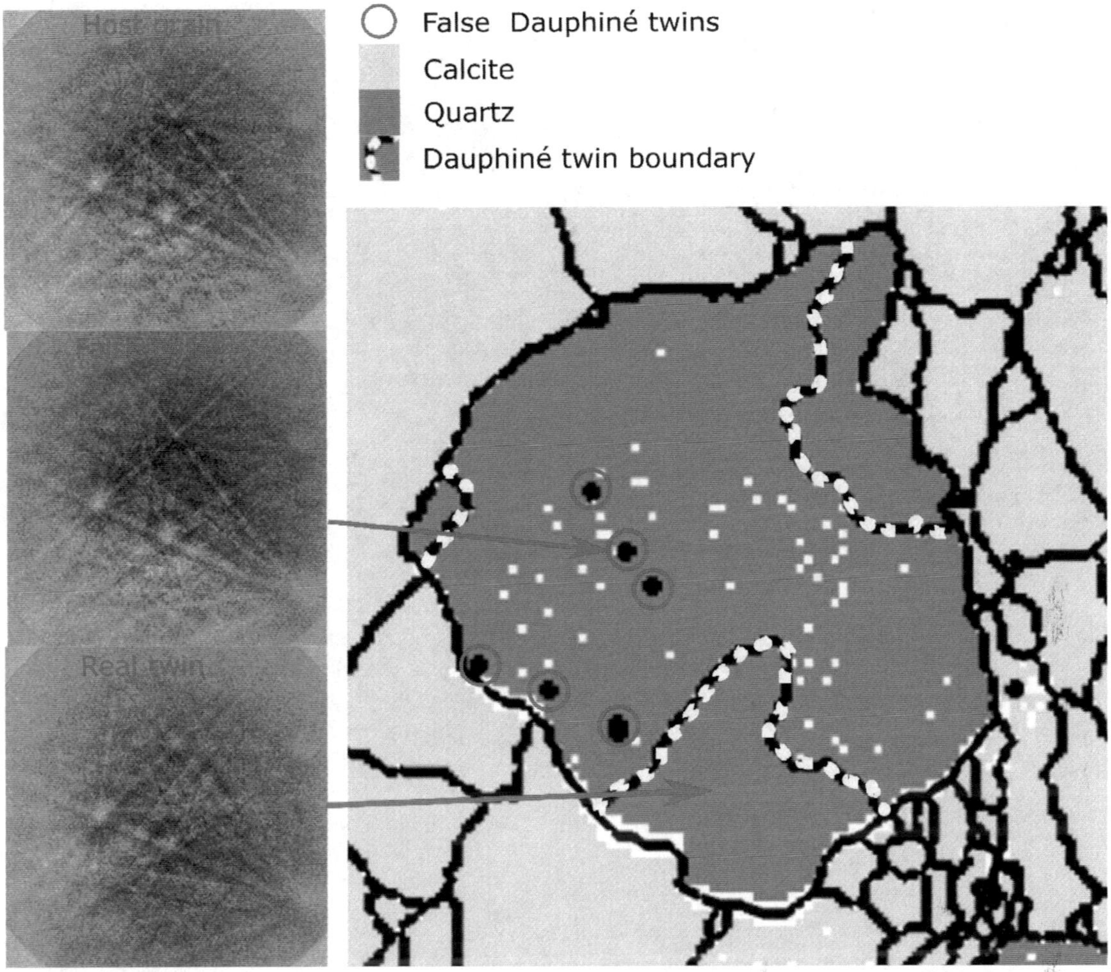

Figure 3. Distinction between false Dauphiné twin and true Dauphiné twin by pattern inspection. False twin occurs when pattern quality reduces and is sensitive to background subtraction and Hough settings.

Coarse-grained materials are hard to handle in EBSD because running at high magnifications only show a few grains or even only parts of one grain, something that makes automatic stich near impossible. Instead samples needs to be mapped at lower magnification. This brings along some artefacts into the raw EBSD dataset. One significant effect is uneven illumination, which is also visible in pattern quality maps at very low magnification. These types of map can be used to differentiate regions that should be removed from the indexed parts, and hence this is very unfortunate. We are currently working on a solution to this using kikuchiPy (https://github.com/kikuchipy), however, so far the indexing results are not better than without pattern adjustments as the tested algorithms introduces other artefacts in the patterns. Another thing that is possible with the offline-data is to use different approaches for indexing, both commercial and freeware such as ASTROEBSD [4] and EMSOFT [5], which may be more suitable in some cases. However, these solutions are not really mature, work only for some data sets and are quite complicated to use. We do expect that these tools will continuously improve and that the possibilities and user friendliness will also do so in the future.

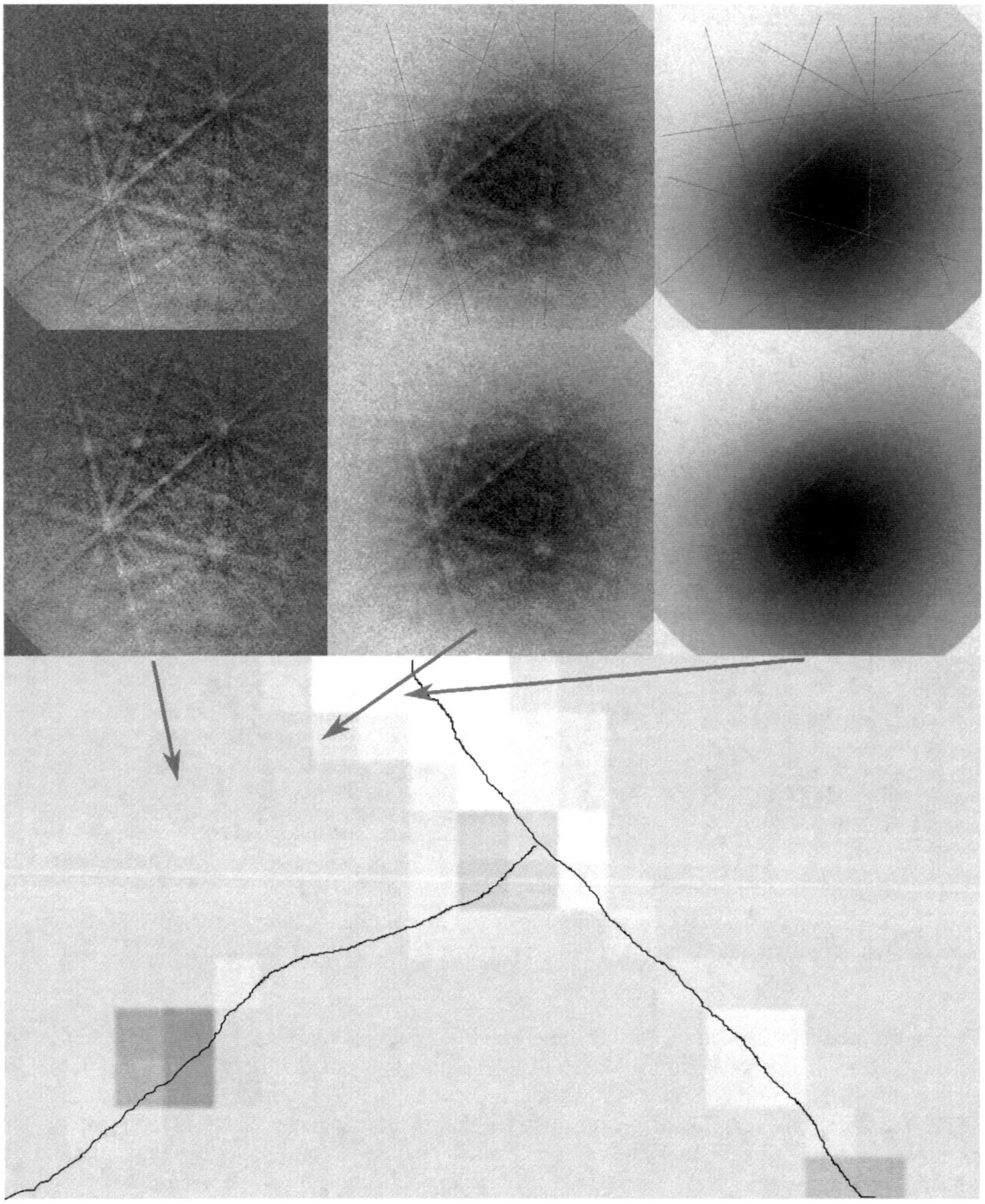

Figure 4. EBSP inspection in unindexed areas. Green is calcite, the lighter the colour the lower confidence index. Reds are false dolomite (see Fig. 5 for discussion). EBSP pattern quality worsen as we enter the unindexed area, but the same diffraction pattern is visible behind all the noise.

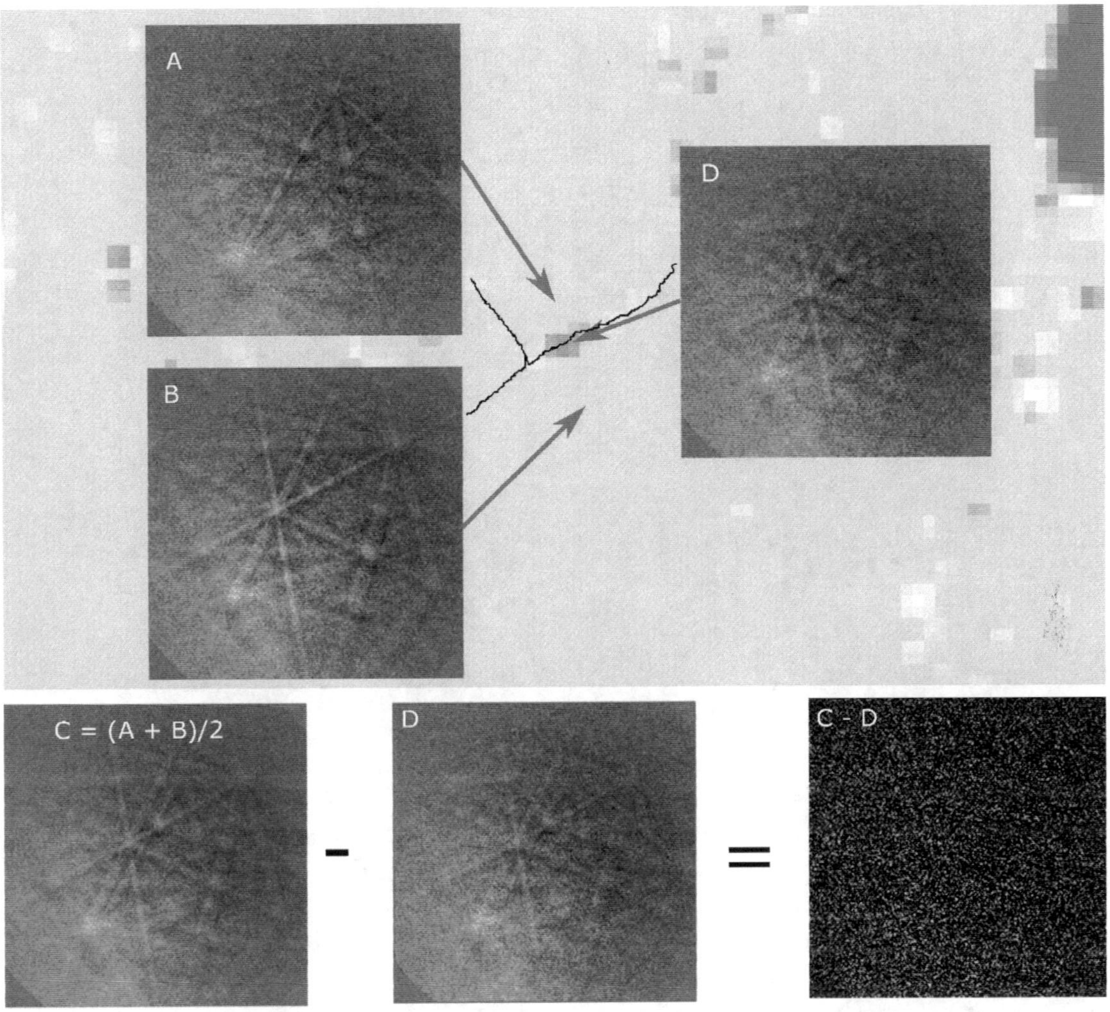

Figure 5. False dolomite recognized as the sum of two neighbouring calcite grain EBSP (A and B) as the average of A and B is the same as the "dolomite pattern" in D.

2.2. EBSD acquisition improvement

Although a lot can be done in offline mode there is still new and unexploited ways of improving the raw data beforehand. The low magnification EBSD maps have big geometric trapezoidal distortions due to the tilt of the sample, that if not corrected will make it impossible to do good stitching. At low magnification, these effects are smaller, but still present. If maps are corrected for this after acquisition then the Y-X coordinates are no longer equally spaced, giving problems with many EBSD processing algorithms. It is therefore much more suitable to have a correction before acquisition. This was implemented in NORDIF version 3 data acquisition software, by correcting the electron beam displacement and making it move in a quadratic grid. A calibration procedure using a square grid on a single crystal silicon standard is applied at different working distance, acceleration voltages and magnifications to fit several future experimental conditions (Fig. 6). If performed correctly this gives undistorted maps even at low magnifications (Fig. 6) and these maps can then be stitched easily without loss of information.

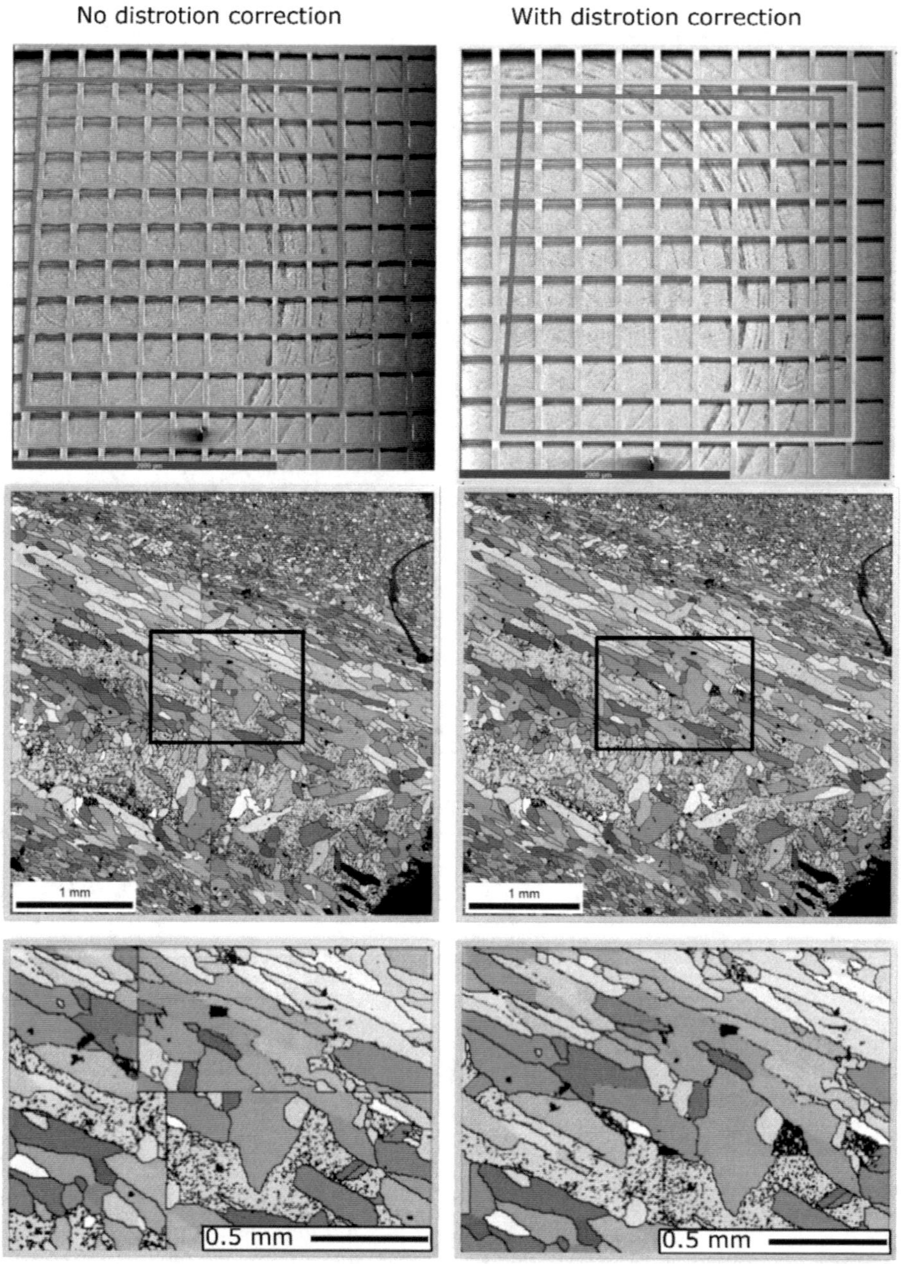

Figure 6. Correction procedure for removing the trapezoidal distortion. Column to the left is without the correction, clear visualised by the square grid not being square in the image. This also gives distorted images of the samples, such that the EBSD maps cannot be stitched properly. In the column to the right distortion correction has been applied and the calibration grid is square and EBSD maps are undistorted and hence stich easily together. It also means that the actual shape and size of the grains are preserved.

Other possibilities related to the acquisition part is to make use of the options that lie in the increasingly advanced EBSD cameras. As an example the auto exposure functionality could reduce some of the uneven illumination problems. In addition this is also interesting for samples with

High variation in phase average z number, where the different phases therefore would give different amount of signal intensity for the same exposure time. In addition, background subtraction schemes could be improved to reduce the background related artefacts in EBSPs for example using better criteria for signal separation into noise and crystallographically relevant parts.

2.3. Processing of indexed EBSD data in MTEX toolbox

MTEX can do most of the functionality implemented in commercial software that, such as grain calculations, grain size and grain boundary statistics. However, MTEX also makes it possible to study problems that are not implemented in commercial packages, simply because there are not industrial customers interested in that problem and hence development of code is not sustainable. The MTEX toolbox is a strong toolbox for EBSD and texture analysis that has a rigorous statistical treatment of orientation and misorientation data and also implements an unbiased orientation colour scheme [6-11]. MTEX is freeware; however, it works from MATLAB so one needs a MATLAB license. Further, the user threshold is higher than the proprietary ESBD software. However, MTEX is much more flexible than any GUI software could ever be and the ODF and MDF functionality is comprehensive. In addition, crystallographic phenomena can be treated both with respect to grain boundaries and physical property calculations within the full realm of point groups. The flexibility also makes things like investigating twins, topo- and epitaxial relations easier.

References

[1] Søfferud M, *et al.* 2008 Development of an ultra-fast EBSD detector system. in: *EMC 2008 14th European Microscopy Congress.* (1-5 September 2008; Aachen, Germany) (Heidelberg, Germany: Springer)

[2] Schwarzer R A and Hjelen J 2010 High-speed orientation microscopy with offline solving sequences of EBSD patterns. in: *Texture and anisotropy of polycrystals III.* (Klein H and Schwarzer R A; Eds.). Solid state phenomena **160** 295-300 (Baech, Switzerland: Trans Tech Publications)

[3] Chen Y, *et al.* 2012 Optimization of EBSD parameters for ultra-fast characterization. *J. Microscopy* **245** 111-118

[4] Britton T B, *et al.* 2018 AstroEBSD: exploring new space in pattern indexing with methods launched from an astronomical approach. *J. Appl. Crystallogr.* **51** 1525-1534

[5] Singh S, Ram F and De Graef M 2017 EMsoft: open source software for electron diffraction/image simulations. *Microsc. Microanal.* **23** (Suppl. 1) 212-213

[6] Bachmann F, Hielscher R and Schaeben H 2010 Texture analysis with MTEX–free and open source software toolbox. in: *Texture and anisotropy of polycrystals III.* (Klein H and Schwarzer R A; Eds.). Solid state phenomena **160** 63-68 (Baech, Switzerland: Trans Tech Publications)

[7] Bachmann F, Hielscher R and Schaeben H 2011 Grain detection from 2d and 3d EBSD data - Specification of the MTEX algorithm. *Ultramicroscopy* **111** 1720-1733

[8] Hielscher R, Schaeben H and Siemes H 2010 Orientation distribution within a single hematite crystal. *Mathem. Geosci.* **42** 359-375

[9] Mainprice D, *et al.* 2015 Descriptive tools for the analysis of texture projects with large datasets using MTEX: strength, symmetry and components. *Geological Society, London, Spec. Publ.* **409** 251-271

[10] Mainprice D, Hielscher R and Schaeben H 2011 Calculating anisotropic physical properties from texture data using the MTEX open-source package. *Geological Society, London, Spec. Publ.* **360** 175-192.

[11] Nolze G and Hielscher R 2016 Orientations - perfectly colored. *J. Appl. Crystallogr.* **49** 1786-1802

IOP Conf. Series: Materials Science and Engineering **891** (2020) 012022 doi:10.1088/1757-899X/891/1/012022

Recent developments in soft X-ray emission spectroscopy microscopy

M Terauchi[1], T Hatano[1], M Koike[2], A S Pirozhkov[2], H Sasai[3], T Nagano[3], M Takakura[4] and T Murano[4]

[1] Tohoku University, Institute of Multidisciplinary Research for Advanced Materials, 2-1-1 Katahira, 980-8577 Sendai, Japan
[2] National Institutes for Quantum and Radiological Science and Technology, Quantum Beam Science Research Directorate, 619-0125 Kizugawa, Japan
[3] Shimadzu Corp., Device Department, 604-8511 Kyoto, Japan
[4] JEOL Ltd., SA Business Unit, 196-8558 Akishima, Japan

E-mail: m_terauchi@tohoku.ac.jp

Abstract. This paper discusses the path to the commercialization of a soft-X-ray emission spectrometer system for EPMA/SEM, its application, how it can be used to investigate new materials, and offers an update on improvements being investigated to further optimise the performance. The ultimate energy resolution of 0.08 eV at Al L- Fermi edge is shown with current optics using a fine pixel detector. The spectral mapping technique can show chemical shift images by using an appropriate region-of-interest energy window. L-emissions of 3d transition metal elements inform one not only of the density of states of bonding but also the number of outer shell or 3d electrons. Furthermore, progress leading to improvements in the detection efficiency has resulted in more than three times increase in the B K-emission peak. Testing and evaluation of new high energy-resolution spectrometer for EPMA, and a new calibration procedure for C K-peak on graphite has resulted in improved energy calibration procedure is presented.

1. Introduction

Observing fine structures of materials and atomic arrangements, examining the composition and the associated chemical state(s) are key issues asked in modern electron microscopy. Especially, understanding the chemical state of element and how it is related to material's function is a core of interest for material's scientists. Transmission electron microscope (TEM) is a powerful and widely used tool for analysing the crystalline order and orientation by electron diffraction pattern, lattice defects by Bragg contrast imaging, atomic arrangements by phase-contrast and/or high-angle dark-field contrast images, the composition by X-ray emission spectroscopy (energy-dispersive X-ray spectrometry, EDS), and chemical state analysis by electron energy-loss spectroscopy (EELS) of nm-scale specimen areas. However, these capabilities of TEM can only be applied for thin specimens. On the other hand, a chemical state analysis method for bulk specimens based on electron microscopy has not been well established in energy resolution compare with those of EELS experiments of 0.1 ~ 1 eV. In the last twenty years, soft X-ray emission spectroscopy (SXES) instruments for electron microscopes have been developed based on TEM to examine its capability [1]. More recently, the SXES was installed on a scanning electron microscope (SEM) and subsequently commercialised as an attachment for electron probe microanalyser (EPMA) and/or SEM [2, 3].

2. X-ray emission due to electronic transition and its information

High energy electron-beam irradiation for a specimen creates a core-hole by exciting a core-electron and this process can lead to X-ray emission as a relaxation process resulting from the transition of an electron to the core-hole state. Figure 1 shows a schematic drawing of electronic transitions among conduction bands (CB), valence bands (VB), and inner-shell levels. The excitation processes, upward arrows, are used in EELS to examine a chemical state of a material. The relaxation processes, downward arrows, are used in X-ray emission spectroscopy (XES) for elemental analysis and/or phase identifications as wavelength-dispersive X-ray spectrometry (WDS) and energy-dispersive X-ray spectrometry (EDS).

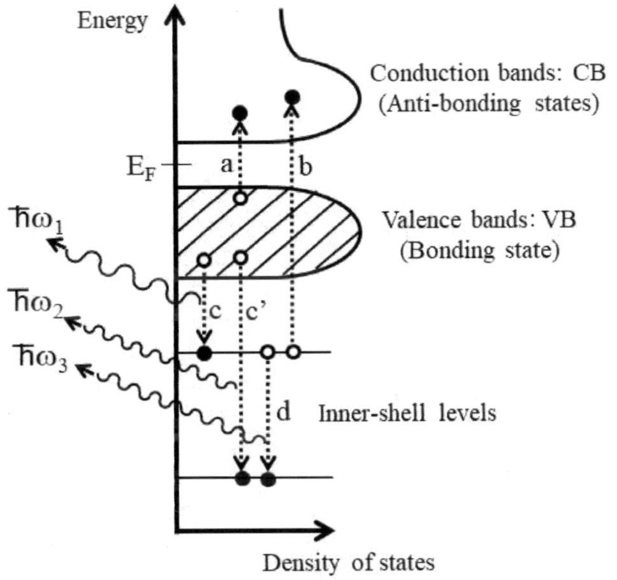

Figure 1. Schematic diagram of electronic structure and electronic transitions related to EELS (a and b), and XES (c, c' and d).

Valence electron excitation EELS spectra (process a) includes information from the band-gap and interband transition energies and refractive indices. Experimental results can be compared with optical experiments by using a Kramers-Kronig analysis. Inner-shell electron excitation EELS spectra (process b) give us information of the density of states (DOS) of CB (anti-bonding states). This inner-shell excitation spectroscopy corresponds to X-ray absorption experiments. Together with the DOS of CB, the DOS of VB (bonding states) it is necessary to understand the whole electronic structure of a material. This is important because the origin of stabilising a material is the energy states of bonding electrons. This information is included in the joint DOS of CB and VB (the imaginary part of ε, ε_2). However, it is not possible to deduce the information from ε_2, because of a lack of information of energy dispersion of bonding states.

Electronic transitions, downward arrows in Fig. 1, from VB to inner-shell core-hole states (processes c, c') and from a shallow core-level to a deeper-lying core-level (process d) accompanied by X-ray emission. Thus, X-ray emissions caused by the process c and c' include information of energy distribution of bonding electrons, DOS of VB. An energy resolution better than 1 eV is necessary for obtaining fine structures of the DOS of VB, because the energy spread of VB is typically from 5 to 10 eV. Since the transitions can happen under a dipole-selection condition ($\Delta l = \pm 1$), the intensity profile of an emission spectrum reflects a partial DOS of VB with a specific symmetry. Furthermore, X-ray emission spectra of different atoms of a compound show us the component of different atoms of the valence DOS. Those emission energies range in the soft X-ray energy region from 15 to 6,000 eV [4]. Since there are many atomic resonances in this energy region, soft X-ray emission spectroscopy (SXES) provides not only the compositional information but also chemical states of elements, which are a sensitive tool for elemental and chemical identification and can be applied to many problems. On the other hand, X-ray emissions caused by the electronic transition d between inner-shell levels does not contain the information of bonding electrons.

SXES is one candidate for valence electron spectroscopy based on electron microscopy. This method does not need an ultra-high-vacuum condition, because the absorption length of soft X-rays is larger than surface thickness of a few nm [5]. It can be combined with conventional electron microscopes (SEM and TEM). One drawback of this method is the low emission efficiency. For example, the yield of X-ray emission of a carbon 1s core-hole state is 1.4×10^{-3} [6]. A longer acquisition time is necessary compared to those of EELS experiments. This is why SXES spectrometer was commercialised for EPMA and SEM which have higher beam currents compared to the TEM and can therefore achieve acceptable signal-to-noise ratios for this technique.

3. Optics of SXES spectrometers

The SXES instrument developed for conventional EPMA/SEM does not have a motorised mechanism to move the spectrometer to satisfy the Rowland circle optics of a traditional EPMA equipped with a wavelength-dispersive X-ray spectrometer. Rather the SXES spectrometer adopts the flat-field grazing-incidence optics by using varied-line-spacing (VLS) gratings [7, 8]. Figure 2 shows a schematic figure of the optics of a spectrometer. T his optics is rather compact, about 50 cm from the specimen to the detector. This compact size results in a narrow energy dispersion on the spectrum plane. Thus, a small pixel sized detector is crucial for this optics to obtain a reasonable energy resolution, typically 13 - 15 μm pixels are used. As the VLS grating forms a spectrum in a certain energy range on a flat plane, a parallel detection by using a CCD detector is employed. The measured energy resolution is typically 0.2 eV or better for Al-L Fermi edge. The grating works with a fixed grazing incidence angle, which ranges from 1.5 to 4 degrees depending on the design of the acceptance energy range of the grating. This small grazing incidence angle improves the reflection efficiency of the grating, but results in a small solid-angle for the spectrometer. To improve detection efficiency, X-ray correction mirrors have been designed and placed between the specimen position and the VLS grating. These mirrors increase the detection solid angle of the SXES instrument by a factor of 2 - 3.

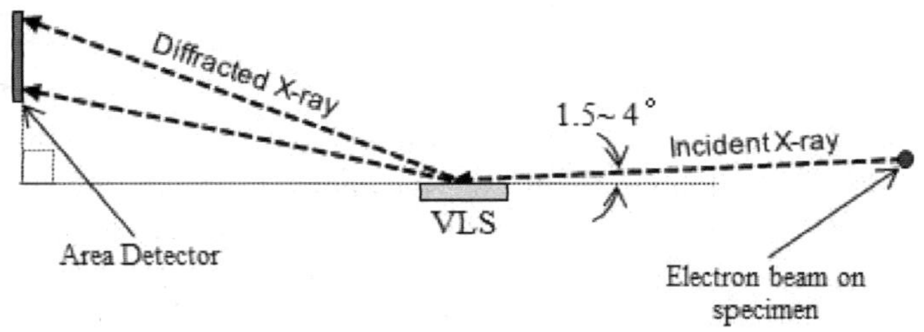

Figure 2. Schematic optics of VLS grating used.

The original optics were designed to cover an energy range of 50 to 3,800 eV by using four VLS gratings, whose optics adopted the flat-field grazing-incidence [9, 10]. Two spectrometers of containing these optics were installed to a TEM [9, 10] and a SEM [2, 11]. At present, the commercial spectrometer accommodates two VLS gratings, which limits the detection energy range compared with the original configuration. Figure 3 shows a photo of the original spectrometer with four VLS gratings attached to a conventional SEM. A fine-pitch micro-channel plate (MCP) optically coupled with a CMOS camera is used as a detector. An aluminium L-emission spectrum obtained by the MCP in photon counting mode is also shown in Fig. 3. The spectrum shows a clear separation of L3- and L2-edges. The energy width corresponds to the intensity increase from 25 % to 75 % of L3-edge is 90 meV. This means that the energy resolution of this spectrum is evaluated as 75 meV by taking into the thermal broadening of Fermi distribution function at room temperature.

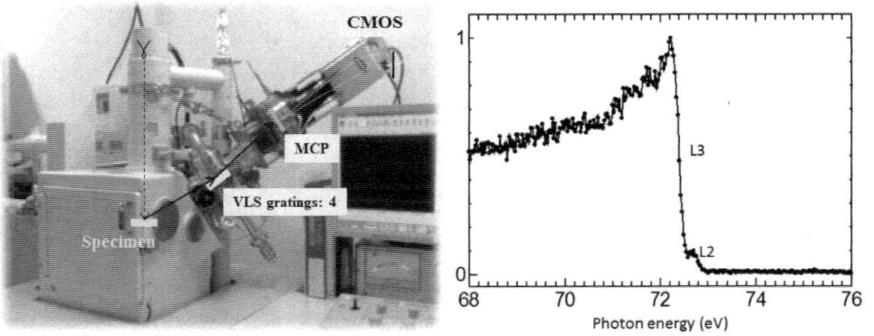

Figure 3. Photo of a SEM-SXES instrument for 50 – 3,800 eV and a spectrum of Al L-emission with an energy resolution of 0.08 eV.

4. SXES spectra reflecting bonding stages and its imaging

Figure 4 shows B K-emission spectra of CaB_6 and CeB_6. Those are composed of a network of B_6 octagonal clusters accommodating a metal atom at the body-centre positon of the unit cell, as shown in an inset figure. A B K-emission spectrum of β-rhombohedral boron (β-r-B), which is classified to a B_{12}-cluster material, is also shown for comparison. Peak intensity at around 185 eV in each spectrum is due to intra-cluster bonding state. The tail intensity in the lower energy side is due to inter-cluster bonding states. In CaB_6, the Ca atom can transfer two valence electrons to the B_6-cluster network. Thus, VB of B_6-cluster network is fully occupied and becomes a semiconductor. The, Ce atom in CeB_6 can transfer 3 or 4 electrons to B_6-cluster network. Thus, excess electrons go into CB making it metallic B_6-material. This can be seen in the intensity profile of CeB_6. On the higher energy side of the main peak, there is an additional peak compared with that of CaB_6. This additional peak is due to electrons occupying the CB. A small onset in the right hand side end of the additional peak could correspond to the Fermi energy position of the material. In addition, the main peak position of CeB_6 is slightly shifted to the lower energy side compared with that of CaB_6. As B K-emission is due to transitions form VB to B1s core-level, this shift is most likely a chemical shift of B1s core-level due to a larger amount of charge transfer of Ce atom compared to the Ca atom.

CaB_6 is a *n*-type semiconductor. However in this case a thermo-electric device using B_6-cluster network materials [12], it is required a *p*-type semiconductor. Sodium was added as the Na-atom has only one valence electron, making Na-substitution to Ca-site of CaB_6 (Na-doped CaB_6) produce hole doping to B_6-cluster network, leading to the material transforming to a *p*-type semiconductor [13]. EDS analysis showed the Na content of the material was a few percent. However, the material may be not a uniform from a production process point of view. A piece of Na-doped CaB_6 bulk specimen was analysed by a EPMA-SXES instrument. Figure 5 shows region-of-interest intensity maps of a) Ca-Lℓ,η intensity (150 - 154 eV), b) B-K intensity (170 - 188 eV), and c) the top part of the B-K intensity (187 - 188 eV). Ca-Lℓ,η emission map shows a low intensity in the central area, due to lower Ca content in this region. B-K emission map shows little intensity change for the corresponding region.

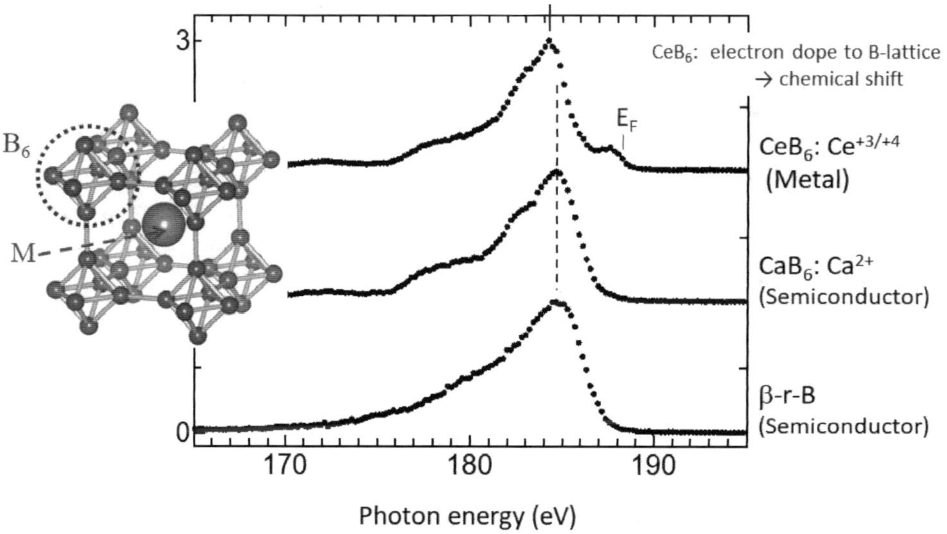

Figure 4. B K-emission spectra of CaB_6, CeB_6 and β-rhombohedral boron (β-r-B). The spectrum of CeB6 shows an additional intensity with Fermi edge in right hand side and a peak shift, chemical shift, due to a larger amount of charge transfer of Ce atom than that of Ca.

An intensity anomaly in the upper left of the image is due to a surface contamination. The composition and chemical bonding state of the central area may be different form the surrounding area. A narrow region-of-interest map for B-K (187 - 188eV) peak, Fig. 5c, shows an apparent intensity increase for the central area. The B-K spectra from the bright region (A) and duller region (B) are shown in Fig. 5c, bottom left. The spectra show a small difference in in their distributions, suggesting some difference of bonding state of B_6-cluster network due to a difference in composition. The peak (arrow) position from region A shows a shift to higher energy of about 0.7 eV compared to that from region B. This shift of the whole spectrum could be due to a chemical shift originated from a larger binding energy of B1s-level of the area. The shift can be explained by a decrease of valence charge of B atoms of point A. It can be understood if the expected Na-substitution was a smaller leading to a smaller charge transfer and/or an impurity of oxygen, which removed two electrons from the B_6-network. Unfortunately, the Na K-peak is not in the energy range of the JS200N grating used (it is noted that the latest commercial system can detect the Na K-emission signal). The impurity oxygen atom cannot be eliminated because a small intensity at 175 eV in the spectrum from region A can be assigned to the 3rd order intensity of O-K peak. A high energy resolution SXES map can provide both an elemental map combined with a chemical bonding state map reflecting, in this case, a valence charge distribution. Chemical state mapping combined with elemental map by an EPMA/SEM-SXES system gives important information for phase identification of bulk specimens.

5. Information of L-emission of 3d transition metal elements

The bonding state and valence charge of 3d transition metal (3d-TM) elements is key for evaluating positive electrode materials used in Li-ion batteries, because a change of valence in 3d-TM elements is the origin of charge and discharge performance of the battery. Information on valence electrons of 3d-TM elements is included in L-emissions of these elements. There are four types of L-emissions of Lα, Lβ, Lℓ and Lη as shown in Fig. 6 [14]. These are classified into two groups having close emission energies group I is Lα,β and group II is Lℓ,η. The Lα,β emission is due to $3d_{5/2,3/2} \rightarrow 2p_{3/2,1/2}$ ($L_{2,3}$) electronic transition, its energy distribution contains an information of DOS of VB (bonding states). While the Lℓ,η emission energy is due to transitions from a shallow inner-shell of $M_1(3s_{1/2})$ to a deeper

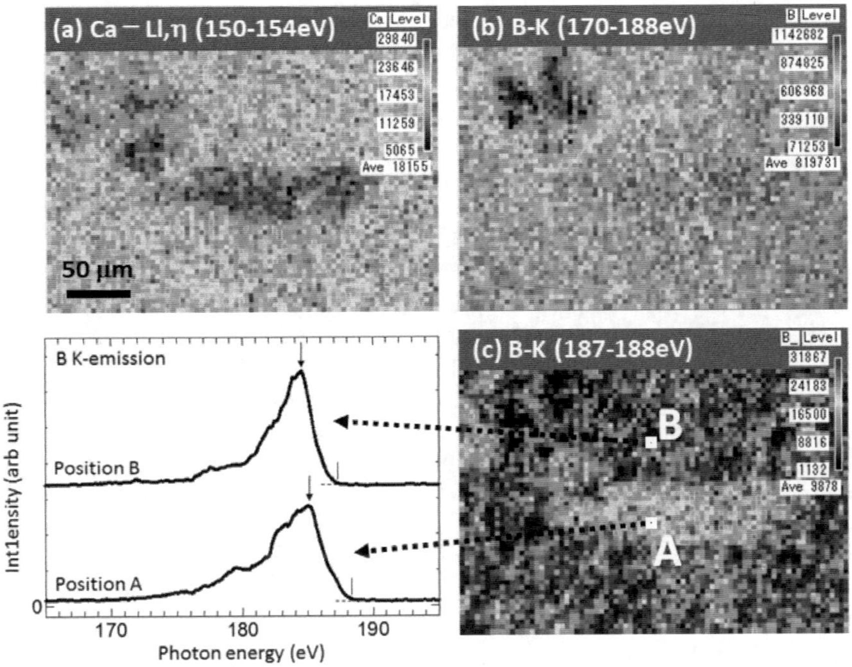

Figure 5. a) Ca-Lℓ,η intensity (150 - 154 eV), b) B-K intensity (170 - 188 eV), c) the top part of the B-K intensity (187 - 188 eV), and B K-emission spectra of the areas A and B in c).

inner-shell of $L_{2,3}$ and reflects the dielectric environment of the transition metal elements [15], which differentiates it from the Lα,β emission.

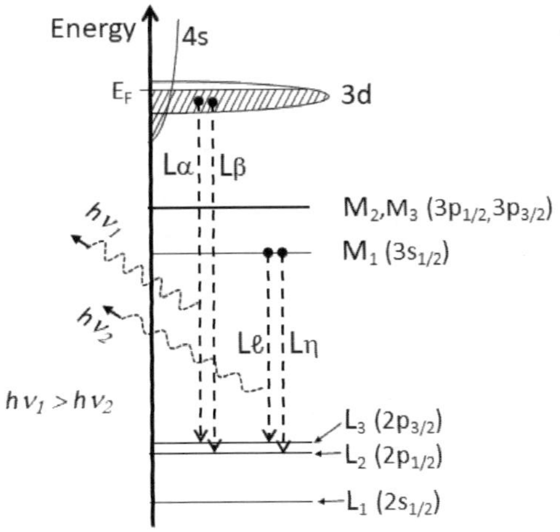

Figure 6. Schematic energy diagram of 3d transition metal elements and electronic transitions of four L-emissions [14].

Figure 7 shows raw L-emission spectra of simple metal materials of Sc (Z = 21, $3d^14s^2$), Ti (Z = 22, $3d^24s^2$), V (Z = 23, $3d^34s^2$), Cr (Z = 24, $3d^54s^1$), Mn (Z = 25, $3d^54s^2$), Fe (Z = 26, $3d^64s^2$), Co (Z = 27, $3d^74s^2$), Ni (Z = 28, $3d^84s^2$), Cu (Z = 29, $3d^{10}4s^1$) and Zn (Z = 30, $3d^{10}4s^2$) acquired by the SEM-SXES instrument shown in Fig. 3 at an accelerating voltage 5 kV [14]. Each spectrum intensity is normalised by the maximum intensity in each spectrum. A JS2000 grating, which covers the energy range from 300 eV to 2,200 eV, was used. To enable a quantitative comparison of spectral intensities recorded by the MCP detector it was set to photon counting mode [14], which results in the integrated intensity of an emission peak corresponding to a number of emitted X-ray photons or related electronic transitions. It can be seen that the L-emission energy increases with an increase of Z, which is caused by an increase of the binding energy of $L_{2,3}$ level according to Z. The intensity ratio of Lα,β/Lℓ,η increases according to an increase of N_{3d}. This trend of the Lα,β/Lℓ,η ratio is expected as the electron number contributed for Lℓ,η emission is a constant of two 3s electrons whereas that for Lα,β emission, the number of 3d electrons (N_{3d}), increases with Z. Thus, the intensity ratio of Lα,β/Lℓ,η can be an indicator for evaluating N_{3d}. Integrated intensities of Lα,β and Lℓ,η are shown in Fig. 7. The intensity ratio of Lα,β/Lℓ,η increases according to N_{3d}, but is not linear. This is due to the two transition of $3d_{5/2,3/2} \rightarrow L_{2,3}$ and $M_1(3s_{1/2}) \rightarrow L_{2,3}$ competing with the L-emission process. The transition probability of $3d_{5/2,3/2} \rightarrow L_{2,3}$ should change according to a change of N_{3d}. The intensity of Lα,β compared with the whole L-emission intensity of Lα,β+Lℓ,η, should reflect not only the number of 3d electrons but also a change of transition probability due to a change of N_{3d}. This is shown in Fig. 8 were the variation of Lα,β/(Lα,β+Lℓ,η) follows to N_{3d}, and is almost linear from Sc to Ni, except for Cr. The deviation of the Cr value from the expected linear trend line may reflect its characteristic electron configuration in 3d orbitals. The linear tendency of Lα,β/(Lα,β+Lℓ,η) was fitted by a linear equation and this was applied to some oxides. In case of 3d metal oxides, the results using the equation led to an over estimate in the number of 3d electrons. This may be due to the charge transfer from ligand oxygen atoms to 3d transition metal elements resulting in a core-hole effect in the intermediate state as reported by Grebennikov *et al.* [16].

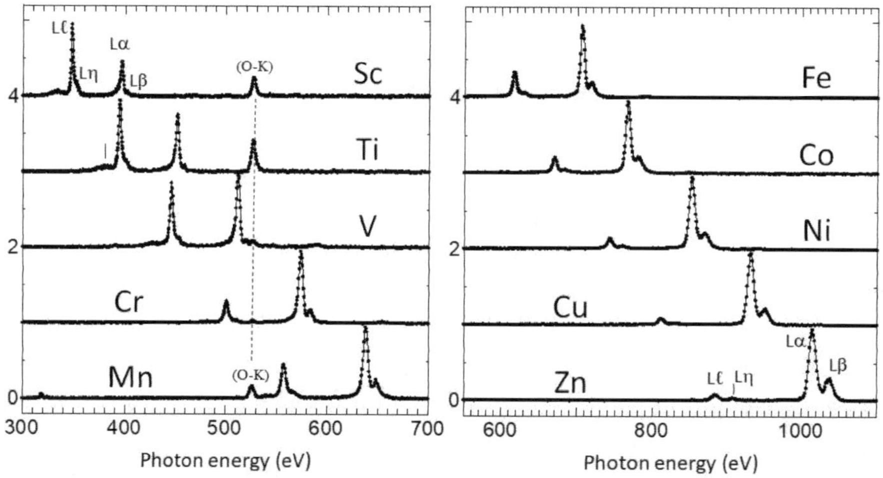

Figure 7. L-emission spectra of Sc, Ti, V, Cr, Mn, Fe, Co, Ni, Cu and Zn. Each spectrum intensity is normalised by its maximum intensity in each spectrum [14].

6. Improvements of SXES instrument

6.1. Improvement of detection efficiency

As mentioned in section 3, the grazing incidence optics intrinsically have a small solid angle for X-ray detection, even though the grazing incidence improves the reflectance of the grating and the optics produce parallel detection within a certain energy range through the use of a flat detector. Further improvement of detection efficiency have been investigated and new optics combined with

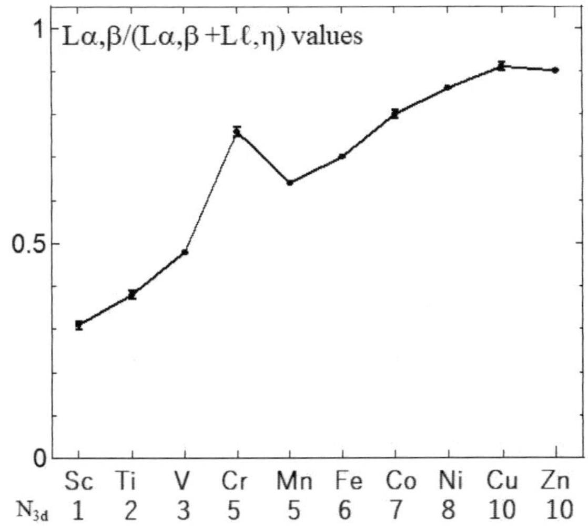

Figure 8. Experimentally obtained Lα,β/(Lα,β+Lℓ,η) values for 3d transition metal materials.

an enhancement of diffraction efficiency of VLS grating by a coating was designed, manufactured and tested [17, 18]. One target was boron, which is important to detect with ppm sensitivity in steel compounds where it is used as a key element to improve structural and mechanical properties.

A conventional JS200N, nickel (Ni) coated VLS grating, used in a commercial SXES instrument has a grazing incidence angle of 3°. In the new optics, this angle was changed to 5.8° keeping the positions of a specimen and a detector unchanged. This improves the solid angle by a factor of two. In addition, the grating surface of Ni layer is coated by La and then carbon (JS200NLC) to improve the reflectivity. The carbon layer was added to minimise oxidation of the La layer. This coating was expected to perform with up to two times better diffraction efficiency than without coating for B K-emission based on a theoretical calculation. In total, ~4 times improvement for B-K intensity was expected compared to that of JS200N grating. A new grating was tested at synchrotron facility and performed about 5 times better [18]. The same grating was tested in a SEM-SXES instrument (Fig. 3), which was modified to accommodate the new optics. Figure 9 the result for B K-emission the observed improvement in spectral intensity was 3.4 times. This is smaller than five measured at synchrotron and is due to poorly aligned X-ray correction mirrors placed between the specimen and the VLS grating. When compared with the new grating, which has only a Ni coat, the enhancement was 2.5 times. This means that about five times enhancement is possible for this NI/La/C grating when the X-ray collecting mirrors work as that for JS200N. Since La is an active material, a stable compound of LaF_3 was coated on a new grating and tested. For this grating, the diffraction efficiency for B-K was 2.1 times compared with that of the Ni only coated surface. Thus, an improved detection efficiency of about 4 times is produced using this configuration.

6.2. High energy-resolution spectrometer for EPMA

Improvement of the energy resolution is also important. A s mentioned in section 3, the energy resolution of the commercial SXES spectrometer is limited by the pixel size of the CCD detector. There are two techniques we have investigated to improve the energy resolution. One is to use a fine pitched new detector. Secondly to construct a spectrometer with a larger energy dispersion. A larger-dispersion spectrometer was originally designed and constructed for TEM [19]. Figure 10 shows the optics of this spectrometer. The distance from the VLS grating to the detector is about twice that of a usual spectrometer. This results in a two times larger energy-dispersion being produced by these optics. However, a spectrometer with a long specimen to detector distance has a smaller solid-angle for X-ray detection and this results in it being difficult to obtain good signal-to-noise ratios when used on a TEM.

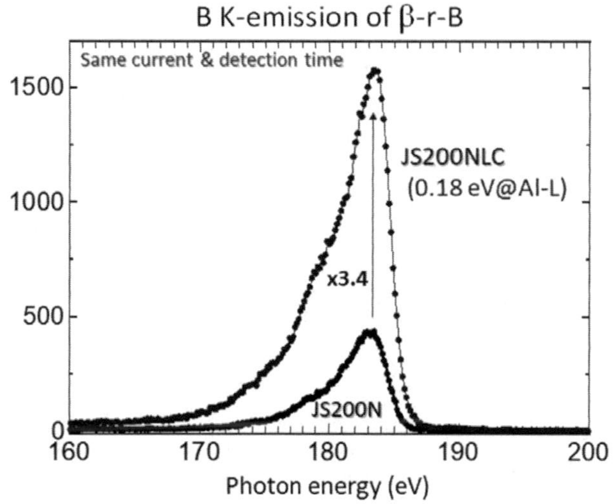

Figure 9. Intensity enhancement of a new grating of JS200NCL compared with a conventional grating of JS200N obtained by SEM-SXES instrument (Fig. 3).

Integrating this larger spectrometer onto EPMA is a reasonable way to improve in the energy resolution with a better signal to noise ratio. Figure 11a shows a photo of the newly constructed SXES spectrometer with the optics of Fig. 10 attached to a EPMA. The EPMA is also equipped with a commercial SXES spectrometer. Two VLS gratings are accommodated in the new SXES instrument and cover the whole energy range of first order L-emissions for the 3d transition metal elements. In this case, the detector is a multichannel plate optically coupled to a CCD camera. The energy resolution at the Fermi energy for B K- peak measured on LaB_6 was 0.3 eV ($E/\Delta E \fallingdotseq 630$) as shown in Fig. 11b. A further improvement can be expected when a finer pixel detection system is utilised. The fine pitched MCP detector already used in SEM-SXES of Fig. 3, resulted in an energy resolution more than three times better than the commercial spectrometer. This performance will be useful for investigating chemical states of 3d transition metal elements together with imaging.

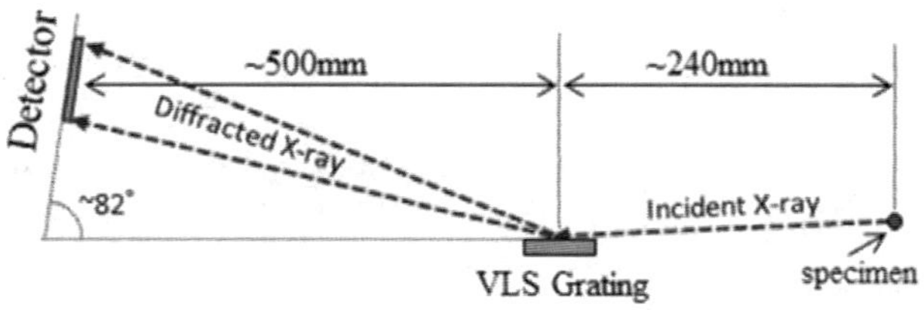

Figure 10. Optics of a larger-dispersion VLS grating.

6.3. Energy determination for improving energy calibration

As K-emissions of light elements of B, C, N are due to electronic transitions of bonding electrons to K-shell core-hole state, those intensity distributions reflect the chemical bonding state of the material examined. This means that the peak energy position depends on the chemical bonding state of the atom in the material. It is difficult to perform an energy calibration on the SXES since there are not a set of suitable metals with known peak positions. Thus, we normally use the sharp peak structure of K-emission from amorphous graphite, which has a peak structure due to sp^2-σ bonding state; however, an alternative calibration can be obtained by using the Al L-Fermi edge and higher-order reflections from Sc. The second and third order sp^2-σ peak energies were consistently explained by the first order energy of 277.0 eV. The peak energy from B K-peak measured on crystalline β-rhombohedral boron was measured to be 185.3 eV. This value is 2 eV higher than the usually reported value of 183.3 eV for B-Kα. This value is close to the peak energy of B K measured on hexagonal BN.

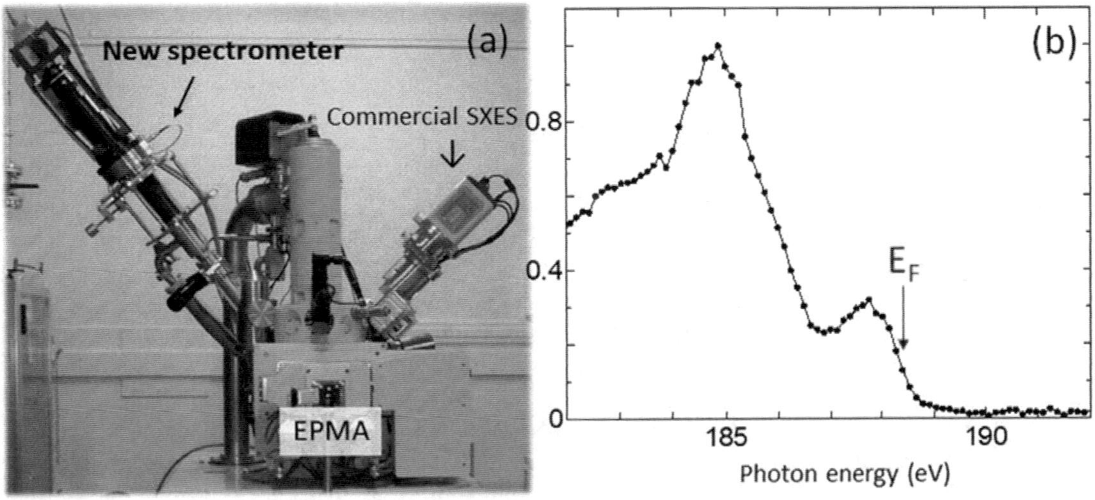

Figure 11. a) Photo of a newly constructed large-dispersion SXES spectrometer attached to EPMA. b) B K-emission spectrum of LaB_6.

7. Summary

Following the developments of SXES spectrometers for TEM, the SXES spectrometer was commercialised as an attachment for EPMA and SEM. This system has a much better energy resolution owing to a high quality VLS gratings combined with a CCD detector and a high detection efficiency owing to a parallel detection. This has enabled compositional analysis and chemical state analysis to be performed in parallel together in either a point analysis mode or an imaging mode. With higher detection efficiencies, new optics and improved grating coatings specifically tuned for a target energy range, we have shown up to five times improvement is possible for B K-emission. A higher energy resolution targeting a chemical state of 3d transition metal elements has just constructed and evaluated.

Acknowledgments
The metal-B_6 materials are part of a collaboration with Prof M. Takeda of Nagaoka University of Technology, Nagaoka, Japan. This work was partly supported by the Research Program of "Dynamic Alliance for Open Innovation Bridging Human, Environment and Materials" in "Network Joint Research Centre for Materials and Devices", and JSPS KAKENHI Grant Number 19K21838.

References
[1] Terauchi M 2014 Valence electron spectroscopy for transmission electron microscopy. in: *Transmission electron microscopy characterization of nanomaterials* (Kumar C S S R; Ed.) (Berlin Heidelberg, Germany: Springer-Verlag) 287-331
[2] Terauchi M, Koshiya S, Satoh F, Takahashi F, Handa N, Murano T, Koike M, Imazono T, Koeda M, Nagano T, Sasai H, Oue Y, Yonezawa Z and Kuramoto S 2014 Chemical state information of bulk specimens obtained by SEM based soft-X-ray emission spectroscopy. *Microsc. Microanal.* **20** 692
[3] Takahashi H, Handa N, Murano T, Terauchi M, Koike M, Kawachi T, Imazono T, Hasegawa N, Koeda M, NaganoT, Sasai H, Oue Y, Yonezawa Z and Kuramoto S 2014 Exciting possibilities of soft X-ray emission spectroscopy as chemical state analysis in EPMA and FESEM. *Microsc. Microanal.* **20** (Suppl. 3) 684
[4] Fabian D J, Watson L M and Marshall C A W 1972 Soft X ray spectroscopy and the electronic structure of solids. *Rep. Prog. Phys.* **34** 601

[5] Attwood D 2000 *Soft X-rays and extreme ultraviolet radiation.* (New York, NY: Cambridge University Press) 3

[6] Krause M O 1979 Atomic radiative and radiationless yields for K and L shells. *J. Phys. Chem. Ref. Data* **8** 307

[7] Harada T and Kita T 1980 Mechanically ruled aberration-corrected concave gratings. *Appl. Optics* **19** 3987

[8] Koike M, Yamazaki T and Harada Y 1999 Design of holographic gratings recorded with aspheric wave-front recording optics for soft X-ray flat-field spectrographs. *J. Electron Spectrosc. Related Phenom.* **101-103** 913

[9] Terauchi M, Takahashi H, Handa N, Murano T, Koike M, Kawachi T, Hasegawa N, Imazono T, Koeda M, Nagano T, Sasai H, Oue Y, Yonezawa Z and Kuramoto S 2012 *J. Electron Microsc.* **61** 1

[10] Terauchi M, Takahashi H, Handa N, Murano T, Koike M, Kawachi T, Imazono T, Koeda M, Nagano T, Sasai H, Oue Y, Yonezawa Z and Kuramoto S 2013 *Microscopy* **62** 391

[11] Terauchi M, Takahashi H, Takakura M, Murano T, Koike M, Imazono T, Nagano T, Sasai H and Koeda M 2016 Chemical state analysis of trace-boron by using an improved SEM-SXES. *Microsc. Microanal.* **22** (Suppl. 3) 414

[12] Takeda M, Terui M, Takahashi N and Ueda N 2006 Improvement of thermoelectric properties of alkaline-earth hexaborides. *J. Solid State Chem.* **179** 2823

[13] Kuribayashi H, Makino S and Takeda M 2014 Synthesis and characterization of p-type CaB_6. *Abstract of ISBB18.* (Honolulu, 31 August -5 September) 148

[14] Terauchi M 2019 Information of valence charge of 3d transition metal elements observed in L-emission spectra. *Microscopy* **68** 330

[15] Terauchi M, Koshiya S and Kimoto K 2017 Information observed in Ti-Lα,β and Ti-Ll,η emission lines of Ti and its oxides. *IOP Conf. Ser.: Mater. Sci. Engng.* **304** 012018

[16] Grebennikov V I, Galakhov V R, Finkelshtein L D, Ovechkina N A and Kurmaev E Z 2003 Effect of atomic magnetic moments on the relative intensity of the Lβ and Lα components in X-ray emission spectra of 3d transition metal oxides. *Phys. Solid State* **45** 1048

[17] Hatano T, Koike M, Nishihara H, Pirozhkov A S, Terauchi M, Sasai H and Nagano T 2018 Enhancement of diffraction efficiency and spectral flux of laminar-type diffraction gratings coated with Ni/La/C layers in soft X-ray region. *Abstract of SRI2018.* (Taipei, 10-15 June) 232

[18] Hatano T, Koike T. Terauchi M, Pirozhkov A S, Hayashi N, Sasai H and Nagano T 2019 Diffraction efficiency measurements of laminar-type diffraction gratings overcoated with lanthanum series layers for boron-K emission spectroscopy. *Photon Fact. Activity Rep.* **36** 2018G012

[19] Terauchi M, Koike M, Fukushima K and Kimura A 2010 Development of wavelength-dispersive soft X-ray emission spectrometer for transmission electron microscopes – An introduction of valence electron spectroscopy for transmission electron microscopy. *J. Electron Microsc.* **59** 251

Advances in electron channelling contrast imaging and electron backscatter diffraction for imaging and analysis of structural defects in the scanning electron microscope

C Trager-Cowan[1], A Alasmari[1], W Avis[1], J Bruckbauer[1], P R Edwards[1], B Hourahine[1], S Kraeusel[1], G Kusch[1], B M Jablon[1], R Johnston[1], R W Martin[1], R McDermott[1], G Naresh-Kumar[1], M Nouf-Allehiani[1], E Pascal[1], D Thomson[1], S Vespucci[1], K Mingard[2], P J Parbrook[3], M D Smith[3], J Enslin[4], F Mehnke[4], M Kneissl[4,5], C Kuhn[4], T Wernicke[4], A Knauer[5], S Hagedorn[5], S Walde[5], M Weyers[5], P-M Coulon[6], P A Shields[6], Y Zhang[7], L Jiu[7], Y Gong[7], R M Smith[7], T Wang[7] and A Winkelmann[1,8]

[1] University of Strathclyde, Department of Physics, SUPA, Glasgow G4 0NG, Great Britain
[2] National Physics Laboratory, Teddington TW11 0LW, Great Britain
[3] University College Cork, Tyndall National Institute, Cork T12 R5CP, Ireland
[4] Technische Universität Berlin, Institut für Festkörperphysik, 10623 Berlin, Germany
[5] Leibniz-Institut für Höchstfrequenztechnik, Ferdinand-Braun-Institut, 12489 Berlin, Germany
[6] University of Bath, Department of Electronic and Electrical Engineering, Bath BA2 7AY, Great Britain
[7] University of Sheffield, Department of Electronic and Electrical Engineering, Sheffield S1 3JD, Great Britain
[8] AGH - University of Science and Technology, Academic Centre for Materials and Nanotechnology (ACMiN), 30059 Krakow, Poland

E-mail: c.trager-cowan@strath.ac.uk

Abstract. In this article we describe the scanning electron microscopy (SEM) techniques of electron channelling contrast imaging and electron backscatter diffraction. These techniques provide information on crystal structure, crystal misorientation, grain boundaries, strain and structural defects on length scales from tens of nanometres to tens of micrometres. Here we report on the imaging and analysis of dislocations and sub-grains in nitride semiconductor thin films (GaN and AlN) and tungsten carbide-cobalt (WC-Co) hard metals. Our aim is to illustrate the capability of these techniques for investigating structural defects in the SEM and the benefits of combining these diffraction-based imaging techniques.

1. Introduction

The scanning electron microscope (SEM) is a very powerful tool for investigating and imaging a wide range of material properties spanning topography, structure, composition and light emission [1-4].

SEMs are extensively used for imaging topography by monitoring the intensity of secondary electrons as a focussed electron beam, with an energy in the range of 100 eV to 30 keV, is rastered over the surface of a sample. Less well known are the techniques of electron channelling contrast imaging (ECCI) [5-14] and electron backscatter diffraction (EBSD) [5-6, 14-20], which exploit diffraction to provide information on crystal structure, crystal misorientation, grain boundaries, strain and structural defects such as dislocations and stacking faults. Here we provide an overview of these techniques and illustrate their application by describing some of our recent investigations of the structural properties of nitride semiconductor thin films (GaN and AlN) and WC-Co hard metals.

To carry out our measurements we use a range of SEMs equipped with both commercial and bespoke detection systems. In the work reported here we have used an FEI Sirion 200 Schottky field emission gun SEM (Sirion SEM) equipped with an in-house developed ECCI system. We have also used an FEI Quanta 250 Schottky field emission gun environmental/variable pressure SEM (Quanta SEM) equipped with gaseous electron detectors for electron detection when the SEM is used in low vacuum or environmental modes. The Quanta SEM is also equipped with an Oxford Instruments Nordlys EBSD detector and forescatter diodes for EBSD and ECCI measurements respectively. The ability to vary the chamber pressure in the Quanta SEM allows the dissipation of charge and therefore imaging of high resistivity materials such as AlN.

2. The electron backscatter diffraction (EBSD) and electron channelling contrast imaging techniques (ECCI)

ECCI and EBSD both exploit diffraction to reveal the structural properties of the crystalline material under investigation. For ECCI it is diffraction of the incident beam which provides the greatest contrast in the resultant images, while for EBSD it is diffraction of backscattered electrons which provide the crystallographic information. The spatial and depth resolution of both techniques is of the order of tens of nanometres. For successful imaging using either ECCI or EBSD, the sample needs to have a reasonably smooth and clean surface. For the imaging of metal surfaces or the surfaces of geological specimens this usually requires careful sample polishing to produce a high quality surface [10], this is also the case for semiconductor wafers cut from bulk crystals. However, for most epitaxially grown semiconductor thin films no surface preparation is required.

2.1. Electron channelling contrast imaging (ECCI)

ECCI micrographs may be produced when a sample is placed so that a plane or planes are at, or close to, the Bragg angle with respect to the incident electron beam. Any deviation in crystallographic orientation or in lattice constant due to local strain will produce a variation in contrast in the resultant ECCI micrograph. The micrograph is constructed by monitoring the intensity of backscattered or forescattered electrons as the electron beam is scanned over the sample. Extremely small changes in orientation and strain are detectable, revealing, for example, low angle tilt and rotation boundaries and atomic steps. Extended defects such as dislocations and stacking faults may also be imaged [5-14, 21-29]. The conditions required to resolve individual dislocations in an electron channelling contrast image are quite stringent: an electron beam with a small spot size (nanometres), high brightness (nanoamps or higher) and low divergence (a few mrad) is required [5, 30]. Such conditions are met in a field emission gun SEM. The ECCI micrographs shown in this paper were acquired at 20 - 30 keV and we show ECCI micrographs where dislocations of order 80 nm apart can be resolved. ECCI micrographs can be acquired at lower electron beam energies, for example see refs. [29] and [31]. In ref. [29] the smallest size of the investigated 3D structures revealed in their ECCI micrographs was 28 nm. While ECCI is an excellent technique for revealing and quantifying defects which reach the surface, and we show some example results in sections 3.1 and 3.3, it is harder to quantify dislocations below the surface due to the surface sensitivity of the ECCI technique and the difficulty in determining its precise sampling depth. Some control of the sampling depth can be achieved by changing the electron beam energy. It is possible to obtain quantitative data on dislocations which lie below the surface, as illustrated for example by the paper by Carnevale *et al.* [22] on misfit dislocations in GaP. They were able to image dislocations lying at up to around 100 nm from the surface, where they acquired ECCI micrographs at a beam energy of 25 keV in the backscatter geometry (see discussion below).

Figure 1 illustrates the two geometries, namely the backscatter and forescatter geometries which are used to acquire ECCI micrographs. The backscatter geometry (Fig. 1a) has the advantage that this geometry does not require a high tilt of the sample and, therefore, a significant correction of the image to account for tilt is not required. This geometry also allows the easiest imaging of large samples, for example full semiconductor wafers. The forescatter geometry (Fig. 1b) has the advantage that images exhibit better signal-to-noise compared to the backscatter geometry due to the increase in intensity of backscattered electrons. The detector used to detect the backscattered electrons is generally an electron-sensitive diode. Key to the acquisition of good quality ECCI micrographs is the use of a good amplification system. An amplifier with a large DC offset and high small signal gain greatly facilitates the acquisition of ECCI micrographs.

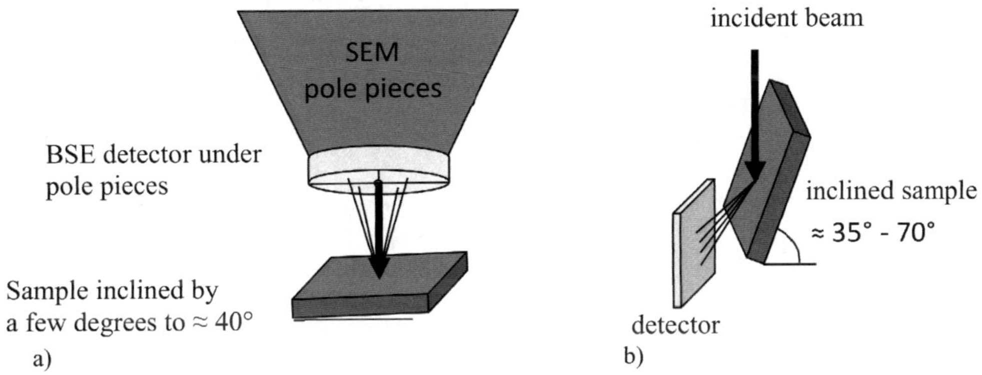

Figure 1. Illustrating the a) backscatter, and b) forescatter geometries for acquisition of ECCI micrographs [reproduced from 32].

2.2. Electron backscatter diffraction (EBSD)

In EBSD the sample is tilted at around 70° to the normal of the incident electron beam. The impinging electrons are scattered inelastically through high angles forming a diverging source of electrons, which can be diffracted. The resultant electron backscatter pattern (EBSP) consists of a large number of overlapping bands, known as Kikuchi bands, which are closely related to a 2D projection of the crystal structure, where each Kikuchi band corresponds to a set of planes, as illustrated in Fig. 2b. EBSPs are generally detected by an electron sensitive phosphor or scintillator screen and a charge-coupled device (CCD) or complementary metal-oxide semiconductor (CMOS) camera [15] (see Fig. 2a), although there have been recent developments of direct electron cameras [33-34]. The EBSP shown in Fig. 2b was acquired at 20 keV from a Si sample using energy-filtered direct electron detection with a collection angle of ≈ 50°. Direct electron detection allows high quality EBSPs to be acquired at low electron beam energies and pattern acquisition down to 3 keV is achievable [30]. EBSD is a well-established technique for texture analysis and for quantifying grain boundaries and crystal phases [5-6, 14-20]. The introduction of cross-correlation based analysis of EBSPs has also made possible measurements of relative strain, geometrically necessary dislocations, and lattice tilt and twist [16, 35], antiphase domains [36] and crystal polarity [37-38].

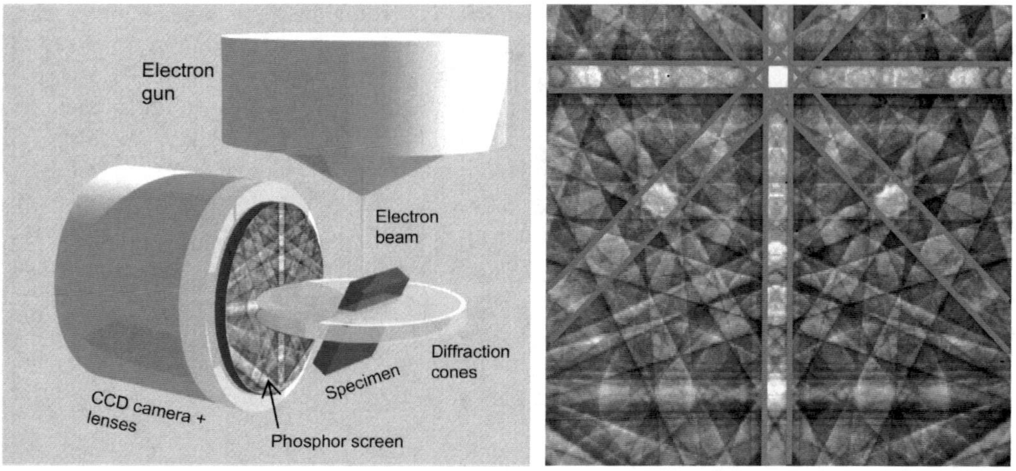

a) b)

Figure 2. a) Illustration of the EBSD detection geometry and a conventional EBSD detector, and b) an EBSP from Si acquired at an energy of 20 keV with a collection angle of ≈ 50°. The red lines outline Kikuchi bands corresponding to {200} planes, the blue lines outline Kikuchi bands corresponding to {220} planes. Adapted from [34] under Creative Commons Attribution (CC BY) license: http://creativecommons.org/licenses/by/4.0/.

3. Example results

3.1. The use of ECCI to image sub-grains and dislocations in a GaN thin film

Figure 3a shows an ECCI micrograph of a GaN thin film grown on *c*-plane sapphire by metalorganic vapour phase epitaxy (MOVPE). The ECCI micrograph was acquired at an electron beam energy of 30 keV in the forescatter geometry in the Sirion SEM. The variation in grey scale in the ECCI micrograph is a result of small differences in orientation of sub-grains in the thin film. This is highlighted in the pseudo-colour image in Fig. 3b. The "spots" in the image, most of which exhibit a black-white contrast (see inset of Fig. 3a for example), are threading dislocations; that is dislocations which thread through the thin film to the surface and are caused by the growth on the lattice-mismatched sapphire substrate. The threading dislocations propagate to the surface of the sample and are revealed due to associated strain fields [39]. For this GaN thin film the average threading dislocation density was determined to be $\approx 3 \times 10^8$ cm^{-2}. A large number of the threading dislocations are seen to lie on sub-grain boundaries. Note that in order to reveal all misorientations, and thus all the sub-grain boundaries, a number of ECCI micrographs need to be acquired under a range of diffraction conditions [40]. To select a particular diffraction condition, that is to select the plane or planes from which the incident electron beam will be diffracted (usually referred to as selecting the g-vector), it is necessary to acquire an electron channelling pattern (ECP). This is discussed in Section 3.2. While the contrast in the ECCI micrograph reveals the presence of sub-grains, it does not provide any quantitative information on their orientation. The magnitude and direction of misorientation can be measured by EBSD, and an EBSD study of sub-grain orientations for an AlN thin film is presented in Section 3.3.

3.2. Selection/determination of the diffraction condition for ECCI, acquiring ECPs

An ECP is obtained when changes in the backscattered electron intensity are recorded as the angle of the incident electron beam is changed relative to the surface of a single crystal area of the sample. When the beam changes its angle with respect to the sample, different planes of the crystal satisfy the Bragg condition, giving rise to the appearance of overlapping Kikuchi bands superimposed on the image of the

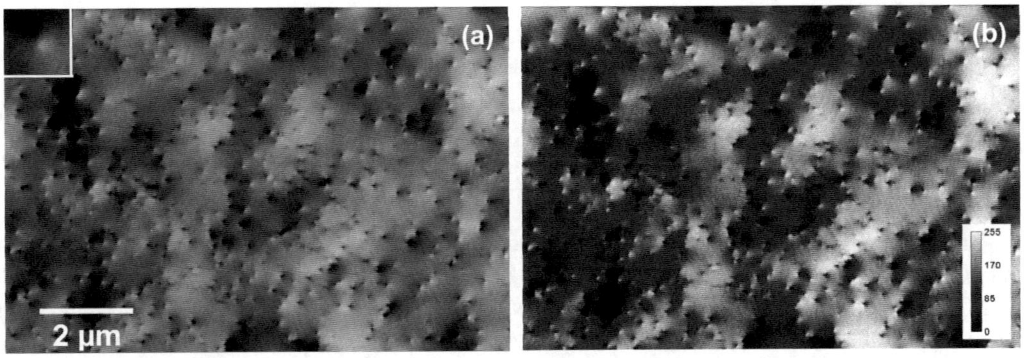

Figure 3. ECCI micrograph from a GaN thin film presented as a) a grey scale image, and b) a pseudo-colour image.

sample; an ECP from a GaN thin film is shown in Fig. 4. The ECP, like an EBSP (the two are related by reciprocity), is closely related to the 2D projection of the crystal structure, with the Kikuchi bands corresponding to different planes in the crystal. Comparing the ECP with kinematical and/or dynamical electron diffraction simulations allows the pattern to be indexed, i.e., the planes in the ECP can be identified. In this case Fig. 4a shows a dynamical simulation [41] and illustrates the detail achievable in such simulations, such detail can greatly facilitate the indexing of the ECP. Figure 4c is a kinematical simulation (produced with ESPRIT DYNAMICS (Bruker Nano) software) and is included to highlight the indices of some of the planes giving rise to the Kikuchi bands observed in the ECP. The plane (or planes) which intersect the centre of the ECP, usually referred to as the pattern centre (PC), are those from which the incident electron beam is diffracted. In the example shown in Fig. 4, the incident electron beam was diffracted from one of the {11-20} planes; so the **g**-vector is one of {11-20}.

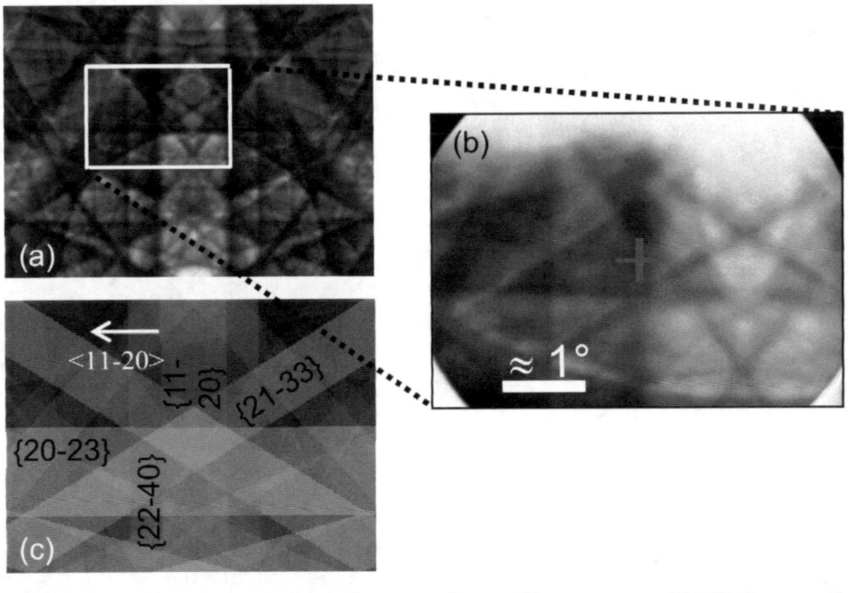

Figure 4. a) Dynamical simulation of an electron channelling pattern (ECP) from a GaN thin film. Electron beam energy is 30 keV, sample tilt ≈ 40°. b) Experimental ECP. The blue cross marks the pattern centre (PC). c) Kinematical simulation of the ECP with some indexed planes highlighted (produced with ESPRIT DYNAMICS (Bruker Nano) software). In this case the PC intersects with the edge one of the {11-20} Kikuchi bands; so the **g**-vector is one of {11-20}. [Reproduced from 32].

An ECP may be obtained by acquiring a backscattered image at low magnification. At low magnification, as the beam is scanned over the sample, it changes its angle with respect to the surface of the sample (in our case, for our Sirion SEM, this is by around ±2.5°), leading to the formation of an ECP [9]. Note that this method of acquiring an ECP is only possible if the scanned area of the sample (of order 5 mm × 5 mm in size) is smooth and all of the same crystallographic orientation. An ECCI micrograph is obtained on zooming in on the PC by increasing the magnification. At higher magnification the beam has a fixed angle with respect to the sample surface as the beam is scanned. The resultant ECCI micrograph will reveal any defects which distort the plane or planes which correspond to the Kikuchi bands intersecting the PC. Alternatively, if 'beam rocking' electron optics are available in the SEM, the angle of the beam with respect to the sample can be changed over a much smaller surface area. Selected area ECPs, referred to as SAECPS or SACPs can be acquired from areas ranging from 10 μm × 10 μm down to of order 500 nm × 500 nm in size [42]. While a lot of useful information can be obtained from ECCI without the acquisition of ECPs, the ability to acquire ECPs makes ECCI far more powerful and easier to use, in particular when it is applied to the identification of unknown extended defects such as dislocations and stacking faults.

3.3. The use of ECCI and EBSD to investigate sub-grains and dislocations in a c-plane AlN thin film

Figure 5 shows ECCI micrographs and EBSD maps from a *c*-plane AlN thin film overgrown by MOVPE on a nano-patterned sapphire substrate (nPSS). For this sample the sapphire substrate has a 0.1° offcut towards the sapphire *m*-plane. The experiments were performed in the Quanta SEM operated in low vacuum mode (0.5 mbar) to avoid charging of the insulating AlN sample.

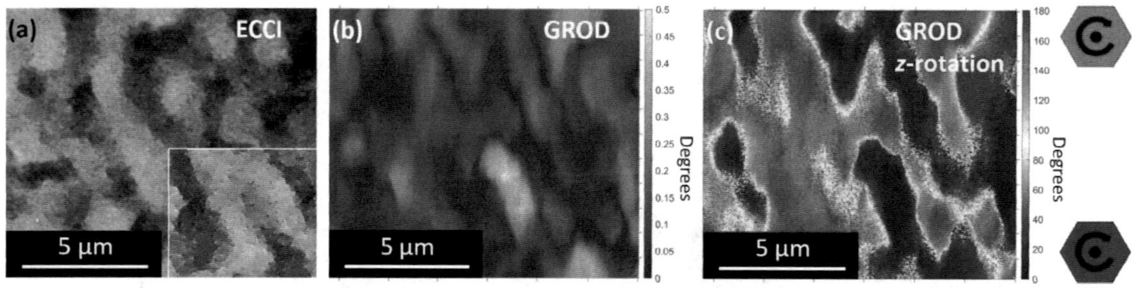

Figure 5. ECCI micrographs and EBSD maps from a *c*-plane AlN thin film. a) ECCI micrograph, acquired at an electron beam energy of 25 keV, revealing sub-grains. The inset shows a higher resolution ECCI micrograph revealing threading dislocations ("spots" exhibiting black-white contrast). b) Grain reference orientation deviation (GROD) map, and c) GROD *z*-rotation map where the colours denote direction of rotation around the *c*-axis. The red regions are rotated in the opposite direction to the blue regions as indicated by the blue and red hexagons. The GROD and GROD *z*-rotation maps were derived from EBSD data acquired at 20 keV. Note that the ECCI micrograph and EBSD maps were acquired from the same sample, but not from the same area of the sample. Adapted from [45] under Creative Commons Attribution (CC BY) license: http://creativecommons.org/licenses/by/4.0/.

The contrast in the ECCI micrograph of Fig. 5a is a result of small differences in orientation of sub-grains in the thin film. The inset shows a higher resolution ECCI micrograph where threading dislocations propagating to the surface of the sample are revealed; the threading dislocation density was determined to be 1.5×10^9 cm^{-2}. A significant number of the threading dislocations are located on the sub-grain boundaries. As stated previously in Section 3.1, while ECCI reveals differences in orientation between sub-grains, it does not provide a measure of the magnitudes and directions of these misorientations. Figure 5b shows a grain reference orientation deviation (GROD) map, the deviation of orientation of the sub-grains relative to a reference orientation [17]. In this case the reference orientation is the *c*-direction; that is the [0001] direction. The GROD map was produced from EBSD data using MTEX [43]. The first step of the analysis involved comparison of each EBSP with dynamical simulations [41]. This pattern matching approach was used in the EBSD analysis as it substantially improves orientation precision [44]. Figure 5c is a GROD *z*-rotation map showing the rotation of the sub-grains with respect to the nominal sample normal, i.e., around the *c*-axis. The colours, blue and red, denote the direction of this in-plane rotation. The red regions are rotated in the opposite direction to the blue regions.

The results presented above illustrate how ECCI and EBSD can provide complementary structural information. ECCI allows fast determination of dislocation densities and their distribution and reveals the presence of sub-grains. Figure 5 and its inset each took around 10 minutes of direct acquisition. EBSD provides quantitative information on the magnitude and direction of the misorientations in the film. However the EBSD data from which the maps of Fig. 5 were derived, took of order three hours to acquire. The EBSD data acquisition was then followed by further data analysis which is also time-consuming. In spite of this, both techniques share the advantages of being non-destructive and can be used to interrogate large areas of a sample.

3.4. The use of ECCI and EBSD to investigate sub-grains in an as-sintered WC-Co hard metal sample
Figures 6a to 6c show an ECCI micrograph, a GROD map and a GROD axis map from WC grains of an as-sintered WC-Co hard metal sample. The sample was first mechanically polished to an optical finish using standard metallographic techniques, finishing with a colloidal silica polish, which leaves a mirror finish and no visible scratches on the surface. It was also further polished in a Hitachi IM4000 argon ion beam polishing system to remove any fine scratches due to the mechanical polishing. This created a polished surface ready for examination in the SEM. While producing a scratch free surface, ion beam polishing induces shallow (of the order of 10 nm) dimples in the WC grains despite rotation under the ion beam to even out variation in milling rate with grain orientation: these dimples need to be taken into account when analysing the acquired data.

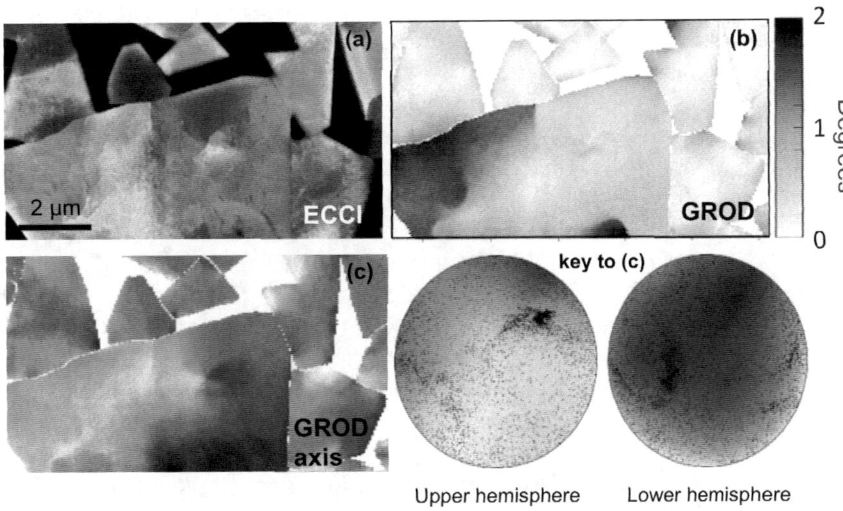

Figure 6. ECCI and EBSD maps from WC grains of an as-sintered WC-Co hard metal. a) ECCI micrograph acquired at 20 keV, b) GROD map, and c) a GROD axis map. The GROD and GROD axis maps were derived from EBSD data acquired at 20 keV. Note in this case the ECCI micrograph and EBSD maps were acquired from the same area of the sample. Reprinted in part from [46] with permission from Elsevier.

For the GROD axis map each pixel has a value denoting the rotation axis with respect to the mean axis for that grain, as defined by the colour key to the right of the image. The ECCI micrograph and EBSD maps reveal sub-grains within nominally single crystal WC grains, indicating how the large grains may have evolved during the sintering process. ECCI also reveals dislocations within the WC grains [46]. In this case the EBSD derived maps again provide quantification of the sub-grain misorientations; the GROD axis map reveals the misorientations are around more than one axis. The colour key indicates the direction of the misorientation axes in real space above (upper hemisphere) or below (lower hemisphere) the sample. The black dots indicate these directions for each point in the map.

4. Summary

To summarise, we have shown that ECCI allows direct, fast imaging of dislocations and sub-grain boundaries, while EBSD provides quantitative information on the magnitude and direction of misorientations in the sample. Both techniques share the advantages of being non-destructive and can be used to interrogate and provide valuable, complementary information on defects and crystallographic properties from large areas of a sample.

Acknowledgements

The authors would like to acknowledge financial support of the EPSRC, UK via Grant No. EP/M015181/1, "Manufacturing nano-engineered III-nitrides"; and Grant No. EP/P015719/1, "Quantitative non-destructive nanoscale characterisation of advanced materials". Data associated with this research are available at https://doi.org/10.15129/b5238863-a088-4f30-8b13-2625260eb73a [45] and https://doi.org/10.15129/c80a3a0c-a718-4941-ac82-8981d21799ce [46], or from the corresponding author.

References

[1] Holt D B and Joy D C (Eds.) 1989 *SEM microcharacterization of semiconductors*. (Cambridge, UK: Academic Press).
[2] Goldstein J, Newbury D, Joy D, Lyman C, Echlin P, Lifshin E, Sawyer L and Michael J 2007 *Scanning electron microscopy and X-ray microanalysis*. (Boston, MA: Springer)
[3] Reimer L 1998 *Scanning electron microscopy: Physics of image formation and microanalysis*. (Berlin-Heidelberg, Germany: Springer)
[4] Zhou W and Wang Z L 2007 *Scanning microscopy for nanotechnology: techniques and applications*. (Boston, MA: Springer Science & Business Media)
[5] Wilkinson A J and Hirsch P B 1997 Electron diffraction based techniques in scanning electron microscopy of bulk materials. *Micron* **28** 279-308
[6] Trager-Cowan C, Sweeney F, Trimby P W, Day A P, Gholinia A, Schmidt N H, Parbrook P J, Wilkinson A J and Watson I M 2007 Electron backscatter diffraction and electron channeling contrast imaging of tilt and dislocations in nitride thin films. *Phys. Rev. B.* **75** 085301
[7] Crimp M A, Simkin B A and Ng B C 2001 Demonstration of the g·bxu=0 edge dislocation invisibility criterion for electron channelling contrast imaging. *Philos. Mag. Lett.* **81** 833-837
[8] Picard Y, Kamaladasa R, De Graef M, Nuhfer N, Mershon W, Owens T, Sedlacek L and Lopour F 2012 Future prospects for defect and strain analysis in the SEM via electron channeling. *Micros. Today* **20** (2) 12-16
[9] Naresh-Kumar G, Hourahine B, Edwards P R, Day A P, Winkelmann A, Wilkinson A J, Parbrook P J, England G and Trager-Cowan C 2012 Rapid nondestructive analysis of threading dislocations in wurtzite materials using the scanning electron microscope. *Phys. Rev .Lett.* **108** 135503
[10] Zaefferer S and Elhami N N 2014 Theory and application of electron channelling contrast imaging under controlled diffraction conditions. *Acta Mater.* **75** 20-50

[11] Deitz J I, Carnevale S D, Ringel S A, McComb D W and Grassman T J 2015 Electron channeling contrast imaging for rapid III-V heteroepitaxial characterization. *J. Vis. Exp.* **101** e52745

[12] Naresh-Kumar G, Thomson D, Nouf-Allehiani M, Bruckbauer J, Edwards P R, Hourahine B, Martin R W and Trager-Cowan C 2016 Electron channelling contrast imaging for III-nitride thin film structures. *Mater. Sci. Semicond. Process.* **47** 44-50

[13] L'Hôte G, Lafond C, Steyer P, Deschanel S, Douillard T, Langlois C and Cazottes S 2019 Rotational-electron channeling contrast imaging analysis of dislocation structure in fatigued copper single crystal. *Scripta Materialia* **162** 103-107

[14] Brodusch N, Demers H and Gauvin R 2018 Electron diffraction techniques in the SEM. in: *Field emission scanning electron microscopy.* (Singapore: Springer) Springer Briefs in Applied Sciences and Technology

[15] Schwartz A J, Kumar M, Adams B L and Field D P 2009 *Electron backscatter diffraction in materials science.* (Berlin, Germany: Springer)

[16] Wilkinson A J and Britton T B 2012 Strains, planes, and EBSD in materials science. *Mater. Today* **15** 366-376

[17] Wright S I, Nowell M M and Field D P 2011 A review of strain analysis using electron backscatter diffraction. *Microsc. Microanal.* **3** 316-329

[18] Wright S I, Nowell M M, de Kloe R, Camus P and Rampton T 2015 Electron imaging with an EBSD detector. *Ultramicroscopy* **148** 132-145

[19] Moussa C, Bernacki M, Besnard R and Bozzolo N 2017 Statistical analysis of dislocations and dislocation boundaries from EBSD data. *Ultramicroscopy* **179** 63-72

[20] Brodusch N, Demers H and Gauvin R 2018 Imaging with a commercial electron backscatter diffraction (EBSD) camera in a scanning electron microscope: A review. *J. Imaging* **4** 88

[21] Naresh-Kumar G, Mauder C, Wang K R, Kraeusel S, Bruckbauer J, Edwards P R, Hourahine B, Kalisch H, Vescan A, Giesen C and Heuken M 2013 Electron channeling contrast imaging studies of nonpolar nitrides using a scanning electron microscope. *Appl. Phys. Lett.* **102** 142103

[22] Carnevale S D, Deitz J I, Carlin J A, Picard Y N, De Graef M, Ringel S A and Grassman T J 2014 Rapid misfit dislocation characterization in heteroepitaxial III-V/Si thin films by electron channeling contrast imaging. *Appl. Phys. Lett.* **104** 232111

[23] Hite J K, Mastro M A and Eddy Jr C R 2010 Approach for dislocation free GaN epitaxy. *J. Cryst. Growth* **312** 3143-3146

[24] Schulze A, Strakos L, Vystavel T, Loo R, Pacco A, Collaert N, Vandervost W and Caymax M 2018 Non-destructive characterization of extended crystalline defects in confined semiconductor device structures. *Nanoscale* **10** 7058-7066

[25] Callahan P G, Haidet B B, Jung D, Seward G G and Mukherjee K 2018 Direct observation of recombination-enhanced dislocation glide in heteroepitaxial GaAs on silicon. *Phys. Rev. Mater.* **2** 081601

[26] Yaung K N, Kirnstoetter S, Faucher J, Gerger A, Lochtefeld A, Barnett A and Minjoo L L 2016 Threading dislocation density characterization in III–V photovoltaic materials by electron channeling contrast imaging. *J. Cryst. Growth* **453** 65-70

[27] Kaboli S and Burnley P C 2018 Direct observations of crystal defects in polycrystalline diamond. *Mater. Characterization* **142** 154-161

[28] Miyajima N, Li Y, Abeykoon S and Heidelbach F 2018 Electron channelling contrast imaging of individual dislocations in geological materials using a field-emission scanning electron microscope equipped with an EBSD system. *Eur. J. Mineral.* **30** 5-15

[29] Han H, Hantschel T, Strakos L, Vystavel T, Baryshnikova M, Mols Y, Kunert B, Langer R, Vandervorst W and Caymax M 2020 Application of electron channeling contrast imaging to 3D semiconductor structures through proper detector configurations. *Ultramicroscopy* **210** 112928

[30] Joy D C, Newbury D E and Davidson D L 1982 Electron channeling patterns in the scanning electron microscope. *J. Appl. Phys.* 53 R81-122

[31] Naresh-Kumar G, Trager-Cowan C and Mingard K P 2019 Electron channelling contrast imaging in a low voltage scanning electron microscope. *Microsc. Microanal.* **25** 504-505
[32] Massabuau F C-P, Bruckbauer J, Trager-Cowan C and Oliver R A 2019 Chapter 8: Microscopy of defects in semiconductors. in: Characterisation and control of defects in semiconductors. (Tuomisto F; Ed.) (London, UK: IET)
[33] Wilkinson A J, Moldovan G, Britton T B, Bewick A, Clough R and Kirkland A I 2013 Direct detection of electron backscatter diffraction patterns. *Phys. Rev. Lett.* **111** 065506
[34] Vespucci S, Winkelmann A, Naresh-Kumar G, Mingard K P, Maneuski D, Edwards P R, Day A P, O'Shea V and Trager-Cowan C 2015 Digital direct electron imaging of energy-filtered electron backscatter diffraction patterns. *Phys. Rev. B* **92** 205301
[35] Wilkinson A J, Meaden G and Dingley D J 2006 High-resolution elastic strain measurement from electron backscatter diffraction patterns: new levels of sensitivity. *Ultramicroscopy* **106** 307-313
[36] Naresh-Kumar G, Vilalta-Clemente A, Jussila H, Winkelmann A, Nolze G, Vespucci S, Nagarajan S, Wilkinson A J and Trager-Cowan C 2017 Quantitative imaging of anti-phase domains by polarity sensitive orientation mapping using electron backscatter diffraction. *Sci. Rep.* **7** 10916
[37] Winkelmann A, Nolze G, Himmerlich M, Lebedev V and Reichmann A 2016 Point-group sensitive orientation mapping using EBSD. in: Proc. 6th Int. Conf. Recrystallization and Grain Growth. (Holm E A, Farjami S, Manohar P, Rohrer G, Rollett A D, Srolovitz D and Weiland H; Eds.) (Berlin-Heidelberg, Germany: Springer) 281-286
[38] Naresh-Kumar G, Bruckbauer J, Winkelmann A, Yu X, Hourahine B, Edwards P R, Wang T, Trager-Cowan C and Martin R W 2019 Determining GaN nanowire polarity and its influence on light emission in the scanning electron microscope. *Nano Lett.* **19** 3863-3870
[39] Pascal E, Hourahine B, Naresh-Kumar G, Mingard K and Trager-Cowan C 2018 Dislocation contrast in electron channelling contrast images as projections of strain-like components. *Mater. Today: Proceedings* **5** 14652-14661
[40] Day A P and Quested T E 1999 A comparison of grain imaging and measurement using horizontal orientation and colour orientation contrast imaging, electron backscatter pattern and optical methods. *J. Microscopy* **195** 186-196
[41] Winkelmann A, Trager-Cowan C, Sweeney F, Day A P and Parbrook P J 2007 Many-beam dynamical simulation of electron backscatter diffraction patterns. *Ultramicroscopy* **107** 414-421
[42] Mansour H, Guyon J, Crimp M A, Gey N, Beausir B and Maloufi A N 2014 Accurate electron channeling contrast analysis of dislocations in fine grained bulk materials. *Scripta Materialia* **84** 11-14
[43] Bachmann F, Hielscher R and Schaeben H 2010 Texture analysis with MTEX - Free and open source software toolbox. *Solid State Phenomena* **160** 63-68
[44] Winkelmann A, Jablon B M, Tong V S, Trager-Cowan C and Mingard K P 2020 Improving EBSD precision by orientation refinement with full pattern matching. *J. Microscopy* **277** 79-92
[45] Trager-Cowan C, Alasmari A, Avis W, Bruckbauer J, Edwards P R, Hourahine B, Kraeusel S, Kusch G, Johnston R, Naresh-Kumar G, Martin R W, Nouf-Allehiani M, Pascal E, Spasevski L, Thomson D, Vespucci S, Parbrook P J, Smith M D, Enslin J, Mehnke F, Kneissl M, Kuhn C, Wernicke T, Hagedorn S, Knauer A, Kueller V, Walde S, Weyers M, Coulon P-M, Shields P A, Zhang Y, Jiu L, Gong Y, Smith R M, Wang T and Winkelmann A 2019 Scanning electron microscope as a flexible tool for investigating the properties of UV-emitting nitride semiconductor thin films. *Photonics Res.* **7** B73-82
[46] Jablon B M, Mingard K, Winkelmann A, Naresh-KumarG , Hourahine B and Trager-Cowan C 2020 Subgrain structure and dislocations in WC-Co hard metals revealed by electron channelling contrast imaging. *Int. J. Refract. Met.* **87** 105159

IOP Conf. Series: Materials Science and Engineering **891** (2020) 012024 doi:10.1088/1757-899X/891/1/012024

Fission product distribution in irradiated safety-tested and non-safety-tested AGR-2 TRISO particles

K E Wright[1] and I J van Rooyen[1]

[1] Idaho National Laboratory, P.O. Box 1625 MS 6157, 83415 Idaho Falls, ID, U.S.A.

E-mail: karen.wright@inl.gov

Abstract. Two similarly-irradiated TRISO nuclear fuel particles, one safety-tested and one not-safety-tested, were examined by electron probe microanalysis (EPMA) as part of the advanced gas reactor (AGR) fuel development and qualification programme. The primary objective of this study was to determine the effect of high temperature safety testing on the fission product quantification and distribution. Concentration profiles for fission products were collected along two radii in each particle. Fission product mass balance distribution for both particles was determined by computing the material mass per traverse step-size and multiplying by the weight fraction analysis at that position. Results of these analyses suggest that the non-safety-tested particle contains more fission product mass in the centre 300 μm of the kernel than the safety-tested particle; however, for mobile fission products such as Cs, Xe, Te, Sr, Ba, and Pd, the mass moves out further from the particle centre in the safety-tested particle than the non-safety-tested particle. This mass balance approach to TRISO particle analysis is beneficial at elucidating fission product migration in irradiated and safety-tested particles and will be beneficial in applying to other particles to determine potential irradiation damage effects on the fission product distribution and mobility.

1. Introduction

The AGR-2 experiment is the second of several fuel irradiation experiments planned for the advanced gas reactor (AGR) fuel development and qualification programme. The purpose of the AGR-2 experiment is to build on lessons learned during the preceding AGR-1 experiment. In contrast to the AGR-1 experiment, which used laboratory-scale fuel fabrication equipment and methods, the AGR-2 experiment used prototype industrial-scale equipment and methods for fuel fabrication.

The AGR-2 UCO tristructural isotropic (TRISO)-coated fuel particle consists of a UCO fuel kernel 425 μm in diameter, surrounded by concentric layers of confining materials (Fig. 1). These include a 100 μm thick porous graphite buffer, followed by a 40 μm thick dense graphite inner pyrolytic carbon layer (IPyC). Outside the IPyC is a 35 μm thick SiC layer. Finally, a 40 μm thick dense graphite outer pyrolytic carbon layer (OPyC) overlies the SiC layer [1]. At time of fabrication, the average kernel composition is $UC_{0.392}O_{1.428}$. The outer portion of the kernel differs from the interior of the kernel in that the kernel periphery (KP) is characterised compositionally by the rapid increase in C concentration and the concomitant rapid decrease in U concentration.

Fission product (FP) distribution in irradiated nuclear fuels has been studied by electron probe microanalysis (EPMA) for several decades. The use of EPMA is advantageous for several reasons. Firstly, the energy resolution of EPMA is typically on the order of 10 eV, compared to ~125 eV for the

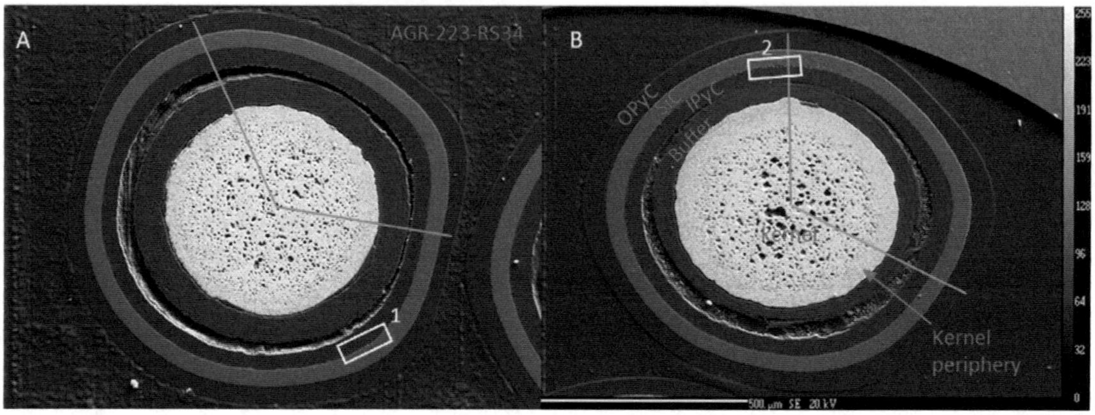

Figure 1. Electron microscope images of the examined particles showing the path of quantitative radial chemistry measurements. A) Secondary electron image of particle AGR2-223-RS34, and b) Secondary electron image of particle AGR2-222-RS19. Particle layer names are shown on the various layers. Boxes 1 and 2 show the location of precipitates occurring along the IPyC-SiC boundary.

modern silicon drift detectors used in scanning electron microscopes [2]. Secondly, the EPMAs used for irradiated specimens are equipped with W-alloy shielding to: 1) reduce detector background noise; 2) reduce the signal to background noise ratio changes that occur when sample-to-detector distance is altered during stage movement; and, 3) preserve the detector life. Finally, the high currents (200 – 300 nA) employed by EPMA in the course of analysing irradiated specimens can improve detection limits by a factor of 10 or more [3].

Numerous earlier EPMA efforts examined FP distribution [4-6], precipitate composition [7], and Cs buffer speciation [8-9]. One previous study examined FP distribution in AGR1 fuel particles [10]. This study expands on these latter efforts by estimating FP product mass in each particle layer.

1.1. Objectives
Several objectives will be achieved by this study. Firstly, the quantitative FP distribution of both irradiated and irradiated- and- safety-tested particles will be characterised. These data are required for modelling fuel behaviour under irradiation conditions, and to date, no such data exists for AGR-2 fuel. Secondly, this study provides the opportunity to separate the effects of irradiation only on FP distribution from those effects resulting from safety-testing similarly irradiated particles. This will lend insight to high temperature effects on FP mobility. Finally, this study demonstrates the utility of using EPMA for such studies. Because of the use of calibration standards and the availability of sophisticated peak deconvolution software, this instrument is essential for quantitative irradiated fuel chemical analysis for the fully coated particle in a time-efficient manner.

2. Methods

2.1. Irradiation and safety testing
TRISO particles in Compacts 2-2-3 and 2-2-2 were fabricated using prototypic-scale equipment and were subjected to irradiation as described by Collin [11]. Table 1 shows the irradiation parameters for both compacts and safety testing parameters for Compact 2-2-2. Following irradiation, Compact 2-2-2 was safety tested using the Oak Ridge National Lab Core Conduction Cooldown Test Facility, which was used to evaluate the release of FPs upon heating. Morris *et al.* [12] discuss the procedures and equipment for this process. Following the safety test, Compact 2-2-2 was subjected to a leach-burn-leach test. Details of the test can be found in Demkowicz [13]. Following the

leach-burn-leach test, both safety-tested and non-safety-tested particles were subjected to gamma counting to determine how the actual FP inventory of each particle compared to the predicted FP inventory. At the conclusion of the gamma counting, one particle each was selected from Compacts 2-2-3 and 2-2-2 for this electron probe microanalysis (EPMA) study. Each particle has exhibited some loss of FPs as evidenced by a measured to calculated ^{110m}Ag ratio of <1 (Table 1).

Table 1. Shows the irradiation and safety-testing parameters for the two examined AGR-2 particles.

Parameter	AGR2-223-RS34	AGR2-222-RS19
^{235}U enrichment	14.03	14.03
% FIMA average burnup	10.8	12.55
Time-average, volume average temperature, °C	1161	1287
Approximate fast fluence (× 10^{25}), n/m^2	2.99	3.39
Measured to calculated ^{110m} Ag ratio	0.84	0.2*
1,600 °C safety test (hours)		300

* measured after safety test

2.2. *Sample preparation*
Each particle was mounted into a metallography mount with Buehler EpoHeat epoxy. The mounts were then polished to mid-plane thickness and <1 µm finish using standard methods. Each mount was then coated with 15 nm of aluminium to ensure sample conductivity. Images of the specimens are shown in Figure 1.

2.3. *Electron probe microanalysis*
Electron probe microanalysis was performed at Idaho National Laboratory's Irradiated Materials Characterization Facility using a CAMECA SX100R electron probe microanalyser, which is specifically designed for the analysis of highly radioactive specimens. To that end, the instrument's electronics and detectors are shielded to 3 Ci of ^{137}Cs.

Various peak and background counting times were employed depending on the phase type and element concentration. In all cases, the accelerating voltage was 20 kV and the beam size was just focussed. The current employed for each element was 200 nA, except for Si in the SiC layer, which was measured using a current of 20 nA to avoid excessive detector dead time.

The choice of X-ray lines, standards, diffracting crystals, and corrected interferences is shown in Table 2. For some elements, an acceptable standard was not available. In this situation, a "virtual" standard was employed. The intensity for a virtual standard is determined by measuring the intensities of the same X-ray lines on the same crystal and spectrometer for elements that are adjacent on the periodic table to the X-ray line of interest. An X-ray intensity can then be interpolated for the line of interest.

Each X-ray line was measured using a peak and two backgrounds. The count time for each X-ray peak varied with the abundance and fluorescence yield of the element, but in general each peak was counted for 300 to 500 seconds, with a background count time approximately one-third of the peak time. For elements with higher count rates such as U in the kernel and Si in the SiC layer, the count rate was 20 - 30 seconds for the peak and 10 - 15 seconds for each background. Note that standards were not counted in the same manner as the sample points; the concentration of the element in the standard is generally so much higher than it is in the unknown that it is not necessary to count for more than 30 seconds for the peak and 10 seconds for the background. Measuring the standards for the shorter time does not negatively affect quantification. In fact, if standards were measured at the same beam currents and for the same time period as unknowns, there would likely be excessive detector counter deadtime, which itself would negatively influence quantification. Finally, some standards, such as phosphates, are beam sensitive and are measured using a defocused 20 μm beam. All standards were measured using a 20 nA current, and the dead time for all analyses was set to 3 μs.

Table 2. Shows the X-ray line, standard used, diffracting crystal, and corrected interferences for each measured element.

Element	X-ray line	Standard	Diffracting crystal	Interferences corrected
U	Mα	U	QTZ	Pd, Ru
Zr	Lα	Zr	PET	
La	Lα	LaP_5O_{14}	LiF	Cs
Ce	Lα	CeP_5O_{14}	LiF	Ba
Nd	Lα	NdP_5O_{14}	LiF	
Pr	Lα	PrP_5O_{14}	LiF	La
Sm	Lα	SmP_5O_{14}	LiF	Ce
Eu	Lα	EuP_5O_{14}	LiF	Pr, Nd
Ba	Lα	barite	QTZ	Pr
Cs	Lα	pollucite	PET	U, Ce, Te
Xe	Lα	virtual	PET	Te, La
Mo	Lα	Mo	PET	
Ru	Lα	Ru	PET	Ba
Ag	Lα	Ag_2Te	QTZ	Cd, U
Pd	Lα	Pd	QTZ	Ru, U
Te	Lα	Ag_2Te	PET	
I	Lα	TlBrI	PET	Cd, Ba
Cd	Lα	Cd	QTZ	
C	Kα	SiC	PC1	U
O	Kα	ThO_2	PC1	
Si	Kα	SiC	TAP	U, Pu
Sr	Lα	$SrTiO_3$	TAP	Si

Quantitative radial traverse data was collected using PROBE FOR EPMA software. The traverse paths measured on each particle are shown in Figure 1. For each particle, one radius was measured in a region in which there was a gap between the buffer and IPyC, and a second radius was measured in a region that lacked a gap, or the gap was small. The distance between points varied from 2 - 5 μm depending on the region of the particle being measured. All X-ray collection programs used the Pouchou and Pichoir (PAP) matrix correction algorithm [14] and the mass-absorption coefficients used were those of Farthing and Walker [15].

2.4. Fission product mass balance calculations

The first approach to evaluating differences between safety-tested and non-safety-tested particles was to estimate the mass of select FPs in the central kernel. This was accomplished by applying the method described in Section 2.5 to the first 150 μm of each radius (from the particle centre extending outward 150 μm. This provides an estimate of 300 μm of the approximate 425 μm kernel diameter. The 300 μm distance was chosen because that portion of the kernel is more homogeneous (based on qualitative micro-porosity distribution) than portions beyond 300 μm. Thus, FP distribution homogeneity in this region is an assumption for this study. The method described in section 2.5 was then applied to the entirety of both radii.

2.5. *Calculation method*

Quantitative weight percent EPMA data from the radial traverse measurements in each particle was converted to an elemental mass in each particle, subdivided according to the particle layer (i.e., kernel, buffer, IPyC, etc.). The procedure and assumptions employed were as follows:

1. The weight percent of the measured element at a given location is assumed to be homogeneous for a slice of the particle equivalent to a hemispheric slice of the measured traverse step-size width.
2. The volume of the hemisphere slice of the measured traverse step-size width is calculated in cubic micrometres.
3. The material density is known (the buffer densifies during irradiation and is assumed to double for this calculation).
4. The mass of the hemispheric slice is computed from volume and density.
5. The weight percent measured at that point is changed to weight fraction and is multiplied by the mass of the hemisphere slice to arrive at a mass in grams of the element, per one hemispheric slice.
6. The procedure is repeated for each hemisphere slice for the first measured radial traverse.
7. The procedure is then repeated for the other radial traverse to account for the inhomogeneity that results from the existence of thermal gradients across the particle.
8. The total mass of the element in the particle is calculated by adding all the slices from both hemispheres together.

3. Results

3.1. *Morphology*

Both studied particles are characterised by a nearly complete delamination of the buffer from the IPyC. In addition, the KP region in both particles exhibits variable thickness, ranging from approximately 30 - 100 µm thick. Finally, the KP is quite distinct from the kernel central region in that the KP is characterised by a large density of small-diameter pores (~ < 1 µm in diameter), in contrast with the centre of the kernel, which is typified by fewer, much larger pores (~1 - 50 µm in diameter).

3.2. *Fission product distribution*

3.2.1. Cs, Xe, I. Figures 2a and 2b show the Cs compositional profiles for both particles. Several observations can be made. Firstly, for both particles, the compositional profile on the gap side is similar to that of the non-gap side. Both show elevated elemental concentrations in the buffer and IPyC as compared to the kernel. For both particles, the elemental concentration begins to rise in the kernel perimeter. In addition, the elemental concentration profile is continuous as it crosses the gap suggesting that the gap formed after the element had migrated from the buffer to the IPyC.

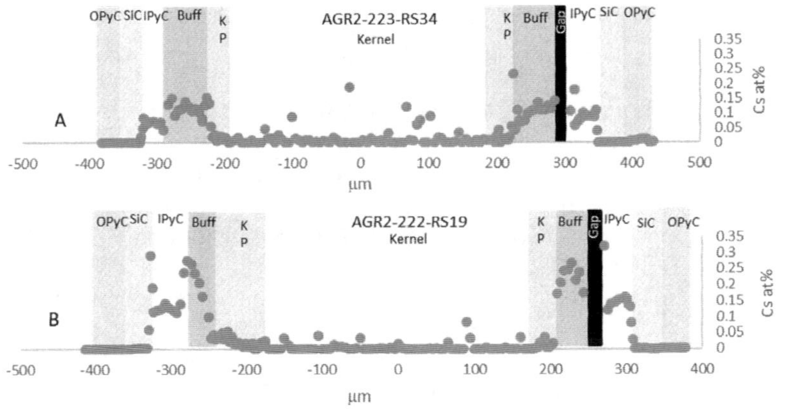

Figure 2. Quantitative radial traverses for Cs for both examined particles. The non-gap side of the particle is shown in blue while the gap side is shown in orange.

There are differences in Cs concentration profiles between safety-tested and non-safety-tested particles. The peak Cs concentration in the buffer of safety-tested particle is on the order of 50 - 100 % greater than in non-safety-tested particle (AGR2-222-RS19 versus AGR2-223-RS34). In addition, elemental kernel concentrations are generally greater in the non-safety-tested particles compared to that of safety-tested particles. Xe and I profiles are similar to Cs (data not shown).

3.2.2. Sr, Ba, Te, Eu. Figures 3a and 3b show the Sr profile for both studied particles. In contrast to Cs, Xe, and I, the Sr concentration profile is asymmetrical across particle AGR2-222-RS19. Specifically, Sr is approximately four times greater in concentration in the KP on the gap side of the particle as compared to the non-gap side. For both particles, the elemental concentration increases in concentration in the KP and reaches its maximum near the KP-buffer boundary. Concentrations then decline precipitously in the buffer. Contrasting the safety-tested particle, concentration profiles on particle AGR2-223-RS34 are more symmetrically distributed. Ba, Te, and Eu profiles are similar to Sr (data not shown).

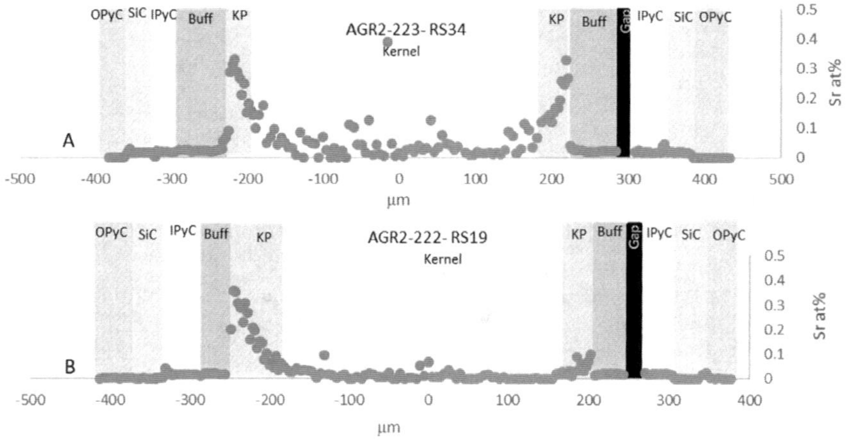

Figure 3. Quantitative radial traverses for Sr for both examined particles. The non-gap side of the particle is shown in blue while the gap side is shown in orange.

3.2.3. SiC-IPyC interface. The SiC-IPyC interface for both particles is shown in Figs. 4a to 4d. With both particles, the Pd and Ag concentration is higher on the non-gap side than the side with the gap. Both particles also show Pd penetration through most of the SiC layer, but little Ag penetration. This may be because the Ag has not penetrated into the SiC layer of the particles or may be because the quantity in the SiC layer is below the detection limit. Although the Ag concentration at the interface is similar for both particles, the Pd concentration is approximately three times higher in the safety-tested particle than the non-safety-tested particle.

3.2.4. Fission product mass balance calculations. The results of the FP mass balance calculation described in section 2.5 for the inner 300 µm kernel are shown in Table 3. The estimated FP mass per particle is shown in Table 4. For Cs, Xe, Te, Ba, Mo, Ru, and Pd, the non-safety-tested particle, RS34 contained a greater FP mass in the inner most 300 µm of the kernel than did the safety-tested particle, RS19. In contrast, the estimated FP mass in the whole particle is greater for particle RS19 than it is for RS34 for all tested elements except Pd and Ru.

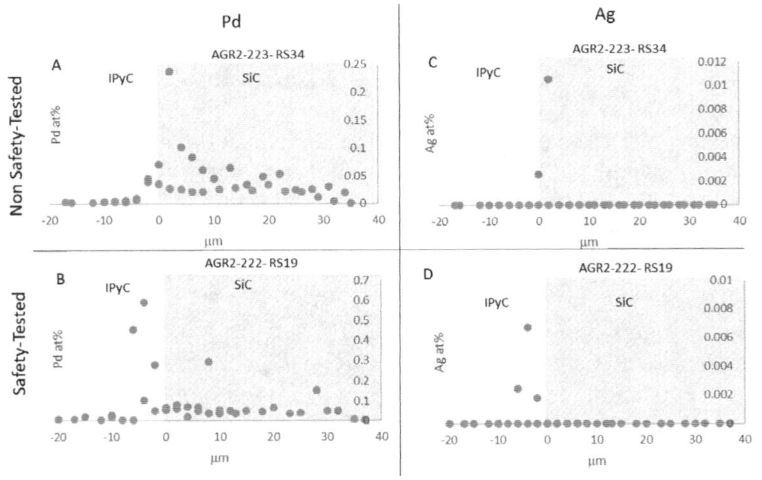

Figure 4. a-b) Pd concentration profile, and c-d) profile of the Ag concentration, across the IPyC-SiC boundary. The gap side of the particle is indicated by the orange symbol and the non-gap side is indicated by the blue symbol.

Table 3. Estimate of the mass of select FPs mass (in gram) located in the inner 300 µm of the kernel diametral sphere. The larger mass is in bold font.

Particle	Cs	Xe	Sr	Te	Ba	Mo	Nd	Ru	Pd
RS34	**5.1E-08**	**1.9E-08**	6.9E-08	**1.8E-07**	**9.6E-08**	**1.7E-06**	1.5E-06	**8.7E-07**	**1.1E-07**
RS19	3.0E-08	7.3E-09	**2.1E-07**	1.4E-07	8.7E-08	1.3E-06	**2.0E-06**	4.6E-07	5.5E-08

Table 4. Estimate of the mass of select FPs mass (in gram) present in each particle. The larger mass is in bold font.

Particle	Cs	Xe	Sr	Te	Ba	Mo	Nd	Ru	Pd
RS34	2.7E-06	3.4E-07	1.5E-06	1.1E-06	2.2E-06	4.6E-06	5.0E-06	**3.1E-06**	**6.1E-06**
RS19	**3.3E-06**	**1.1E-06**	**2.0E-06**	**2.4E-06**	**3.7E-06**	**5.5E-06**	**7.0E-06**	3.0E-06	1.2E-06

4. Discussion

4.1. Particle morphology and fission product distribution

For both particles AGR2-222-RS19 and AGR2-223-RS34, two radii were examined, one on the side of the particle where the buffer remained bonded to the IPyC, and one on the side where the buffer became delaminated from the IPyC, creating a substantial gap between the buffer and IPyC. This morphology is quite common among irradiated AGR1 particles (AGR1 particles preceded the AGR2 particles in the AGR sets of experiments) and is thought to be the result of the inward buffer densification [16]. Assuming that this hypothesis is correct (as opposed to the gap having formed during sample preparation), the FP distribution evidence suggests that the gap formed after Cs, I and Xe had migrated into the IPyC, or that the gap did not alter the transport properties of these elements. Figure 2 shows that Cs has crossed the gap and appears to maintain the same concentration profile as the other side of the particle that has no gap.

While most FPs have similar concentration profiles on both sides of the particle (gap side versus non-gap side), the exceptions include Sr, Te, Ba, and to a lesser extent, Eu. Sr, Ba and Te concentrations are up to four times higher in the KP of the non-gap side of safety-tested particle than on the side with the gap (Figus. 3a and 3b). However, this difference cannot be due to the gap as the gap is beyond the buffer, outside of this concentration spike. For non-safety-tested particle AGR2-223-RS34, the concentration asymmetry is not present; however, this asymmetry has been observed in other non-safety-tested particles (data not shown). Perhaps concentration gradient differences based on incipient asymmetry are enhanced through the safety-testing process.

Several researchers describe the effects of temperature gradients across TRISO particles. Tiegs [5] and Minato *et al.* [7] both describe the transport of Pd and rare earth elements down the temperature gradient. Grubmeier *et al.* [17] noted the diffusion of rare earth and platinum group elements to the cooler side of the particle but stated that they were not necessarily co-located. Kleykamp [4] described the presence of an "azimuthal temperature gradient" across the particle that influenced FP transport and precipitation.

Figures 4a to 4d show that on the non-gap side of both particles, Pd and Ag are on the order of 3 – 4 times higher in concentration than it is on the gap side of the particle. Figure 1 (a-b, boxes 1 and 2) shows visible precipitates at the SiC-IPyC interface, whereas no such obvious precipitates are present at the SiC-IPyC interface where the buffer has been separated from the IPyC. Given that the precipitates are located where the buffer-IPyC remains intact, this suggests that this is the cooler side of the particle and that delamination occurs preferentially where the particle is hotter.

Kleykamp [4] showed that for TRISO with UO_2 kernels, Te and Pd migrated to the cold side of the particle. As Te, Sr, Ba, and Eu all exhibit similar behaviour, and Sr, Ba, and Eu are predicted to exist as carbides [18], this suggests that transport properties for these species are similar and that they accumulate on the colder side of the particle.

4.2. Fission product distribution in safety-tested versus non-safety-tested particles

Section 3 provided data concerning differing concentration gradients in the safety-tested versus non-safety-tested particles as a two-dimensional representation of the FP behaviour as a concentration traverse across a plane. By employing the method describe in Section 2.5, it is possible to approximate FP masses in different particle layers across three dimensions, thus arriving both an estimated FP inventory for the whole particle, and the mass inventory for each particle layer. This allows a more thorough comparison of FP migration between the two particle treatments.

Tables 3 and 4 show that although AGR2-223-RS34 has a higher FP mass for Cs, Xe, Ba, Te, Mo, Ru, Pd in the 300 μm kernel centre, AGR2-222-RS19 has overall higher masses of Cs, Xe, Sr, Te, Ba, Mo, and Nd in the whole particle. However, the distribution of some of these elements amongst the particle layers is different in the high-temperature safety-tested particle compared to the non-safety-tested particle.

Figures 5a and 5c show a larger quantity of Cs and Xe in the KP, buffer, and IPyC of AGR2-222-RS19 compared to what is contained in these layers in AGR2-223-RS34 (Figs. 5b and 5d), while AGR2-223-RS34 contains more Cs and Xe in the inner kernel than what is observed in AGR2-222-RS19.

Figures 6a and 6b show that for Sr, most of the mass is located in the KP for both particles. Figures 7a and 7b show distinctively different behaviours between the safety-tested particle and the non-safety-tested particle. In the safety-tested particle, Nd is most abundant in the KP, while in the non-safety-tested particle, most of the mass remains in the inner kernel. Particle AGR2-222-RS19 experienced slightly higher burn up, so it would have produced more nuclide mass, but the data in Fig. 7 suggests that for this element, the safety-test promotes diffusion from the kernel to the more peripheral KP region.

Similarly, Figs. 8a and 8b show that Pd behaves quite differently in the safety-tested particle in that much of the mass has moved to the IPyC and SiC, whereas in particle AGR2-223-RS34, most Pd resides in the KP. Palladium vapour pressure increases from 0.006 Pa at 1,161 °C (the approximate time average volume average irradiation temperature) to 7.4 Pa at 1,600 °C, the temperature of the safety test. This increased vapour pressure will likely result in greater diffusion, thus explaining this observation.

IOP Conf. Series: Materials Science and Engineering **891** (2020) 012024 doi:10.1088/1757-899X/891/1/012024

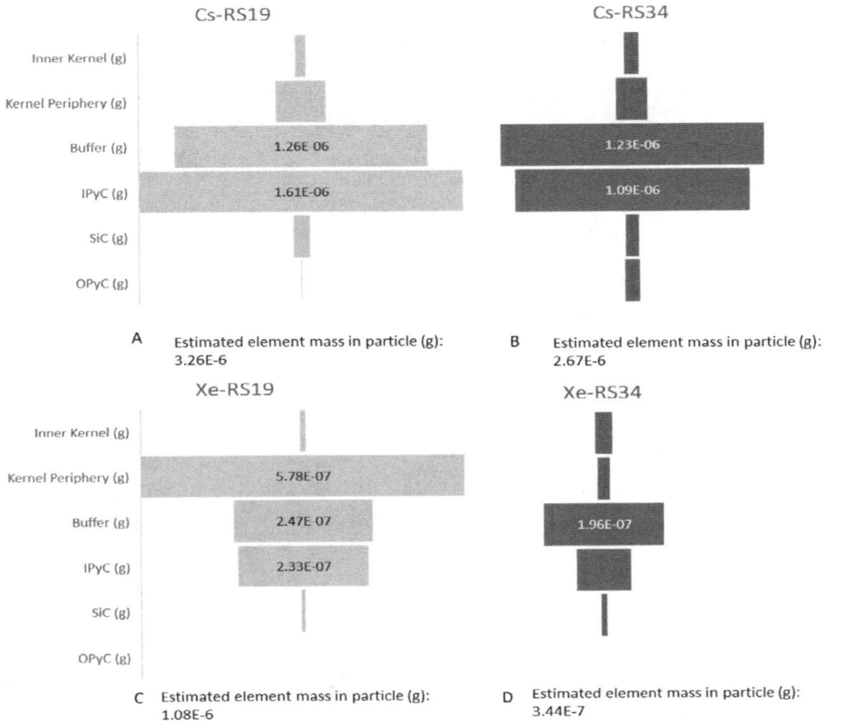

Figure 5. Cs and Xe total elemental mass estimate (in gram) and distribution within particles AGR2-222-RS19 and AGR2-223-RS34.

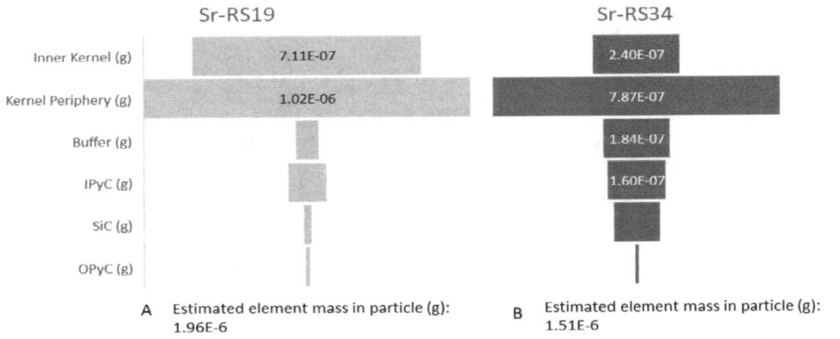

Figure 6. Sr total elemental mass estimate (in gram) and distribution within particles AGR2-222-RS19 and AGR2-223-RS34.

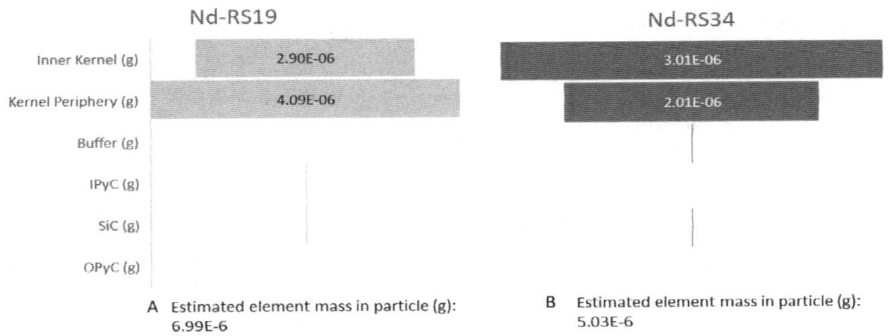

Figure 7. Nd total elemental mass estimate (in gram) and distribution within particles AGR2-222-RS19 and AGR2-223-RS34.

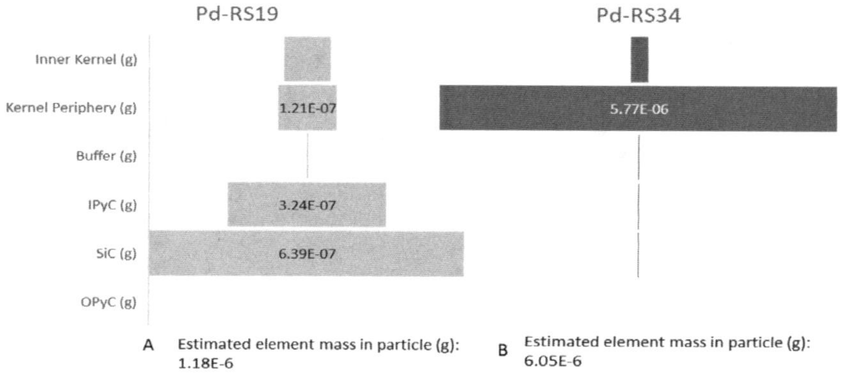

Figure 8. Pd total elemental mass estimate (in gram) and distribution within particles AGR2-222-RS19 and AGR2-223-RS34.

4.3. Sample size

Due to the high cost of analysing radioactive samples, only two radii on one plane were measured. Observations and resulting conclusions are based on the data obtained from these radial traverses and the assumption of hemispheric homogeneity. However, these particles could be much more heterogeneous than assumed, which could alter the data and subsequent conclusions.

5. Conclusions

- Sr, Te, Ba, and Eu are up to four times higher in concentration on the non-gap side of safety-tested particle AGR2-222-RS19 compared to the gap side. Asymmetry is absent in the non-safety-tested particle AGR2-223-RS34.
- Pd and Ag are up to four times higher in concentration on the non-gap side of the SiC-IPyC boundary of both the safety-tested and non-safety-tested particles, suggesting that the gap forms on the hotter side of the particle.
- Although the safety-tested particle contains a greater mass of FPs Cs, Xe, Sr, Te, Ba, Mo, and Nd than the non-safety-tested particle, the non-safety-tested particle retains more FP mass of Cs, Xe, Ba, Te, Mo, Ru, and Pd in the central kernel compared to the safety-tested particle.
- The FP transport as determined by three-dimensional mass calculations shows that for mobile FPs such as Cs, Xe, Te, Sr, Ba, and Pd, the mass moves out further from the particle centre in the safety-tested particle than the non-safety-tested particle. Palladium is much more abundant in the IPyC and SiC layers of the safety-tested particle than the non-safety-tested particle.
- Computing the FP mass by determining the electron interaction volume and extrapolating the measured concentration layer by layer appears to offer a method to more carefully explore FP transport and mass balance.
- It will be beneficial to re-evaluate this method using additional radial line scan data and mass balance calculations to better evaluate sample heterogeneity along the central plane. In addition, it would be beneficial to compare these results with results obtained from FP production modelling (e.g., ORIGEN) for each particle to evaluate how this FP mass as estimated by EPMA measurements compares to that calculated by the FP model.

Acknowledgements
This work was sponsored by the U.S. Department of Energy, Office of Nuclear Energy, under U.S. Department of Energy Idaho Operations Office Contract DE-AC07-05ID14517, as part of the Advanced Reactor Technology Program. The United States Government retains and the publisher, by accepting the article for publication, acknowledges that the United States Government retains a nonexclusive, paid-up, irrevocable, world-wide license to publish or reproduce the published form of this manuscript, or allow others to do so, for United States Government purposes.

References

[1] Collin B 2018 *AGR-2 Irradiation test final ss-run report.* (Idaho Falls, ID: Idaho National Laboratory) Report INL/EXT-14-32277 Rev. 3

[2] Ritchie N, Newbury D and Davis J 2012 EDS measurements of X-ray intensity at WDS precision and accuracy using a silicon drift detMector. *Micros. Microanal.* **18** 892-904

[3] Perrot M 1995 Microanalyse d'echantillons irradies par la microsonde de Castaing, application au combustible Nncleaire MOX. Diplome Universitaire Supèrieur de Recherche thesis. (Orsay, France: Université Paris Sud, UFR Scientifique d'Orsay)

[4] Kleykamp H 1975 *Mikrosondenuntersuchungen zum Verhalten der Spalprodukte in hoch abgebrannten HTR-brennstoffen.* (Karlsruhe, Germany: Kernforschungszentrum Karlsruhe) Report KfK 2213.

[5] Tiegs T 1980 *Fission product Pd-SiC interaction in irradiated coated particle fuels.* (Oak Ridge, TN: Oak Ridge National Laboratory) Report ORNL/TM-7203

[6] Schenk W, Pott G and Nabielek H 1990 Fuel accident performance testing for small HTRs. *J. Nucl Mater.* **171** 19-30

[7] Minato K, Ogawa T, Fukuda K, Shimizu M, Tayama Y and Takahashi I 1994 Fission product behavior in Triso-coated UO_2 fuel particles. *J. Nucl. Mater.* **208** 266-281

[8] Barrachin M, Dubourg R, Kissane M and Ozrin V 2009 Progress in understanding fission-product behaviour in coated uranium-dioxide fuel particles. *J. Nucl. Mater.* **385** 372-386

[9] Barrachin M, Dubourg R, DeGroot S, Kissane M and Bakker K 2011 Fission-product behaviour in irradiated TRISO-coated particles: Results of the HFR-EUbis experiment and their interpretation. *J. Nucl. Mater.* **415** 104-116

[10] Wright K and Van Rooyen I 2016 Electron probe microanalysis of irradiated and 1600 °C safety-tested AGR-1 fuel particles with low and high retained ^{110m}Ag. in: *Proc. 2016 Int. Top. Meeting High Temperature Reactor Technology (HTR 2016).* (Las Vegas, NV)

[11] Collin B 2018 *AGR-2 irradiation experiment test plan.* (Idaho Falls, ID: Idaho National Laboratory) Report INL/MIS-11-21682

[12] Morris R, Hunn J, Baldwin C, Montgomery F, Gerczak T and Demkowicz P 2018 Initial results from safety testing of US AGR-2 irradiation test fuel. *Nucl. Eng. Des.* **329** 124-133

[13] Demkowicz P 2013 *AGR-2 post irradiation examination plan.* (Idaho Falls, ID: Idaho National Laboratory) Report INL/MIS-13-30534

[14] Pouchou J and Pichoir F 1991 Quantitative analysis of homogeneous or stratified microvolumes applying the model "PAP". in: *Electron probe quantitation.* (Heinrich K F J and Newbury D E; Eds.) (New York, NY: Plenum Press) 31-76

[15] Farthing I and Walker C 1990 *Heinrich's mass absorption coefficients for the K, L, and M lines.* (Karlsruhe, Germany: European Commission, Institute for Transuranium Elements) Report K0290140

[16] Ploger S, Demkowicz P, Hunn J and Kehn J 2014 Microscopic analysis of irradiated AGR-1 coated fuel compacts. *Nucl. Eng. Des.* **271** 221-230

[17] Grubmeier H, Naoumidis A and Thiele B 1977 Silicon carbide corrosion in high-temperature gas-cooled reactor fuel particles. *Nucl. Tech.* **35** 413-427

[18] Homan F, Lindemer T, Long Jr. E, Tiegs T and Beatty R 1977 Stoichiometric effects on performance of high-temperature gas-cooled reactor fuels from the U-C-O System. *Nucl. Tech.* **35** 428-441

AUTHOR INDEX